Too Hot to Touch

The Problem of High-Level Nuclear Waste

When the nuclear energy industry was launched in the 1950s, Robert Oppenheimer dismissed the waste problem as "unimportant." Over a half-century later, the waste issue is as prominent as reactor safety in the international controversies surrounding nuclear power. It is particularly topical in the US since the 2010 closure of the Yucca Mountain repository project. With no long-term plan in sight, high-level radioactive waste remains scattered across 121 sites in 39 states.

William and Rosemarie Alley provide an engaging and authoritative account of the controversies and possibilities surrounding disposal of nuclear waste in the US, with reference also to the difficulties and progress of other countries around the world. The book tells the full history from the early days after World War II up to the present time, with an insightful perspective drawn from William Alley's expertise in the field, including leading the US Geological Survey study of Yucca Mountain. Stories of key players bring to life the pioneering science, the political wrangling and media drama, and the not-in-my-backyard communities fighting to put the waste somewhere else.

Written in down-to-earth language, this is a fascinating book for public interest groups, affected communities, and anyone interested in finding out more about this issue. The timely and important subject also makes it a valuable resource for policymakers, political staff, environmentalists, and research scientists working in related fields.

WILLIAM AND ROSEMARIE ALLEY are a husband and wife team, writing for the general public on Earth Science issues confronting society. As a leading expert in the field of hydrogeology, Dr. William M. Alley has won numerous awards for his work, including the US Geological Survey (USGS) Shoemaker Award for Lifetime Achievement in Communication and the Meritorious Presidential Rank Award. Dr. Alley served as Chief of the Office of Groundwater for the USGS for almost two decades and oversaw the Yucca Mountain project from 2002 to 2010. Rosemarie Alley has a Master's Degree in special education. As a literacy specialist, she has taught young adults with language delays and conducted numerous reading workshops for teachers, administrators, and parents. Currently, Rosemarie is a writer, sculptor, potter, and gardener. The Alleys live in the foothills above San Diego, California.

Too Hot to Touch

The Problem of High-Level Nuclear Waste

WILLIAM M. ALLEY
AND
ROSEMARIE ALLEY

CAMBRIDGE
UNIVERSITY PRESS

CAMBRIDGE UNIVERSITY PRESS
Cambridge, New York, Melbourne, Madrid, Cape Town,
Singapore, São Paulo, Delhi, Mexico City

Cambridge University Press
The Edinburgh Building, Cambridge CB2 8RU, UK

Published in the United States of America by Cambridge University Press,
New York

www.cambridge.org
Information on this title: www.cambridge.org/9781107030114

First published 2013

Printed and bound in the United Kingdom by the MPG Books Group

A catalog record for this publication is available from the British Library

Library of Congress Cataloging in Publication data
Alley, William M.
Too hot to touch : the problem of high-level nuclear waste /
William M. Alley and Rosemarie Alley.
pages   cm
Includes bibliographical references and index.
ISBN 978-1-107-03011-4 (hardback)
1. Radioactive waste disposal – United States.
2. Radioactive wastes – United States.   I. Alley, Rosemarie.   II. Title.
TD898.118.A45   2013
363.72'890973 – dc23   2012021832

ISBN 978-1-107-03011-4 Hardback

Additional resources for this publication at
www.cambridge.org/alleyalley

This book is dedicated to all the scientists who have devoted their working lives to trying to solve the problem of high-level nuclear waste.

Contents

Acknowledgments

We are grateful to Laura Clark, our editor at Cambridge University Press, for her unflagging interest, insights, and support in helping to bring this book to completion. We also would like to thank Dr. Isaac Winograd, at the US Geological Survey, for his thoughtful and generous review of an early draft of Part II. A special thanks also goes to Dr. Deserai Crow, at the University of Colorado, for her help in framing the discussion questions. Finally, we are indebted to five anonymous reviewers, who provided fresh perspectives on the strengths and weaknesses of the manuscript.

Units

We use a combination of American and metric units, and show both where it is important for comprehension. Conversion factors are listed below for some commonly used units.

1 foot (ft) = 0.3048 meters (m)
1 inch (in) = 2.54 centimeters (cm)
1 mile (mi) = 1.609 kilometers (km)
1 square mile = 2.59 square kilometers
1 acre = 0.4047 hectares
1 gallon = 3.785 liters
1 pound = 0.45 kilograms (kg)
1 rem = 0.01 sieverts
1 metric ton = 1000 kilograms = 1.1 tons

Abbreviations

AEC	Atomic Energy Commission
AFR	Away-from-reactor
AFSWP	Armed Forces Special Weapons Project
ANEC	American Nuclear Energy Council
BRC	Blue Ribbon Commission on America's Nuclear Future
DOE	US Department of Energy
EBR-I	Experimental Breeder Reactor-I
EEG	Environmental Evaluation Group
EIS	Environmental Impact Statement
EPA	US Environmental Protection Agency
EPRI	Electric Power Research Institute
ERDA	Energy Research and Development Administration
ESF	Exploratory Studies Facility
ET	Evapotranspiration
FEPs	Features, events, and processes
GAO	Government Accountability Office
GNEP	Global Nuclear Energy Partnership
INL	Idaho National Laboratory
IRG	Interagency Review Group
ISFSI	Independent spent fuel storage installation
LLRW	Low-level radioactive waste
MRS	Monitored retrievable storage
MOX	Mixed oxide
NAS	US National Academy of Sciences
NASA	National Aeronautics and Space Administration
NIMBY	Not in my backyard
NIMS	Not in my State
NRC	US Nuclear Regulatory Commission
NRTS	National Reactor Testing Station (now INL)

NTS	Nevada Test Site (now named Nevada National Security Site)
NWPA	Nuclear Waste Policy Act
NWTRB	Nuclear Waste Technical Review Board
PUREX	Plutonium URanium EXtraction reprocessing
R&D	Research and development
RCRA	Resource Conservation and Recovery Act
RMEI	Reasonably maximally exposed individual
SLAC	Stanford Linear Accelerator Center
TSPA	Total system performance assessment
TVA	Tennessee Valley Authority
UNLV	University of Nevada-Las Vegas
USGS	US Geological Survey
UZ	Unsaturated zone
WIPP	Waste Isolation Pilot Plant

Introduction

On Labor Day in 1954, President Eisenhower appeared on television to announce the start of construction of the Nation's first nuclear power plant. The plant was to be built in the small town of Shippingport, Pennsylvania on the Ohio River, about 25 miles (40 km) west of Pittsburgh. The site was not far from Titusville, Pennsylvania, the birthplace of the petroleum industry, and was almost on top of one of the world's greatest coal fields. The Pittsburgh utility had come onboard the demonstration project for one basic reason – pollution control. Pittsburgh, once the "Smoky City," had instituted strict air pollution requirements, and the local citizens were resisting plans for a coal-fired power plant. [1]

President Eisenhower, vacationing in Denver, Colorado, made good use of the new television medium to dramatize the event. The President held up a large neutron-generating "magic wand" for the viewers to behold. As he waved the atomic-age magic wand over a neutron counter, an electronic signal traveled 1,200 miles (1,900 km) from Denver to Shippingport. The signal activated an unmanned, remote-controlled bulldozer which began to break ground for the new plant. It was an impressive feat. The local crowd rose to their feet and applauded. Two weeks later, Lewis L. Strauss, Chairman of the Atomic Energy Commission, delivered his oft-cited speech that "Our children will enjoy in their homes electrical energy too cheap to meter." [2–3]

Signed into law a few months earlier, the Atomic Energy Act of 1954 opened the way for peaceful uses of the atom. For the first time, civilians were allowed to join the elite nuclear club by building and operating privately owned nuclear power plants. Soon thereafter, the Atomic Energy Commission announced its Power Demonstration Reactor Program. The program offered free nuclear fuel for up to seven years, money for research and development and, in some cases, a large part of the capital needed to build nuclear plants. One member of Congress characterized it as an attempt to "force-feed atomic development" with tax dollars. [4]

In the euphoria of abundant and cheap energy just around the corner, the question of what to do with the radioactive waste from nuclear power plants rarely came up. Nor were there many concerns about the high-level nuclear waste that had been accumulating at Hanford and other atomic weapons plants since the start of the Manhattan Project. For many years, the problem of nuclear waste would go unrecognized or be considered trivial. Predictions that the problem would almost solve itself would prove miserably wrong.

By the late 1960s, a general optimism prevailed that there would be a permanent site for high-level nuclear waste by 1980. Time and again, the date would be revised. In 1982, the Nuclear Waste Policy Act mandated that the federal government begin accepting high-level wastes for burial in a geologic repository by January 31, 1998. The government was so sure of itself that it signed binding agreements with the utilities to accept responsibility for the waste by this date. A few years later, the Office of Technology Assessment for the US Congress suggested that the 1998 goal was unrealistic, but expressed "considerable confidence" that a repository would be operating by 2008 and a second by 2012, "even if significant delays are encountered." Newly revised predictions of the date for an operational repository continue to be pushed further and further into the future, as the dates of previous predictions recede into the past. [5]

Meanwhile, high-level radioactive waste from electricity generation and weapons production remains scattered among 121 sites in 39 States. As of 2012, the United States had accumulated almost 70,000 tons of spent fuel from nuclear reactors. In addition, about 20,000 giant canisters of defense-related high-level radioactive waste will need a final resting place. [6]

More than a half-century has passed since Eisenhower waved his magic wand. The Shippingport plant has come and gone. It shut down in 1982, after 25 years of successful operation. By any measure, it is well past the time to tackle the waste problem. No magic wand will make it go away.

Part I   The problem

1

The awakening

I can't think about that right now... I'll think about that tomorrow.
Scarlett O'Hara, *Gone with the Wind* [1]

In January 1949, the Atomic Energy Commission (AEC) held a seminar on radioactive waste. In his opening remarks, AEC Chairman David Lilienthal cast the problem of waste disposal as part of "learning to live with radiation." According to Lilienthal, this learning curve was the same as how we humans learn to live with anything else unfamiliar. The Chairman of the AEC acknowledged that radioactive wastes could become "a subject of emotion and hysteria and fear... [but] we do not believe those fears are justified provided technology applies itself to eliminating the troubles." The previous year, Robert Oppenheimer, Chairman of the AEC's General Advisory Committee, had dismissed the waste problem as "unimportant." [2–3]

In spite of these pronouncements, dealing with radioactive waste gained greater urgency upon passage of the Atomic Energy Act of 1954, making possible the widespread use of nuclear energy for civilian purposes. As such, the nuclear industry would now be close to major cities and towns. And dilution was not the solution. Given the anticipated size of the US nuclear industry by the year 2000, it would require a volume equal to about five percent of the world's oceans to dilute the dangerous waste to recommended safe levels. This exceeded the volume of freshwater stored worldwide in lakes, rivers, groundwater, glaciers, and polar ice caps. [4]

In February 1955, the AEC signed a contract with the National Academy of Sciences (NAS) to establish a committee of leading scientists to study the problem of geological disposal of radioactive waste. Five months earlier, Eisenhower had waved his magic wand to start groundbreaking for the Nation's first commercial nuclear power plant. Arthur E. Gorman, a champion within the AEC for dealing responsibly with the waste, spoke to the newly formed committee. The remote locations of the agency's atomic weapons plants had made it possible, he said, "to sweep the problem under the rug, [but now] we must face up to the fact that we are confronted by a real problem." Finding satisfactory methods for radioactive waste disposal had to be accomplished if the nuclear industry was to reach its full potential. [5]

Chartered by Congress during the Civil War, the NAS serves as the federal government's premier scientific advisor. The NAS Committee included several distinguished scientists. Hydrogeologist, C.V. Theis, had developed a mathematical equation for predicting the response of groundwater levels to pumping. Published in 1935, the *Theis equation* revolutionized the science of groundwater hydrology. Another member, John C. Frye, is credited with creating the field of environmental geology. And then there were Harry H. Hess, chair of the committee, and M. King Hubbert. Hess and Hubbert were widely respected in the scientific community, but were quite different personalities.

Harry Hammond Hess, head of the geology department at Princeton University, was a giant in the world of marine geology. While serving in the US Navy during World War II, Hess managed to map the Pacific Ocean floor while cruising from one battle to the next. His understanding of the seafloor later played a pivotal role in a revolution in geologic understanding of the Earth. [6]

For centuries, mapmakers had observed the parallelism of the opposing coasts of the Atlantic Ocean, suggesting that the continents had drifted apart. How this might have happened remained a mystery. Hess proposed that the seafloor is created as magma rises up from the Earth's interior along mid-oceanic ridges. The new seafloor spreads outward and eventually sinks into deep oceanic trenches in a conveyor-belt motion. The continents are carried along as part of large, rigid plates. As evidence for seafloor spreading gained credibility, this hypothesis became accepted as the theory of plate tectonics.

Hess also played a prominent role in designing the Nation's space program. In 1962, he was appointed by President John F. Kennedy to the prestigious position of Chairman of the Space Science Board of the National Academy of Sciences. Hess suffered a fatal heart attack while chairing a meeting of the Board in 1969. In spite of his considerable fame, Hess is remembered as a humble man who was sought after throughout his life for his fairness and open-minded nature.

Figure 1.1 Harry Hess commanding the *USS Cape Johnson* in 1945. On this troop transport ship, Hess mapped the Pacific Ocean floor with sonar during World War II. Photograph courtesy Department of Geosciences, Princeton University.

M. King Hubbert, the other key member of the NAS Committee, was a widely regarded geoscientist who worked for Shell Oil. King, as he was known, came from a family with a tradition of unusual names. King's great grandfather named many of his 15 children after his heroes. There was David Crockett Hubbert, Benjamin Franklin Hubbert, and Andrew Jackson Hubbert. Educated in a one-room schoolhouse in the hill country of Texas, King Hubbert managed to work his way to the University of Chicago. There he became the first graduate with a triple major in physics, mathematics, and geology. He refused to pick just one. King wanted an *education*, not a *major*. [7]

Figure 1.2 M. King Hubbert (on the left) setting up a resistivity instrument for a geophysical survey in Franklin Co., Alabama, September 22, 1934. Photograph by E.F. Burchard; courtesy US Geological Survey.

M. King Hubbert could be abrasive and was often a lightning rod for controversy. He was on the NAS Committee because of his intellectual depth and breadth in the earth sciences. Among his contributions, Hubbert's theoretical work laid the foundation for the study of regional groundwater flow systems. Although he made landmark contributions to earth science, King Hubbert would become most identified with the concept of *peak oil*.

In 1956, as the NAS Committee continued to deliberate on nuclear waste, Hubbert was invited to give a keynote address on the world energy situation at a meeting of the American Petroleum Institute. The public relations representative of Shell Oil pleaded with Hubbert to tone down the "sensational parts" of his speech. Not one to heed such advice, Hubbert informed the oil men in attendance that oil production in the lower 48 States would peak between the late 1960s and the early 1970s. Production would then decline sharply on the downward side of the classic bell-shaped curve. As the reserves of oil and other fossil fuels diminished, Hubbert predicted that they would be replaced by nuclear energy, "an energy supply adequate for our needs for at least the next few centuries." [8]

Hubbert's conclusions about peak oil were widely criticized. Following his pronouncement, Morgan Davis, president of Humble Oil (the largest domestic oil producer at the time), had King Hubbert followed from meeting to meeting to refute his arguments. There were also limits to his methodology. Hubbert focused on "easy oil" that was recoverable by the methods practised at the time. He also did not directly consider the effects of price on supply and demand. Nonetheless, Hubbert's prediction of peak oil in the lower 48 was realized in 1970, about the same time that the Arab oil embargo temporarily brought the Western world close to a stand-still. In essence, Hubbert predicted an oil crisis some 20 years before it actually occurred.

The first meeting of the NAS Committee was held in April 1955 at the Johns Hopkins University. The university was a logical choice. Abel Wolman, chairman and founder of the Sanitary Engineering Department at Johns Hopkins, was a pioneer in public health and an outspoken, though constructive, critic of the AEC's waste practices. Once, while dining at the home of Robert Oppenheimer, Wolman told his distinguished host: "[I have] tremendous respect for your field of activity and your views," *but*, he added, "When you enter my field . . . your ideas as to how we manage this 'unimportant' problem are characterized by a total ignorance of the nature of disposal." Wolman later recalled that, despite the strong differences of opinion, they "parted friends." In 1950, Abel Wolman and Arthur Gorman called it "highly questionable" whether the important task of dealing with radioactive waste should be left entirely to the AEC – a premonition that became increasingly hard to refute. [3, 9]

One problem faced by the NAS Committee was that outside the hallowed halls of the AEC, and apart from a few individuals like Abel Wolman, little was known about the problems associated with nuclear waste. The only member of the Committee who had any direct knowledge of the AEC facilities was C.V. Theis, from coordinating work between the US Geological Survey and the AEC. Most of the information on the Nation's nuclear waste legacy was classified.

The meeting at Johns Hopkins was followed by a conference at Princeton University. Two years later, in 1957, the NAS Committee summed up their findings in a report. Like many to follow, the Committee viewed nuclear waste disposal as essentially a technical problem. Three key conclusions stood out that would significantly influence events over the next couple of decades [5]:

(1)   Wastes may be disposed of safely at many sites in the United States, but conversely, there are many large areas in which it is unlikely that disposal sites can be found.

(2)   Disposal in cavities mined in salt beds and salt domes is suggested as the possibility which promises the most immediate solution to the problem.

(3)   Disposal could be greatly simplified if the waste could be made into a solid form, relatively insoluble in character.

The first conclusion, though reasonable in the light of knowledge at the time, contributed to a false sense of confidence about the ease of finding a suitable site. The second conclusion – burying the waste in salt beds – became the cornerstone of waste disposal policy for the next 20 years. The third conclusion – converting radioactive waste into solid form to reduce its mobility – has stood the test of time and has become the focus of considerable worldwide research and development (R&D). The report also represents the beginning of the international consensus to pursue on-land, subsurface disposal of high-level radioactive wastes.

The NAS Committee was very forthright about the limitations of their knowledge. They were the first to admit that certain assumptions needed to be verified. For example, would the extreme heat from the radioactive material reduce the ability of salt to contain it? The Committee also warned, "The hazard related to radioactive waste is so great that *no element of doubt* [emphasis added] should be allowed to exist regarding safety." Over time, it has become clear that such absolute assurances are not possible. [5]

To put the Committee's report into context, in 1957 the first practical computer models of hydrologic systems were a decade away and very few studies had been undertaken of subsurface chemical processes – both critical areas for nuclear waste management. The Committee also assumed that the waste would be reprocessed and would require isolation for only 600 years or less. The idea that tens of thousands to hundreds of thousands of years of safe containment might be required was simply not on the radar screen. In the same way that plate tectonics required revision of many principles of geologic processes, the Committee's conclusions needed re-evaluation as new insights were gained. It would be decades before this reality sunk in.

The Committee report appeared to have little effect on the AEC. While the Academy scientists suggested searching for a repository site among the best possible geologic locations, the AEC, with an eye to cost and convenience, had other ideas. They favored establishment of repositories at existing atomic weapons facilities where virtually all of the wastes were located at the time. "They pressured us right from the start that they wanted a disposal site at each of these plants," recalled M. King Hubbert. "They never let up on this." [10]

The NAS Committee became increasingly frustrated by the AEC's lack of responsiveness to their recommendations. In 1960, Harry Hess wrote a letter to AEC Chairman, John A. McCone. As spokesman for the Committee, Hess recommended that urgent action be taken to establish facilities at suitable geologic sites, instead of taking the path of convenience. Hess also urged that approved plans for safe disposal of radioactive wastes be a prerequisite before any new nuclear power plants could be built. The AEC responded that the Committee's proposals were costly and unnecessary. Having "practically no further duties except for trivialities," fumed Hubbert, the NAS Committee might as well have been disbanded. [10–11]

AEC Chairman McCone, a California industrialist, had a history of steadfast support for established AEC positions. He also had a will to match that of Hubbert. A few years earlier, during the 1956 presidential election, ten California Institute of Technology (Caltech) scientists concerned about radioactive fallout had issued a statement supporting Adlai Stevenson's proposal to ban atmospheric nuclear weapons tests. McCone, a member of the Caltech board of trustees and campaigner for Eisenhower, accused the scientists of being taken in by Soviet propaganda. According to the scientists, he tried to get them fired. When they weren't fired, McCone resigned from the board. Two years later, Eisenhower appointed him Chairman of the AEC. [12]

In 1963, when M. King Hubbert became chairman of the earth sciences division of the Academy's National Research Council, he promptly confronted the AEC. "I told them I didn't propose to keep any committee standing around twiddling its thumbs...they should either discharge it or give it something worthwhile to do." The AEC reluctantly agreed to let the Committee review their R&D program on waste disposal for nuclear power. For perhaps the first time, the AEC conducted a fieldtrip for an outside group. The NAS Committee visited Hanford and other weapons sites, as well as two salt mines near Lyons, Kansas where preliminary experiments on waste disposal were taking place. After being briefed on research involving waste solidification in glass and ceramics, the Committee reported that it was "favorably impressed with the whole solidification program." However, they were definitely not impressed with the waste disposal practices at the weapons sites, and said so in a draft report to the AEC. [10, 13]

A tug of war ensued. The AEC argued that the NAS scientists had overstepped the scope of their study and pressured the Committee to delete their criticisms. The Committee refused to delete anything. As responsible citizens, they felt a duty to raise these concerns. When the NAS Committee submitted their final report to the AEC in May 1966, they again stated their conviction that "none of the major sites at which radioactive wastes are being stored or disposed of is geologically suited for safe disposal of any manner of radioactive waste other than very dilute, very low-level liquids." The Committee acknowledged one possible exception – some intermediate waste might be safely disposed by mixing with grout and injecting it into fractured shale at Oak Ridge. [10, 13]

The AEC fought back. First, they prepared a 15-page response to the NAS report, taking the position that the scientists had been misguided in their major conclusions. Second, they suppressed the report, arguing that it had already been made available to pertinent personnel. Finally, the Commission disbanded the NAS Committee and replaced it with a new group that had "a broader spectrum of scientific discipline." [10, 14]

With this accomplished, the Atomic Energy Commission turned its attention to salt deposits as the preferred medium for disposal. Within a few years, the next phase of confrontation would begin.

THE SALT OF THE EARTH

Salt beds are formed by the evaporation of inland seas or enclosed coastal bays. Under the right conditions salt can really pile up, eventually forming beds hundreds, even thousands, of feet thick under vast areas. For a number of reasons, salt beds are a good choice for disposing of radioactive waste. Given that salt dissolves easily in water, thick salt deposits mean that groundwater has been absent for the many millions of years required to form them. The physical and mechanical properties of salt are also favorable. Salt is approximately equal to concrete in its ability to shield harmful radiation. And because of its plastic nature, particularly at depth, fractures tend to seal up. Salt also conducts heat better than other types of rocks, alleviating localized over-heating. Finally, salt deposits usually occur in areas of low seismic activity.

Yet, there are drawbacks. When water does appear, it forms highly corrosive saline brines – bad news for the waste containers. Brine also tends to migrate toward heat sources such as hot waste.

Another disadvantage is that if radionuclides somehow escape from the container, salt is particularly bad at grabbing and holding onto them. For many other rock types, attachment (sorption) on mineral surfaces can significantly delay or even stop subsurface contaminant movement. Salt is the Teflon of minerals.

The first in-depth study of salt for high-level nuclear waste disposal was conducted between 1965 and 1968 in an abandoned salt mine near Lyons, Kansas. Dubbed Project Salt Vault, scientists from Oak Ridge National Laboratory inserted 14 spent fuel assemblies into the floor of the mine. Heaters were installed to determine how salt would respond to high temperatures. Nineteen months later the fuel and heaters were removed. The experiment was deemed "successful in all respects." In early 1970, the Oak Ridge scientists declared that "most of the major technical problems pertinent to the disposal of highly radioactive waste in salt have been resolved." [3, 15]

This breakthrough could not have come at a better time. Having finally ridden themselves of the NAS Committee and King Hubbert's incessant badgering, the AEC was now under intense pressure from Senator Frank Church. The Democratic Senator from Idaho was raising the roof about a serious problem in his State.

The problem began in 1969, when a fire gutted the Rocky Flats plutonium weapons plant 16 miles northwest of Denver. The *New York Times* ran a story on the fire, and how tons of plutonium-contaminated debris were being shipped to the Idaho National Reactor Testing Station (NRTS) for burial. The NRTS had been accepting radioactive waste from Rocky Flats for years with little controversy. Thanks to the *Times*, this time it was different.

Robert Erkins, an Idaho resident and owner of the world's largest trout farms, would have none of it. A prospective buyer of one of his farms had clipped the *New York Times* story and sent it to Erkins, saying he didn't want to buy a trout farm next to a nuclear waste burial site. Erkins was alarmed. The NRTS sat above the prolific Snake River Plain aquifer – the source of water supplying his business. This aquifer was a major reason the NAS Committee had advised against using the Idaho site for radioactive waste disposal in their ill-fated report. [16–18]

After visiting the NRTS and seeing atomic waste in cardboard containers dumped into trenches, Erkins complained to the Governor and NRTS management. Dissatisfied with the blanket assurances he received, Erkins mailed letters to newspaper editors throughout the State and soon had plenty of media attention. The public outcry

Figure 1.3 Stacks of nuclear waste storage containers at a shallow land
burial site at Idaho National Laboratory, 1976.
Source: US Department of Energy.

reached Senator Church, who also learned about the suppressed NAS
report. Church demanded to know why the report had not been made
public.

After several months of foot dragging, a reluctant AEC finally
released the suppressed NAS report to a blitz of media attention. The
Idaho-Falls Post-Register headlined the event, "Science Academy Doubts
Safety of Waste Disposal at NRTS." This public relations nightmare for
the AEC was compounded when the Federal Water Quality Administration concluded that the NRTS waste-burial practices were "a potential
threat to the water resources of the State of Idaho." The Water Quality
Administration recommended not only stopping the waste-burial practice, but also removing previously buried wastes. [3, 19]

Ironically, Glenn Seaborg, the current AEC Chairman, had co-
discovered plutonium – the primary culprit of this mess. Seaborg
promised Senator Church that all of the plutonium-contaminated waste,
including what had been buried in the past, would be recovered and sent
to a geologic repository. Seaborg was quite sure that this could be underway by the end of the decade.

The pressure was on, but Chairman Seaborg had every reason to believe that the Lyons salt mine would save the day. The project offered much needed jobs for the citizens, who were generally receptive to the idea. Project Salt Vault had been conducted in an overall atmosphere of goodwill among federal, State, and local officials. Public tours of the mine had even been given. [20]

Naturally, there were a few skeptics. From the outset of early studies, an editorial in the nearby *Great Bend Daily Tribune* spoke of people "just plain scared of anything that has to do with nuclear fission" and "workers glowing in the dark like watch dials." Concerned about public opinion, Frank C. Foley, Director of the Kansas Geological Survey, suggested that the term "atomic waste disposal" should be replaced by the more reassuring term "atomic by-products storage." Sure to boomerang on their agency, the AEC wisely rejected Foley's word-smithing. [3]

Kansas politicians held mixed opinions about the prospect of a geologic repository for radioactive waste in their State. Governor Richard B. Docking, a Democrat in a heavily Republican State was, in 1970, seeking a third term. He decided that the best course of action would be to take a wait-and-see approach. Then there was Congressman Joe Skubitz, an ardent and outspoken opponent from the outset. Like many Republicans in Kansas, Skubitz was deeply suspicious of the federal government. [3]

As opposition began to grow, the AEC once again became embroiled with its old nemesis – the NAS Committee. This time the controversy was in public view right from the start. The NAS Committee had established a special panel of scientific experts to study the feasibility of burying wastes in salt beds. On June 17, 1970, they held a meeting in Lawrence, Kansas. Unbeknownst to these scientists, John Erlewine, their point-of-contact with the AEC, was simultaneously holding a press conference 30 miles (50 km) down the road in Topeka. While the scientists were discussing the scientific complexities in Lawrence, Erlewine announced the "tentative selection" of the Lyons salt mine to demonstrate the feasibility of burying high-level wastes in salt. Erlewine's surprise announcement came after astonishingly little consultation with State officials. Congressman Skubitz immediately sent a letter to the Governor expressing his "grave doubts about the safety of this project" and followed with a 12-page, single-spaced letter to AEC Chairman Seaborg. [3]

Erlewine's announcement put the NAS Committee under pressure to address the suitability of the Lyons site, with little time for thoughtful evaluation. As a result, their report concluded that the site was "satisfactory, subject to the development of certain additional confirmatory data and evaluation." In essence, the report was a qualified let's-take-a-closer-look endorsement, yet some saw it in black and white terms. Proponents of the site immediately proclaimed that the NAS Committee had given the green light to proceed. Milton Shaw, Director of the AEC Division of Reactor Development and Technology and a strong proponent of nuclear power, announced that the site "had been recommended" by none other than the prestigious National Academy of Sciences. The trade journal, *Nucleonics Week*, once described Shaw as "probably without peer in convincing someone that nuclear power is to be embraced with little or no reservation." [3, 21]

William Hambleton, a member of the NAS expert panel, was skeptical. Hambleton had become Director of the Kansas Geological Survey in 1970, just as the Lyons issue was gaining prominence. Among his concerns was the presence of oil and gas drilling holes in the immediate vicinity of the mine. He also wanted to take a closer look at possible structural weaknesses in the geologic formation, as well as to obtain a clearer understanding of the effects of heat and radiation on salt. The June 17 meeting in Lawrence was conducted, in large part, to discuss some of Hambleton's concerns. [3]

With the AEC all but ready to begin construction, Hambleton was becoming increasingly frustrated by the lack of meaningful attention to the serious issues he had raised. The newly re-elected Governor Docking relied heavily on Hambleton's expertise in guiding his position. Hambleton and Docking's frustrations came to a head during congressional hearings of the Joint Committee on Atomic Energy in March 1971. The purpose of the hearings was to address the AEC's request for federal funds to purchase the land around Lyons, so that preliminary design and engineering work could begin.

Appearing before the Joint Committee as the Governor's spokesman, Hambleton announced that Docking had *reluctantly* concluded that the AEC's efforts to minimize the problems raised by scientists in Kansas "support fears of many Kansans that if funds are appropriated for design and site acquisition the project cannot be stopped at a later date if it is ... found to be unsafe." He urged that funding be deferred until scientific studies were completed and the results evaluated. Milton Shaw of the AEC rebutted that "another year's work of research and development in this area, on top of fifteen years of work, will not be particularly productive ... We need the project and are ready to proceed with it." [3, 16]

The debates continued, with both Docking and Hambleton denouncing the high-handed and patronizing manner of the AEC. Five months later, in August 1971, with the help of Senator Robert Dole

(R-KS), compromise legislation was passed. Funds were allocated to lease the land, rather than purchase it. The legislation also specified that any radioactive materials must be fully retrievable – although no one had a clue how to do this in self-sealing salt beyond a few years' time.

The compromise was short-lived. A consultant hired by Oak Ridge National Laboratory discovered that 29 exploratory oil and gas wells had been drilled into or below the salt formation near the site. The consultant's best guess was that 26 of these wells could *probably* be plugged. The likelihood of plugging the other three was *very low*. Additionally, these discovered wells opened up the possibility of more wells, some of which might never be found. [10]

Worse news soon followed when it was learned that 175,000 gallons (650,000 liters) of water had mysteriously disappeared during hydraulic fracturing at a nearby mine. No one knew where the missing water had gone. The Lyons site was beginning to look, in Hambleton's words, "a bit like a piece of Swiss cheese." A few weeks later, Congressman Skubitz declared that the "Lyons site is dead as a dodo for waste disposal." The AEC held on for a while, but officially abandoned the project in 1972. [3]

As noted by J. Samuel Walker, historian for the Nuclear Regulatory Commission, "The AEC's first effort to identify a suitable site for disposing of high-level radioactive waste from commercial nuclear power failed spectacularly. In its haste to fulfill its pledge to Senator Church and to build a repository for the growing quantities of commercial reactor wastes, it not only selected a location that proved unsuitable but also offended political leaders and scientists whose backing for the project was essential . . . it tended to view critics of the Lyons proposal as a monolithic whole. It failed to distinguish between the reservations that Hambleton cited and the much more strident and intractable position that Skubitz adopted." [3]

The public had been given a first-hand look at the AEC in action on radioactive waste disposal, and many were not pleased with what they saw. As the consequences played out, Congressman Skubitz's words were not only applicable for the Lyons site. In a few years' time, the AEC also would be as dead as a dodo.

A TECHNOLOGY AHEAD OF ITSELF

A quarter of a century had now passed since the establishment of the Atomic Energy Commission, in 1946. The commitment of budget and personnel to waste disposal had been woefully inadequate and the complexity of the problem greatly underestimated. "How," asks Luther Carter in his 1987 landmark book on radioactive waste, "could the U.S. Atomic Energy Commission, the agency behind the start-up of nuclear power in the 1950s and 1960s, have launched this important new commercial enterprise without first knowing what was to be done with the radioactive waste?" Nuclear power, Carter concludes, got ahead of itself as an important new technology. [22]

Four years before Eisenhower officially launched commercial generation of electricity by nuclear fission, James B. Conant, President of Harvard University and a key advisor to President Roosevelt on the atomic bomb, predicted that the world would eventually turn away from nuclear power in large part because of problems with waste disposal. At the time, Conant appears to have been alone in such an assessment. [23]

The Atomic Energy Commission's failure to put together a coherent and effective waste policy has been linked to a variety of causes. The Atomic Energy Act of 1946 primarily focused on how to preserve the US monopoly on nuclear weapons and technology. The Act is silent on the subject of radioactive waste. The revised Atomic Energy Act of 1954 allowed private industry to enter the nuclear business, but generally ignored safety issues, including waste disposal. Harold P. Green, a former AEC attorney, noted that there were only 31 references to the health and safety of the public among the 4,000 pages of reports, testimony, and debates before Congress relating to the Act. "Nobody really ever thought that safety was a problem," he says. "They assumed that if you just wrote the requirement into the Act that it be done properly, it would be done properly." [24]

Concerns about waste disposal were even less likely to come from other quarters. The nuclear industry had its beginnings under conditions that allowed no possibility of meaningful public discussion. "There was secrecy, promotional hype, lack of open debate, and a profound lack of knowledge and understanding on the part of the public about nuclear energy," notes Carter. [22]

Politicians were not exempt from this lack of knowledge. Former Secretary of the Interior, Stewart Udall recalled that "the atomic scientists had such great prestige that if you were an ordinary congressman, and not skilled in science, and certainly the whole thing about the atomic bomb was secret, what kind of questions could you ask? The questions began about the time I left office in the early 1970s, when people like Senator Muskie and others began asking serious questions." [25]

The complex technology and high-level security classification peculiar to atomic energy led Congress to make special provisions. One of these was the creation of the Joint Committee on Atomic Energy, established as a congressional oversight committee by the Atomic Energy Act of 1946. The Joint Committee had jurisdiction over "all bills, reso- lutions, and other matters" related to civilian and military aspects of nuclear power. It was perhaps the most powerful congressional com- mittee in history, and the only congressional committee ever to have legislative veto power – later found to be unconstitutional by the US Supreme Court.

In theory, the Joint Committee should have raised concerns about waste disposal. Yet, as Harold Green notes, the Joint Committee was "almost always more aggressive and expansionist than the Atomic Energy Commission and the Executive Branch, and it constantly pressed for larger and more ambitious atomic energy projects." The AEC and the Joint Committee established a track record of downplaying health and safety issues, including the dangers of atmospheric nuclear testing, uranium mill tailings, and improper ventilation of uranium mines. This tradition continued with high-level waste. [16, 26]

In extensive hearings held in 1959, the Joint Committee heard one expert after another from the AEC, the national laboratories, as well as academia and industry, testify that a technological solution to the waste management problem was possible. The conclusion by the 1957 NAS Committee report, that many sites might be suitable for waste disposal, contributed to this overall complacency. With these assurances in place, Congress largely dropped the matter until the AEC was disbanded in 1975. [27]

In addressing the complex issues of radioactive waste disposal, the government and policymakers have followed an all too human course. Pressing problems receive priority while anything that can wait, does wait. The United States has not been alone in this dilemma.

The United Kingdom built the world's first commercial nuclear power plant (Calder Hall) near Sellafield on the coast of the Irish Sea. Officially opened by Queen Elizabeth II on October 17, 1956, the UK plant was up and running while the first US nuclear power plant at Shipping- port was still under construction. Little or nothing had been done about waste disposal in the United Kingdom when the Royal Commission on Environmental Pollution was created, 14 years later in 1970, to advise the Queen, the British government, and the public on environmental issues. It was a time when the environmental movement was at full steam worldwide. A few months later, the United States established the Environmental Protection Agency (EPA).

In 1976, the Royal Commission issued a report on nuclear power and the environment that attracted a great deal of attention. Although

reflecting many of the views of the nuclear industry, the report condemned the "conspicuously backward" progress on nuclear waste in the
UK. There should be no commitment to a large program of nuclear fission
power "until it has been demonstrated beyond reasonable doubt that a
method exists to ensure safe containment [of the wastes] ... for the indefinite future." This recommendation carried considerable weight in that
the commission chair, Sir (later Lord) Brian Flowers, was a distinguished
nuclear physicist. [28]

In response to the Flowers report, the government promised to
"ensure that waste management problems are dealt with before any large
nuclear programme is undertaken." The British nuclear industry fought
back. Donald Avery, deputy managing director of British Nuclear Fuels,
the primary nuclear facility operator, argued, "What we have been trying
to do is to persuade the government that Brian Flowers was wrong ... We
should not commit ourselves now to a particular course which leaves the
poor wretches fifty years from now with no option." In 1981, the search
for a geologic repository was abandoned and official government policy
became aboveground storage for a minimum of 50 years. [22]

For the next 16 years, British plans for burying nuclear waste went
in fits and starts. Sites were selected largely in secret, unveiled, and then
defended under fire. This pattern became ridiculed as *decide, announce,
defend*. In 1997, all plans for a geologic repository were abandoned and
the House of Lords declared the UK nuclear waste program had "stopped
dead in its tracks." The government spent the next decade exploring
other options. Finally, in 2006, the government returned to geologic
disposal as the only long-term solution, this time seeking a volunteer
using an open process. [29]

The situation is no further along in Germany where efforts to
address high-level waste date back to the 1960s. From the beginning, the
Germans limited their focus to salt domes as the host rock. In 1973, after
successful experiments in an underground salt laboratory, the search
began for a site to build a one-stop nuclear waste management center.
The center would store, reprocess, and dispose of nuclear waste from the
equivalent of about 45 planned large reactors. [22]

In 1976, the announcement of three potential sites in the northern plains of Lower Saxony, where most salt domes in Germany occur,
immediately triggered strong local opposition. Ernst Albrecht, the Prime
Minister of Lower Saxony, used his considerable power and rejected all
three sites. Instead, he designated a salt dome near the village of Gorleben as the sole candidate. Located along the River Elbe separating West
and East Germany, Gorleben had a small population and a depressed

economy. Albrecht's decision came more or less out of the blue – there had been no testing or surface investigations at the site. [22, 30]

From 1979 to 2000, site investigations proceeded haltingly, while fierce public opposition and protests put Gorleben on the map. Stop-work orders were repeatedly issued by the courts, and then reversed by higher courts. In 1980, in an early rendition of "Occupy Wall Street," thousands of protestors occupied a key borehole site, denying access to drilling crews. The protestors erected wooden huts and tents calling their village the Free Republic of Wendland, after the Wends, an ancient Slavic tribe of the region. Flags were hoisted and Wendland "passports" were issued as evidence of sovereignty. Three months after the occupation, a force of 3,000 police ousted the protestors and bulldozed their village. Periodic protests continued over the next two decades as the site investigations muddled along. [22, 30]

In 1998, the political environment in Germany changed dramatically with the election of a coalition government of the Social Democrats and Green Party. The government formed an independent advisory group (AkEnd) to develop an equitable site selection process that would include active public participation. The AkEnd released its recommendations in 2002 to national and international acclaim, yet key politicians and the electric utilities (who insisted on Gorleben as a viable solution) refused to back the recommendations. [30]

In 2000, exploration of the Gorleben salt dome was suspended when the new government and the energy suppliers reached an agreement to phase-out all nuclear power plants within two to three decades. Later that decade, German Chancellor Angela Merkel, a trained scientist with a Ph.D. in physics, shifted German policy back toward nuclear power. However, Merkel reversed her position soon after the 2011 Fukushima disaster – eight of the country's nuclear reactors were promptly shut down, with the remaining nine reactors to close by 2022, as part of a bold transition to wind and solar energy. [31]

Regardless of Germany's nuclear future, a repository is still needed for the country's high-level waste. Site investigations at Gorleben started back up in October 2010, but remain controversial. Meanwhile, some key players have suggested that the planned shutdown of Germany's nuclear industry may have opened up some "political space" for environmentalists and industry to finally come together to address the waste issue. Time will tell. [32]

A paralysis in decision-making about nuclear waste has not been restricted to the USA, United Kingdom, and Germany. Even Sweden and France, two countries making progress today toward a geologic

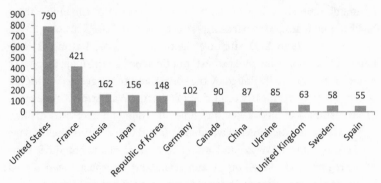

Figure 1.4 Nuclear energy production (net) in the top 12 countries
generating electricity from nuclear power in 2011. Data are in billion
kilowatt-hours from the International Atomic Energy Agency
(http://pris.iaea.org/public).

repository, were stymied for decades by political opposition. Over a
half-century since the birth of the commercial nuclear power industry,
there are some 440 nuclear power plant units in 31 countries. More are
on the way. Yet, no country on Earth has an operating high-level waste
disposal facility. [33]

2

Brainstorming

The general public can be divided into two parts: those that think science can do everything, and those who are afraid it will.

Dixy Lee Ray [1]

There is a strong worldwide consensus that disposal in a geologic repository is the only feasible, permanent solution to the high-level waste problem. The concept seems simple and comes to mind almost instinctively – bury the waste in a hole excavated in rocks deep underground. Burial in deep geological formations provides more than adequate shielding from the waste's radiation and decreases the likelihood of inadvertent or malicious intrusion by humans. Yet, this international consensus favoring geologic burial is the product of decades of brainstorming. No matter how outlandish or seemingly impossible, one by one, each conceivable solution was given its turn at the table.

In these early discussions, one idea stood apart from the rest. In April 1951, Representative Albert Gore, Sr. (D-TN), father of Nobel Peace Prize winning Al Gore, suggested using radioactive waste from the Nation's plutonium production to "dehumanize" a belt across the entire 38th parallel of the Korean peninsula. The contamination would prevent further attacks by Communist forces invading from the North. Yet, the most important result would be "its psychological effect as a mystery weapon, analogous to the initial use of poison gas and of tanks in World War I." Representative Gore argued that such widespread use of radioactive waste was morally justified as a deterrent, rather than an attack strategy. The enemy would be duly forewarned. Gore had in mind relatively short-lived isotopes. He added that the belt would need to be replenished periodically with new wastes, until a "satisfactory solution to the whole Korean problem" was reached. [2–3]

Albert Gore, Sr. was not your typical war hawk. He was a staunch liberal and an opponent of McCarthyism and later the Vietnam War. The day after Gore's suggestion was made public, Brien McMahon (D-CT), chairman of the Joint Committee on Atomic Energy, downplayed the idea as impractical. McMahon was a hawk. He considered the atomic bomb "the most important thing in history since the birth of Jesus Christ" and viewed the scientists who had created it to be "secular saints." [4–6]

The concept of dropping radioactive products over enemy territory went back almost to the beginning of the atomic age. In 1941, a National Academy of Sciences committee ranked radiological weapons number one among the three military applications proposed for atomic energy. The other two were developing a fission bomb and use of nuclear reactors for submarine and ship propulsion. [7–8]

In the late 1940s and early 1950s, the use of radiological weapons was discussed behind closed doors, as well as in popular magazines and newspapers. Major General J.L. Homer caught the attention of the *Los Angeles Times* in 1947 when he suggested that guided missiles of the future might carry radioactive waste to the enemy. Homer was envisioning what later became cruise missile technology, but with radioactive waste used in place of an atomic warhead. In 1950, the Atomic Energy Commission reported to Congress on continuing studies to develop weapons that would contain concentrated radioactive waste "great enough to kill large populations if dropped on big cities." This report influenced Albert Gore, Sr. to suggest its use in the Korean War. [9–10]

Gore's proposed use of radiological weapons became popularly known as an "atomic Maginot Line," harking back to the supposedly impenetrable line of fortifications built by France along their border to deter German aggression. The Germans easily circumvented it in World War II by attacking Belgium first. In 1954, the *New York Times* called the idea of an atomic Maginot Line one of the "dreams of the atomic age." *Colliers* magazine highlighted it as an "intriguing plan" by military men about what to do with radioactive waste. [7, 11]

By the late 1950s, using radioactive waste as weapons was seen as the mad solution it, in fact, was. The idea disappeared from public discourse, as the world turned to more peaceful approaches to dealing with radioactive waste. No place was off limits. Shoot it into space. Sink it into polar ice caps. Bury it beneath a remote island. All were considered.

Launching the waste into space could be viewed as the ultimate in permanent solutions. The idea surfaced as early as 1954 when Professor Ira Freeman, a physicist at Rutgers University, suggested concentrating radioactive wastes and loading them on expendable "tanker rockets," which would carry them to Mars or Venus. If there were objections to this deep space solution, Professor Freeman saw no reason why we should not "throw the wastes overboard somewhere between the Earth and the moon, whereupon they would revolve around the Earth like satellites." [12]

During the 1970s, the newly created Department of Energy (DOE) and NASA jointly studied several options for extraterrestrial disposal of nuclear waste. These included sending the wastes to the moon, to the Sun, or into orbit midway between Earth and Venus. Given the enormous expense of this approach (nuclear waste is heavy), the idea was pursued only for disposal of some reprocessed wastes. Unreprocessed spent fuel from nuclear reactors would remain earthbound.

The idea was complicated. A space shuttle would be launched into low Earth orbit. During the early stage, the orbiter would be able to abort and return to the launch site or ditch into the ocean. The nuclear "payload" also could be ejected and make its own return to Earth with a re-entry vehicle. Once in orbit around Earth, the waste would be transferred to an orbit transfer vehicle which would shoot it into space. After release of the nuclear waste, the transfer vehicle would rendezvous with the space shuttle and return to the launch site for refurbishment and use on a later flight. According to DOE, this complex space disposal could be operational by the year 2000. Never mind that the first space shuttle had yet to be launched. In an obvious understatement, the government noted, "While the space option appears technically feasible, there are engineering problems that would require resolution." [13]

It did not take long for this idea to come down to earth. Potential failure of a space launch and the ensuing catastrophic explosion of the waste packages was an obvious concern. Fool-proof packaging would be required. This is not what Floridians had in mind when they welcomed the Kennedy Space Center to Florida. In 1982, the Department of Energy and NASA discontinued further study of the space disposal concept. Four years later, the space shuttle Challenger explosion clearly illustrated the dangers of the space disposal concept in a hard-learned lesson in technical overconfidence.

Ironically, permanent space disposal of nuclear waste has preceded permanent disposal of similar wastes on Earth. In dark parts of space where solar power is not practicable, special generators convert the heat from radioactive decay of plutonium-238 (a much shorter-lived isotope of plutonium than that used for nuclear weapons) to small amounts of energy. The electricity from plutonium-238 allowed cameras on the spacecraft Galileo to capture images of the ice-covered surface of Europa, one of Jupiter's moons. Europa is possibly the most likely place in the solar system to find life, so there was considerable interest in taking a peek. As planned, the Galileo spacecraft plunged into Jupiter to end its 14-year voyage. Likewise, the Mars rovers, Spirit and Opportunity, have tiny nuclear-powered heaters to keep their axles and instruments warm through the frigid Martian nights. Each heater contains about two grams of plutonium. The rovers and their heaters performed admirably. Their final resting place is on Mars. [14]

In recent years, another space-dumping possibility has been suggested. Ground-based laser systems would fire "bullets" containing radioactive waste into space. Writing about the possibilities in *The Space Review*, Jonathan Coopersmith suggests that geologists are "professionally inclined to look down, not up. That's shortsighted," he laments. [15]

For those with a more down-to-earth perspective, there were still six other alternatives to a geologic repository. Perhaps the most outrageous idea was *ice-sheet disposal* in Antarctica or the Greenland ice cap. The political impediments to disposal, in either place, were formidable. The Antarctica Treaty of 1959 explicitly prohibits disposal of nuclear waste on the continent of Antarctica. Greenland is a Danish territory. Nonetheless, the idea was briefly pursued.

This time the concept was simple. A container with radioactive waste would be placed in a shallow hole or on the surface of the ice. The heat from radioactive decay would cause the container to melt its way to the bottom of the ice sheet. A variation on this theme would attach an anchor cable to the canister, limiting its descent and permitting retrieval.

There were several arguments for ice-sheet disposal. Large ice sheets can be very thick; in some places they are several miles deep. Ice tends to behave as a plastic and would "flow" to seal fractures and close openings. Antarctica and Greenland also have few inhabitants to protest about disposal in their backyards. On the other hand, the technical impediments to ice-sheet disposal are huge. High costs and safety concerns with transporting the wastes to these remote and forbidding environments would be significant challenges. Future effects of climate change and inadequate understanding of ice dynamics have also entered the debate. Given current concerns about climate change, the idea of ice-sheet disposal would be considered preposterous today.

A third alternative, *deep well injection*, works like an oil or water well, except in reverse. Instead of pumping fluids out of the ground, they are injected into the Earth, usually under high pressure. The technique was developed by the oil industry as a way to dispose of unwanted brine (salty water) that is pumped to the surface along with the oil. The brines are pumped back into the petroleum reservoir, often helping drive more oil toward the producing well. Well injection technology soon expanded outside the oil patch as an approach for disposal of industrial chemical wastes. In recent years, deep well injection has seen a resurgence of interest as part of plans for carbon sequestration to reduce greenhouse emissions. The idea is to capture carbon dioxide from coal-fired power plants, cement plants, and other large point sources of carbon dioxide and inject (sequester) it deep below the Earth's surface.

Over the years, experience with deep well injection has proven that isolation of wastes cannot be assured with absolute certainty. Wastes can migrate back toward the surface through unforeseen geologic disconti-nuities or faults, or if not properly sealed, back up through the well bore itself. If the surrounding area has been drilled before, abandoned and forgotten wells can provide a fast track back to the surface.

Deep well injection of radioactive wastes was used until the 1980s at Oak Ridge National Laboratory in Tennessee, but the lion's share took place in the former Soviet Union. The practice remained a State secret until 1994. By then, half of all Russian nuclear waste had been injected into the ground near three major nuclear facilities. [16–18]

Deep well injection became the Soviet method of choice after suf-fering serious setbacks with the use of surface storage. In 1957, the explo-sion of a large concrete storage tank containing high-level nuclear waste at Kyshtym, near the Ural Mountains, contaminated over 5,000 square miles (13,000 square kilometers), caused evacuation of more than 10,000 inhabitants, and killed an undisclosed number of people. Denied for decades, the Soviet government finally acknowledged the disaster in 1989. The long denial of this catastrophe was in stark contrast to the fan-fare surrounding the launch of Sputnik, just five days after the Kyshtym explosion. [19]

A major drawback to deep well injection is that it requires liquid wastes. Because liquids are much more mobile than solids, there is now almost universal agreement that radioactive waste must be solidified before disposal. This leads to a fourth option – dispose of the wastes in a *very deep hole*, far enough beneath the land surface to preclude *any* possible opportunity for re-emergence. The concept sounds simple. Drill a very deep hole, drop the wastes down it, fill it back in, and forget about it. But how deep is deep enough? Suggestions ranged from a few miles to six miles (10 km).

Aside from drilling a hole in the Earth about the same distance as commercial jets fly above the Earth, the logistics were daunting. A couple of thousand holes might be needed to accommodate all the waste, with each hole possibly requiring several years to drill. The drilling operation would have to withstand intense subterranean pressures and temperatures, as well as the hostile chemical environment deep in the earth. Once in place, retrieval of the wastes would be all but impossible. There were too many unknowns and a staggering price tag. The idea was dropped, until recently.

Some scientists have advocated resurrecting the very deep borehole concept as a result of advances in deep drilling technology. Scientists at Sandia National Laboratories have proposed that the Nation's current inventory of spent nuclear fuel could be disposed of in a few hundred boreholes drilled three miles (5 km) into granite or other crystalline basement rocks. Because crystalline rocks are relatively common at this depth, regional disposal facilities might be constructed to share the burden and minimize transportation distances. A substantial research and pilot program would be needed to test the approach. The concept might be most useful for disposal of reprocessing and plutonium wastes that presumably no one ever wants back (see Table 2.1 for definitions of different types of waste). [20–21]

A fifth alternative to a mined geologic repository was *rock melting*, also known as DUMP, for deep underground melt process. The rock-melting concept would be implemented in two steps. First, hot radioactive wastes, along with water to cool the wastes, would be placed in deep underground cavities. Once a cavity was filled, the water would be boiled off. In the absence of water, the cavity temperature would rise rapidly, melting the surrounding rock and trapping the radioactive material deep underground. The original proposal, by Lawrence Livermore Laboratory in California, involved setting off a nuclear bomb to create the underground cavity. The rock-melting proposal was riddled with problems, and it was soon abandoned. [22–23]

The final two alternatives to a geologic repository turned toward the sea. One of these, *disposal beneath an uninhabited island*, is basically a geologic repository with the addition of an over-water transportation route. Greater uncertainty comes as part of the bargain. Many islands are near tectonic plate boundaries and vulnerable to seismic activity. Possibilities for tsunamis were not considered but are, of course, self-evident now. Rising sea levels caused by climate change is a further complication. Disposal beneath an uninhabited island seemingly avoids protests of *not in my backyard*; however, residents of any nearby islands, or those near shipping routes, might see it differently. In the 1990s, a group of private investors unsuccessfully sought congressional approval

Table 2.1 *Categories of radioactive waste.*

Category	Description
High-level radioactive waste	Commonly referred to simply as *high-level waste*. In its most general sense, high-level waste includes highly radioactive waste from nuclear fuel reprocessing and weapons production along with spent fuel from nuclear reactors. There is debate among many people involved in the nuclear industry about including spent fuel in the high-level waste category. To some, spent fuel is a future resource that can be recycled (reprocessed), and therefore, is not a "waste." In this book, we use the term high-level waste in its most general sense to include spent fuel as well as reprocessing waste. We use more specific terms, such as spent fuel and reprocessing waste, as needed to distinguish particular high-level waste types.
Intermediate-level waste	This term, rarely used in the United States, has various definitions around the world to describe radioactive waste that is not quite as dangerous as high-level waste. We only use this term to maintain accuracy in citing sources.
Mill tailings	This term refers to slightly radioactive materials left over after uranium extraction from ore.
Transuranic waste	These wastes include materials contaminated with enough plutonium and other heavy radioactive elements (i.e. beyond uranium) to require some form of long-term isolation, but which do not have the radioactivity and heat output of high-level waste. Transuranic waste comes mostly from reprocessing and the use of plutonium in making nuclear weapons.
Low-level waste	This term is a catch-all category for almost everything else. Low-level waste can be solid or liquid. The radioactivity of low-level waste is generally low enough that the materials can be handled when in containers. Examples are filters, contaminated clothing and rags from nuclear power plants, and radioactive waste from medical and research laboratories. Most of the radioactive elements in low-level waste are relatively short-lived and the waste has low radioactivity. However, the waste may contain some small amounts of longer-lived radionuclides.

to build a repository for US and Russian nuclear waste on Wake Island, a US military base in the South Pacific. [24]

The United Kingdom has been particularly interested in disposal beneath uninhabited islands. In 1976, a report by the Royal Commission on Environmental Pollution noted that a disposal facility on a small uninhabited island would be particularly advantageous as "any leakage of radioactivity into the island's ground water would be easily detected and . . . dilution of seawater would provide a further line of defence." The British Geological Survey postulated that more than 100 small offshore islands might be suitable for nuclear waste disposal. Two tiny, uninhabited Scottish islands appeared on a list of 12 potential sites drawn up in the late 1980s. [25–27]

Vast stretches of the seafloor also received considerable attention. This approach, known as *subseabed disposal*, is addressed in the next chapter as part of an extended discussion of using the oceans for all types of radioactive waste disposal.

3

The ocean as a dumping ground

Roll on, thou deep and dark blue Ocean – roll!
Ten thousand fleets sweep over thee in vain;
Man marks the earth with ruin – his control
Stops with the shore

Lord Byron [1]

In the early days of atomic energy, the oceans were viewed as a convenient place for getting rid of all kinds of radioactive waste. The volume of water is huge and dilution would quickly reduce concentrations to minuscule levels. There was a catch, though. Toxic materials can be ingested by microorganisms and concentrated as they are passed up the food chain to higher organisms, including human beings. Moreover, ocean circulation can carry waste over large distances in short times.

Any doubts about the potential for long-distance transport of waste, particularly in the ocean's surface, should be dispelled by the studies of Dr. Charles Ebbesmeyer, a retired oceanographer living in Seattle. Ebbesmeyer originally monitored ocean currents by tracking buoys and markers dropped at sea. When his mother heard about 80,000 Nike shoes floating at sea in 1990, she brought it to her son's attention. Thus began Ebbesmeyer's second career, tracking objects accidentally spilled at sea using a worldwide network of enthusiastic beachcombers. In 1992, Ebbesmeyer and his network began to track 29,000 bathtub toys washed overboard from a container ship during a storm in the eastern Pacific. Adrift on the open sea, many of the castaways floated south where some beached on the far-flung coasts of Australia, Indonesia, and western South America. An armada of about 10,000 rubber ducks, beavers, turtles, and frogs headed north toward Canada and Alaska, propelled by wind and sea. Some circled back to Japan. Others got caught in the Arctic and hitched their way eastward with the floating ice. By 2000, some of those caught in the Arctic ice had reached the North Atlantic, where they were freed from the thawing ice. In 2003, after an epic 11-year journey halfway around the world, a rubber duck was found in Maine and a rubber frog in Scotland. [2–3]

Evidence for ocean dispersal is not limited to bathtub toys. Two days after the 1954 nuclear tests at Bikini Atoll in the Pacific, the radioactivity of the surface waters near Bikini was observed to be one million times greater than normal. Four months later, after transport and dilution by ocean currents, concentrations three times the natural radiation level were found 1,500 miles (2,400 km) from the test area. A year after the blasts, the contaminated water mass had spread over 1 million square miles (2.6 million square kilometers). The ocean's potential for long-distance transport of radionuclides was clearly evident. [4]

Concern about the effects of widespread dispersal of atomic radiation on humans and other living organisms quickly grew among scientists in the 1950s, and spurred the US National Academy of Sciences to form six committees to assess the Biological Effects of Atomic Radiation (known as the BEAR committees). The BEAR committee on oceanography and fisheries was chaired by Roger Revelle of the Scripps Institution of Oceanography. Previously, Revelle had organized the scientific program to study radionuclide movement and its effects on marine life after the first postwar atomic test on Bikini Atoll.

Dr. Roger Randall Dougan Revelle was one of the most influential environmental scientists of the twentieth century. His interests and intellectual reach spanned the physical, biological, and social sciences. Revelle would serve as science advisor to Interior Secretary Stewart Udall during the Kennedy Administration, and as president of the American Association for the Advancement of Science in 1974. In a seminal scientific paper in 1957, Revelle and fellow Scripps scientist Hans Suess rekindled debates about greenhouse gas emissions, when they debunked a long-standing counter argument that the immense mass of the oceans would quickly absorb excess carbon dioxide (CO_2) from human activities. According to Revelle and Suess, humanity was carrying out a large and unprecedented global experiment by rapidly returning to the world's biosphere vast amounts of carbon that had taken hundreds of millions of years to accumulate in sedimentary rocks. Revelle later introduced a young college student named Al Gore to the dangers of global warming, when Gore took Revelle's course on climate science. [5]

Revelle knew that it was essential to establish precise measurements of the CO_2 content of the atmosphere to gain insights into the processes affecting climate. He assigned a postdoctoral student, Charles Keeling, to develop the approach. Keeling reasoned that continuous

Figure 3.1 Roger Revelle in his laboratory. Courtesy Scripps Institution of Oceanography Library Archives.

monitoring of CO_2 at the newly established observatory at Mauna Loa, Hawaii, far from any large human source, might provide a good representation of temporal trends in carbon dioxide in the Earth's atmosphere. Revelle was skeptical that an instrument in one location could provide a meaningful global record. He recommended that priority be given instead to a one-time global survey of carbon dioxide, to be repeated a decade or so later. Keeling persisted. The Mauna Loa record, now known as the *Keeling curve*, is the world's longest continuous record of atmospheric concentrations of CO_2, celebrating its 50th anniversary in 2008. The record provides the most compelling evidence that the concentration of carbon dioxide in the world's atmosphere has been steadily rising. Revelle later became one of the Mauna Loa record's strongest champions. [6]

The BEAR committee on oceanography and fisheries, which Revelle chaired, concluded that there is no location where very large amounts of radioactive materials can be introduced into the *surface* waters of the ocean, without the possibility of their eventual appearance endangering human activities. The committee emphasized the vulnerability of coastal waters and the upper layers of the ocean that are home to commercially important fish. They warned that, in spite of the vast area and volume of the world's oceans, the lessons of human and industrial waste disposal in nineteenth-century cities should be remembered. During the early stages of urban growth, these wastes were dumped in nearby lakes and rivers and released into the air in what seemed innocuous quantities. As cities grew, it wasn't long before toxic wastes overwhelmed the natural cleansing processes, creating serious pollution and disease problems. [4]

While the BEAR committee report gave a clear message about potential problems with ocean disposal of radioactive waste, it did not oppose the practice. The committee emphasized the need for further study, to "find places in the ocean where the rate of transfer of radioactive materials to the surface waters would be slow, or where great dilution would occur before radioactive materials came in contact with marine food products or human beings." Hopefully, there were places where both conditions could be met. In addition to the dangers of ocean-dispersed radioactivity, the committee was intrigued by the scientific possibilities of using radioactive isotopes, introduced from atmospheric weapons testing and waste disposal, as tracers to study ocean circulation and track the flow of materials through food chains. [4]

Like many others at the time, the BEAR committee assumed that nuclear power was the way of the future. Revelle, in particular, was caught up in the optimism for peaceful uses of the atom. He argued that humans would soon deplete the world's mineral deposits and other natural resources. Nuclear energy would be needed to create new substances and provide the vast power needed for future generations. Otherwise, Revelle wondered, "Will our children's children look forward only to a slow decline into misery and fear?" [7]

In *Poison in the Well*, a history of radioactive waste disposal in the sea, Jacob Hamblin argues that the conclusions of the BEAR committee largely reflected the tendencies of Revelle and other oceanographers to promote research interests. Nevertheless, the scientists had many unresolved questions. Little was known about how upward circulation might affect containment of radioactive wastes dropped to the ocean floor, as well as the effects of biological uptake of radionuclides and transfer through the food chain. Even in the absence of physical circulation, would vertical migration of organisms transfer dangerous material from the ocean floor to shallow waters? Was there a safety threshold for the amount of waste that could be put into the oceans, or was the threshold concept an illusion? No one really knew. Allyn Vine, of Woods Hole Oceanographic Institution and a member of the BEAR committee, summarized the degree of uncertainty as "somewhere between nonexistent and insolvable." [7]

A better understanding of the world's ocean currents unfolded as part of the International Geophysical Year of 1957–8, with more than 60 Nations participating. For the first time in history, scientists from around the world took part in a series of coordinated observations of the Earth, sea, and atmosphere. Many scientific discoveries followed. Major discoveries included the Van Allen radiation belts surrounding Earth (a complicating factor for future space travelers), and the mid-oceanic ridges deep beneath the sea (a precursor to understanding plate tectonics). French and Japanese scientists probed the ocean depths with Jules Verne like bathyscaphs. They found deep-sea currents and living organisms. These discoveries called into question whether dead zones for long-term isolation of high-level wastes could be found even in the deep ocean. The Soviets excelled in exploring both the sky and ocean. They commemorated the International Geophysical Year by launching Sputnik, the world's first artificial satellite – an event that shocked the free world. Using echo sounders, they also found the deepest part of the ocean in the Mariana Trench. [8]

The Mariana Trench, lying almost seven miles below the ocean's surface, is an alien and forbidding world of cold, total darkness, and immense pressure of almost 17,000 pounds per square inch. In January 1960, Jacques Piccard, the Swiss deep-sea explorer, and his associate, US Navy Lieutenant Donald Walsh, piloted the bathyscaph Trieste to the deepest hole in the trench. To their surprise, they found white flat fish and small red shrimp where no life was expected to exist. Oxygen and currents, it turned out, also exist in the deepest part of the ocean. Prior to the voyage of the Trieste, it was believed that the water in the trench had been essentially static for a million years or more. According to Piccard, the chief significance of his findings was that "radioactive material dumped there would eventually find its way back to the surface." Their discovery was a further nail in the coffin for deep-sea disposal of high-level radioactive waste. [9]

In the previous year, 1959, ocean disposal was a centerpiece of discussions at the first international scientific conference on radioactive waste disposal held in Monaco. Less than a square mile in size, Monaco has neither nuclear power plants nor room for waste disposal sites. But Prince Rainier III, perhaps best known for his marriage to Hollywood actress Grace Kelly, came from a royal family long interested in the marine environment. The Prince hosted 280 experts from around the world to discuss radioactive waste disposal on land and sea. The conference was sponsored by the newly formed International Atomic Energy Agency.

The Monaco meeting opened with a heady speech by Sterling Cole, the first Director General of the International Atomic Energy Agency. Cole was a leading spokesman for commercial nuclear development. He had served for more than two decades as a Republican congressman from upstate New York, and as chairman of the Joint Committee on Atomic Energy from 1953–4. In his presentation, Cole emphasized that the problem of radioactive waste was actually one of temporary storage for later use, rather than disposal. Even if there was no use for this material now, this did not mean a use would not be discovered in the future. In spite of Cole's assurances, the participants discussed at length the radioactive waste disposal problem. Oceanographers emerged from the Monaco conference with a consensus that scientific knowledge of the oceans did not support *large-scale dumping* of radioactive waste anywhere in the open seas. More research was needed. [10]

While scientists at the conference were saying one thing, the atomic energy establishments of Britain, France, and the United States were finding common ground in their belief that they already knew enough to make estimates of what could be dumped safely in the sea. The gulf between marine scientists and the nuclear industry had widened. [7]

In spite of the BEAR committee report, the Monaco conference, and the Trieste discoveries, dumping radioactive waste in the world's oceans continued for decades. Some of this disposal took place legally, some when no one was looking, and some by accident. With a few exceptions, the disposal was mostly low-level wastes.

RADIOACTIVE WASTE DISPOSAL AT SEA

From 1946 to 1970, the United States dumped about 87,000 containers in offshore waters under license by the Atomic Energy Commission (AEC). These were mostly low-level radioactive wastes packed in steel drums. The first wastes were dropped about 50 miles (80 km) off the coast of California. Most were dumped in the Atlantic Ocean. By the late 1950s, US disposal of radioactive waste had become fairly routine along the Atlantic coast. For example, in September 1957 the *New York Times* featured a story about a 65-foot vessel, the *Irene May*. The vessel, operating out of Boston Harbor, had recently made her 500th voyage with "hot cargo." Many European countries likewise dumped in deep offshore waters. All told, 14 countries disposed of radioactive wastes at sea. [11–13]

Initially, nuclear waste was dumped at sea without public controversy. By the late 1950s and early 1960s, rising fears about atmospheric fallout and greater recognition of the need to preserve the resources of the sea began to attract the public's attention to ocean dumping. Jacob Hamblin notes, "in a few short years, radioactive waste disposal at sea became a rallying point for an assortment of strange bedfellows: marine scientists, antinuclear activists, politicians, and a vicious propaganda campaign from the Soviet Union." At the Monaco meeting and every opportunity thereafter, the Soviets took Western countries to task for dumping radioactive waste indiscriminately into a shared resource. [7]

Operations at sea did not always run smoothly. Some packages were dropped outside the designated sites. Some floated rather than sank. Others reappeared in undesirable places. In 1957, the US Coast Guard sighted a giant canister containing 20 tons of low-grade radioactive sodium floating in the Atlantic Ocean, about 200 miles (320 km) southeast of New York City. The drum, along with 24 others like it, had been rigged with explosives. The fuse on the runaway drum had failed to ignite. The AEC claimed the radioactivity posed no risk, but the canister was a navigational hazard due to its size. There was also the possibility of explosion if the sodium came into contact with water. The drum was subsequently lost in the darkness. The next day, Navy planes sighted the canister and sank it after a day-long strafing mission. [14]

The incident made front-page headlines and motivated *New York Times* reporter James Reston to draw attention to the pervasiveness of all types of radioactive waste. He observed that in addition to government agencies, there were more than 4,000 separate establishments throughout the country using radioisotopes. Reston expressed the general view that this need not hamper rapid development of peace-time atomic energy, but it did require extra vigilance and cooperation among communities and governments. [15]

A growing concern with ocean dumping was that the waste packages might end up on beaches or in fishermen's nets. In 1960, a 30-gallon (110 liter) drum of radioactive waste was netted by fishermen 12 miles (19 km) off the coast of Massachusetts. In June 1962, the French announced that fishing trawlers had twice discovered drums of radioactive waste in their nets, marked as property of the UK Atomic Energy

Authority. In 1964, drums of British waste again showed up in the nets of French fishing boats trawling off the coast of Brittany. [7]

According to AEC guidelines, radioactive waste should be dumped no closer than 100 miles (160 km) from shore and in waters at least 1,000 fathoms (6,000 feet) deep. In actual practice, however, the AEC had been allowing waste disposal as close as 15 miles offshore and in waters as shallow as 300 feet (90 m). To justify these practices, and to address requests by commercial firms for permission to dump at sites more conveniently located near ports, in January 1958, the AEC sought advice from the National Academy of Sciences Committee on Oceanography. After deliberations, the committee identified 28 sites that might be suitable.

The sites "lie off every major seaport region from Boston to Corpus Christi, Tex." announced the *New York Times* in June 1959, which also published a map. Politicians in coastal States from Massachusetts to Texas were soon inflamed. A storm of media attention ensued when it came out in congressional hearings that the AEC had already allowed disposal of radioactive waste in near-shore, shallow areas. Protests from the California legislature about dumping waste in the Pacific, and from Mexico about a proposed site in the Gulf of Mexico, added to the controversies. An AEC spokesman later indicated that the furor "in effect stopped all ocean dumping for all practical purposes." The United States also lost interest in ocean dumping of low-level radioactive wastes when the nuclear industry decided that disposal on land was much cheaper. By 1962, more than 95 percent of low-level radioactive wastes were being buried on land. [7, 16–17]

Ocean dumping continued in Europe, as did the controversies. In 1960, the French atomic energy agency decided to "experiment" by dumping about 2,000 tons of low-level liquid and solid radioactive waste packed in drums into the Mediterranean Sea, about 50 miles (80 km) off the coast between Nice, France and Corsica. The site seemed ideal. It was in relatively deep water, in an area with minimal surface currents, and not near any known fishing waters. The French atomic energy agency claimed that even in the highly unlikely event that all of the drums burst, the surrounding water would dilute the material so much that the danger to human health would be "completely negligible." [7]

When French newspapers picked up on the story, oceanographers in France and Monaco harshly criticized the proposal. Prince Rainier intervened as did Jacques-Yves Cousteau, the legendary underwater photographer. City councils and mayors of towns all along the French Riviera joined the fray. The conflict soon became worldwide news. Cousteau, then Director of the Musée Océanographique de Monaco, was dubbed by the *New York Times* as the "unofficial leader of the anti-dumping campaign." The incident contributed to Cousteau's rise to worldwide fame. The French atomic energy agency backed down. [18]

The ocean is a particularly convenient place to dispose of old nuclear submarines and their reactors. In 1980, the US Navy announced that it was considering disposal of decommissioned nuclear submarines by scuttling them in the deep ocean. According to the Navy, preliminary information indicated that the submarines and their reactors would remain intact after landing on the ocean bottom. The Navy argued that a submarine hull was far more durable as a waste container than the steel drums used by European nations in their continued ocean dumping of radioactive waste. [11]

This would not be the first disposal of a nuclear submarine, or its reactor, at sea. In 1959, with no public announcement, a barge bearing the nuclear reactor from the submarine Seawolf was towed 120 miles (200 km) off the shore of Delaware. The reactor was unceremoniously dropped in an area commonly used for radioactive waste disposal. When later revealed, the Navy insisted that the Seawolf reactor should decay rapidly within decades, and much of the radioactivity would be contained as an "integral part" of the corrosion-resistant reactor vessel. The reactor's predominant radionuclide is cobalt-60, which has about a 5-year half-life. Nonetheless, purposefully scuttling nuclear submarines and their reactors at sea was soon rejected by the United States. [11]

In the late 1970s, Japan was greatly interested in the possibility of using the deep ocean for disposal of low-level radioactive waste. A site in the deep ocean, just over 500 miles (800 km) southeast of Tokyo Bay, was selected for experimentation. If successful, 60 percent of Japan's low-level radioactive waste would be dumped at this site. The plans were dropped after fierce opposition by South Pacific island nations, whose early traumatic experiences with the nuclear age had caused them to band together. The South Pacific island nations had good reasons to be distrustful. Inhabitants of the Rongelap and Utirik atolls had not been evacuated before the Bravo thermonuclear test in 1954, resulting in heavy radiation exposure from fallout. In another blunder, the people of Bikini, who had been evacuated in 1946 when the first atomic weapons test was conducted, were told in 1968 it was safe to return home. Ten years later, they were evacuated once again when their local foods were found to be contaminated by radionuclides. [19]

Britain was long considered to be the primary contributor of radioactive waste to the sea. Donald Avery, a key British nuclear industry spokesman, noted in 1980, "One of the most important things for us in this country is to keep open, and if possible increase, the use of the deep oceans for dumping low- and intermediate-level wastes . . . When you've got a small, highly populated island, you don't have to go through much science to come to that conclusion." Perhaps, the largest *single* source of radioactive pollution to the world's seas and oceans has come from British nuclear facilities at Sellafield, on the west coast of England. After the United States refused to share its atomic weapons technology with Britain after World War II, the British established two massive reactors at Sellafield in order to produce plutonium for a "British Bomb." Sellafield is also the location of Calder Hall, the world's first commercial nuclear power station and other nuclear reprocessing facilities. [19]

Since 1952, liquid radioactive waste has been discharged from Sellafield into the Irish Sea. The discharges which (at least officially) have rarely exceeded regulatory limits, have nonetheless led to considerable public concern. The British House of Commons once referred to the Irish Sea as "the most radioactive sea in the world." Ireland, which shares in the environmental risks but receives none of the benefits, has understandably been a long-time critic of the Sellafield practices. Findings of radioactive cesium and technetium in fish and lobster in the Irish and North Seas have unleashed concerns by the area's fishing industry about the safety, not to mention *reputation*, of their product. [20]

The most embarrassing episode of discharges to the sea at Sellafield took place in November 1983. Greenpeace divers, who were surreptitiously attempting to plug the underwater discharge pipes from Sellafield, suddenly found their Geiger counters "going crazy," while big globs of "oil" spewed from the pipes. Contaminated flotsam washed up along 20 miles (30 km) of beach. Exposure of the incident by Greenpeace became a public relations nightmare for the Sellafield operator. Both the plant operator and Greenpeace were fined, although the incident almost assuredly brought Greenpeace more than enough contributions to pay off their fine. In response to continuing public concerns and controversies, Sellafield has substantially ratcheted down the allowable discharge limits over time. [19]

In 1989, the Soviet government declared that it "did not dump, does not dump, nor plans to dump radioactive waste at sea." In truth, the Soviet Union had relied on ocean dumping for decades. In the early 1990s, Russian President Boris Yeltsin decided to come clean about these practices. As part of Russian–American negotiations, Russian authorities admitted to dumping large amounts of radioactive waste into the ocean. In addition to liquid and packaged wastes, the Soviets had dumped 16 nuclear reactors (several with their radioactive fuel) from submarines and an icebreaker into the Arctic Ocean, in waters less than 300 feet (90 m) deep. The admission of these practices revealed not only the previous lies, but also massive hypocrisy. The Soviets had long used the issue of radioactive waste disposal at sea in their propaganda war against the West. They repeatedly spoke of the West as "poisoners of wells," as though the sea was a folksy village well. [7, 21]

In spite of their earlier self-righteous proclamations, it turned out that the Soviet Union and subsequent Russian Federation had dumped twice as much radioactivity into the ocean as all other countries combined. The major culprit was the Russian nuclear submarine fleet and merchant icebreakers. Plans for building facilities to handle radioactive waste generated by the fleet had been halted in 1972. Another plan to build special storage facilities for nuclear submarine reactors was never implemented. Waste facilities on land were overfilled. Money was in short supply. Dumping the waste at sea just seemed a whole lot cheaper and easier. Moreover, the collapse of the Soviet Union left numerous aging nuclear submarines rusting away in harbor. The situation was a mess. In 1996, the local power station twice cut power to the submarine fleet for not paying its bills, even though docked submarines need power to cool their reactors. To pay their utility and other bills, one of the submarines transported potatoes in its missile tubes. [22–24]

The Russians sought and received help from the West to address these issues, as part of a global partnership to reduce proliferation and environmental threats from the Soviet Cold War legacy. However, the process has been complicated by reluctance to allow their civilians or any outsiders near the military facilities.

A bizarre aftermath of the Russian nuclear submarine story involved a retired naval captain. Aleksandr Nikitin was arrested by the Russian Security Service (the successor of the KGB) in 1996. Nikitin had co-authored a report with Bellona, a Norwegian environmental group, documenting the hazards posed by the Russian Northern Fleet's nuclear waste dumping. In spite of the fact that the information in his report came entirely from public sources, Nikitin was placed in solitary confinement and charged with treason for releasing state secrets – punishable by death. Under pressure from the West, the charges were temporarily reduced to misusing his identity card while gathering the information, but were later reinstated. Finally, in 2000, Aleksandr Nikitin became the first person in history charged with high treason by the KGB and its successor to be fully acquitted. [25–26]

The short- and long-term environmental effects of dumping radioactive wastes into the ocean are unknown. Monitoring the radioactivity and its effects is extremely difficult, even in situations where the dumping was carefully documented. Monitoring is impossible for wastes disposed of at undisclosed locations. The incidences that make the headlines are not always the largest sources. For example, dumping from ships has received considerable attention, yet accounts for only a tiny share of the total amount of radioactivity put into the ocean. In the early 1960s, British health officials pointed out that dumping from ships in the first decade or so of Britain's operations was equivalent to about a week's discharge from the Sellafield pipeline. [7]

A more recent episode further illustrates the public's sensitivity to any sort of radioactive discharge at sea. In July 2008, a valve leak aboard the USS Houston, a nuclear-powered attack submarine, went undetected for months, leaking while the ship was in port in Guam, Hawaii, and Japan. The total amount of radioactivity released was less than that contained in a common bag of lawn and garden fertilizer. Nonetheless, it created a brief diplomatic stir. [27]

Regardless of the history of ocean dumping, the international community has worked long and hard to first regulate, and then ban, the practice of dumping radioactive waste at sea. In 1958, the United Nations Conference on the Law of the Sea concluded that "every State shall take measures to prevent pollution of the sea from dumping of radioactive wastes," but provided little further guidance. The scientific conference held in Monaco the following year had been designed to help countries conform to this agreement. [12]

The London Dumping Convention of 1972 took things a step further. It was at the height of the environmental movement and the United Nations Conference on the Human Environment had just been held in Stockholm. The purpose of the London Dumping Convention was to build upon the Stockholm conference, with a focus on marine pollution of all types. For regulatory purposes, the participants drew up "black" and "grey" lists. Disposal of substances on the "black" list was prohibited except in trace quantities. Substances on the "grey" list could be dumped, but only under the provisions of a special permit. High-level radioactive wastes were on the "black" list. Low- and intermediate-level radioactive wastes made it to the "grey" list. [12]

In 1972, shortly before the presidential election, Richard Nixon signed into law a prohibition on ocean dumping of high-level radioactive wastes (as well as materials for chemical, biological, and radiological weapons). This was largely a symbolic gesture, as these decisions had been made many years before. International agreements prohibited disposal of any radioactive waste in the Baltic Sea in 1974 and the Mediterranean Sea in 1976. In 1983, a nonbinding moratorium was placed on dumping any form of radioactive waste in the ocean by signatories of the London Dumping Convention. Ironically, Britain opposed the voluntary ban, but ceased dumping nuclear waste from ships into ocean waters after members of the National Union of Seamen refused to handle the wastes. As the London Dumping Convention only applied to disposal at sea, radioactive wastes continued to be piped into the ocean at facilities like Sellafield. [7, 12]

In 1993, 37 nations voted to make the ban on ocean dumping permanent and legally binding. Several key countries like Britain, France, and Russia abstained, although each later accepted the decision. Japan at first refused to sign-on to the ban. Its position quickly changed when public outcry arose from a Greenpeace documentary. The film was released just before the meeting, and showed Russians illicitly dumping liquid radioactive wastes into the Sea of Japan. Although progress has been slow and enforcement difficult, using the ocean for dumping radioactive wastes is no longer tolerated by the world community. [28]

SUBSEABED DISPOSAL

During the 1970s and 1980s, another idea for how to dispose of radioactive waste gained momentum – burial beneath the ocean floor. The proposal, which focused on high-level waste, became known as *subseabed disposal*. In this scenario, the waste would not be tossed overboard but, rather, buried in ocean floor sediments. Although this idea has large public perception and international barriers to overcome, for many years it was considered to be the primary alternative to a land-based geologic repository for high-level waste. [29]

For those familiar with plate tectonics, the first place beneath the seabed that might come to mind might be in a deep oceanic trench along a subduction zone. Presumably, plate motions at subduction zones would carry the wastes down into the Earth's mantle. This would remove them not only from the biosphere but from the entire Earth's crust. Such a solution seems as permanent as shooting the wastes into the Sun. However, subduction zones are some of the most geologically unpredictable places on Earth. There is abundant evidence to show that not all the sediment in a subduction zone is carried downward. Some is pushed up against the adjacent continent or island arc (like Japan), potentially transporting some of the wastes to shallow water or land. [30]

Figure 3.2 Charles Davis Hollister. Courtesy Woods Hole Oceanographic Institution Archives.

In contrast, vast regions lying within the interior parts of the oceans are probably the most geologically stable environments anywhere on the Earth's surface. These areas are remote from human activities, have few resources known to humans, are relatively biologically unproductive, and have weak bottom currents. These vast "abyssal plains" are blanketed by clays hundreds of feet thick, created over the eons by the slow deposition of particles blown far out to sea from land masses. The plates move slowly, at a rate of a few inches per year. A site located hundreds of miles or more from a plate margin would not reach the edge of the plate for at least a few million years.

The idea of burying high-level waste below the abyssal clays was pioneered by Charles Hollister, a senior scientist of marine geology at Woods Hole Oceanographic Institution. Hollister was affectionately known to friends and colleagues as "the cowboy oceanographer," a sobriquet that fit him well. Having grown up on his family's cattle ranch in Santa Barbara, California – one of the largest in the State – his love for the great outdoors was matched with passion, daring, and (sometimes unorthodox) vision. Hollister excelled as a horseman, skier, fly fisherman, and hunter, but it was the Earth's heights and depths that captivated him. [31]

As a youth, he began with the traditional climbs in the Cascades and Sierras. Then, in 1962, he was part of the climbing team on the first ascent of the southeast side of Denali. The month-long expedition was featured in *The New York Times* and *Look* magazine. He joined the expedition that made the first ascents of Antarctica's highest mountains, including Vinson Massif, the continent's highest peak. He climbed in Europe, the Himalayas, and Asia. Quite fittingly, he served as President of the American Alpine Club – a post once held by John Muir. [31]

In 1967, Charles Hollister completed his Ph.D. in geology from Columbia University and joined Woods Hole Oceanographic Institution. He was among the first oceanographers to document that large areas of the seafloor, long believed to be tranquil, are swept by strong currents known as benthic storms. As an expert in deep-sea ocean current, he led the American delegation that negotiated with the Russians over monitoring of their dumped radioactive material in the Arctic Sea, and how to prevent its escape. Hollister also helped develop a giant piston corer, dubbed "Super Straw," that broke the core record. The 100-foot-long (30 m) core contained 65 million years of ocean basin history. Over the course of his career, he participated in 27 ocean research cruises, for 21 of them as chief scientist. "Ocean research is a lot like climbing a new route to the top of a mountain," he once said. [31]

Hollister proposed the idea of subseabed disposal in 1973, after talking with a chemist from Sandia National Laboratories about the problems with the proposed salt repository for high-level waste at Lyons, Kansas. He immediately thought of the clays in the deep-sea floor. From previous studies, Hollister knew that radioactive particles that had settled there as a result of atmospheric nuclear testing clung strongly (sorbed) to the clays. Hollister noted that the abyssal clays cover nearly 20 percent of the Earth. "So one thing is clear," he said, "although other factors may mitigate against subseabed disposal, it will not be constrained by a lack of space." As an added plus, the abyssal plains "appear to constitute some of the least valuable real estate on the planet." [32]

Hollister followed up with a meeting of biologists, physicists, and oceanographers at Sandia to see if they could "destroy" his idea, in what he called the "biggest shootout since the OK Corral." The idea was to look for problems with the concept. Instead, many of the scientists told Hollister they wanted to work with him. A subseabed research program was initiated in 1974. It soon grew into an international research program, involving ten countries and 200 scientists. They called themselves the Seabed Working Group. [33]

The clay-rich abyssal mudflats, with the consistency of peanut butter, have several desirable characteristics for high-level waste disposal: low permeability to water, a strong capacity to sorb dangerous elements like plutonium, and a natural plasticity to seal any cracks that might develop around a waste container. Laboratory experiments, conducted by the Seabed Working Group from 1976 to 1986, concluded that plutonium and other radioactive elements buried in the clays would not migrate more than a few meters from a breached canister after 100,000 years. In 1993, Hollister examined a Soviet submarine that had sunk years before in an active fishing ground in the Norwegian Sea. He concluded that the scientific evidence suggested "zero impact if the nuclear material sits beneath the bottom of the sea or even on the bottom." [33]

Many research questions remained to be answered. These were well recognized by Hollister. How would the heat affect movement of water and chemicals in the clays? Could organisms deeper in the clays transport radioactive substances upward to the seafloor? Are there currents strong enough to bring clay-bound radionuclides to the ocean surface?

The most obvious approach for burying radioactive waste in deep-sea sediment is analogous to offshore drilling – lower a steel pipe several miles to the seabed and drill an emplacement hole. A less costly approach was also considered – drop the wastes overboard in projectile-shaped *penetrometers*. Calculations and an experiment in 1984 suggested such projectiles would achieve speeds adequate to penetrate the soft sediment to depths of about 100 feet (30 m). Alternatively, a boosting system might be used to fire the penetrometer downward as it approached the sea bottom. Initial laboratory work suggested that the hole would close rapidly after penetration, yet these results needed confirmation by real-world testing.

As demonstrated by earlier experiences with sea disposal, transportation would pose an additional challenge for subseabed disposal. The waste would have to be moved long distances overland, then transferred to an ocean-going vessel and carried hundreds or thousands of miles to the disposal area. The probability of accidents is greater at sea than on land. In the event of an accident, the waste could be impossible to retrieve.

In 1986, the United States terminated its participation in the international subseabed disposal research program, in order to concentrate efforts on land-based disposal. The following year, as the federal government proposed to focus exclusively on Yucca Mountain in Nevada, the Subseabed Nuclear Waste Disposal Research Act of 1987 was sponsored by Senator Chic Hecht (D-NV). From the viewpoint of Nevada, the idea of burying wastes below the sea seemed like a good one. Hecht's bill failed to pass. Soon thereafter, the Office of Subseabed Disposal Research was created within the Department of Energy, but was never adequately funded. Other bills were later introduced that would ban research on subseabed disposal.

The 1972 London Dumping Convention prohibited "dumping" high-level waste or spent fuel into the waters of the ocean. In 1996, parties to the London Dumping Convention voted to classify the disposal of nuclear material below the seabed as "ocean dumping," and it became prohibited by international law. The bylaws allowed for subseabed disposal to be reviewed in 25 years. Not to be deterred, Hollister suggested that this interval "would provide sufficient time to complete a comprehensive appraisal of this disposal method." Hollister had no problem with a ban on subseabed disposal until more experiments had been conducted. What troubled him were people who were trying to ban *research* on the subject. [32]

Charles Hollister died in 1999 while climbing in the mountains of Wyoming. With his passing, the possibility of subseabed disposal most likely died also.

4

Radioactivity and atomic energy

"The Italian navigator has just landed in the new world."

"Is that so; Were the natives friendly?"

"Everyone landed safe and happy."

> Coded dialog between Arthur Compton and James Conant that
> Enrico Fermi's nuclear reactor had worked. [1]

In 1789, the German chemist Martin Heinrich Klaproth discovered that pitchblende, a black mineral with a dull, pitch-like luster from the mountains of central Europe, contained an unknown element. He named it uranium after the newly discovered planet Uranus, Greek for *Titan of the Gods*. The substance that Klaproth identified, however, was not pure uranium; it was uranium oxide. In 1841, Eugène M. Péligot, a French chemist, was the first to isolate elemental uranium.

The first known use of uranium dates to antiquity. When archaeologists excavated the Imperial Roman Villa near Naples, they found pale yellow–green glass from a first century AD mosaic mural. Roman artisans had used a uranium-bearing mineral to obtain the color. When Rome fell, the technique disappeared with it. [2]

Eventually, people rediscovered that uranium produces lustrous hues of orange or yellow to glassware and ceramic glazes. Photographers used it to tone photographs. Outside of these trades, demand for uranium remained low. Uranium ore was considered mostly a worthless byproduct. Miners would toss it aside in their search for more valuable elements like steel-hardening vanadium. Then, at the end of the nineteenth century, a few major scientific discoveries changed everything.

In 1896, French physicist Henri Becquerel found that uranium compounds emit mysterious rays that fogged photographic plates he had stored in a dark drawer. Quite by accident, Becquerel had discovered radioactivity. Becquerel's discovery is part of a long list of chance discoveries in the history of science, including synthetic rubber, insulin, penicillin, Teflon, Post-it notes, corn flakes, and even Viagra (originally intended to treat cardiovascular ailments).

Two years later, through careful detective work, Marie and Pierre Curie discovered that only certain elements in uranium ore emit the radiation. Radioactivity, as Marie Curie named it, was a property built into the structure of certain atoms. Becquerel and the Curies shared the 1903 Nobel Prize in Physics for their discoveries. In 1911, as a result of painstaking chemical analyses, Marie Curie received a second Nobel Prize; this time for her discoveries of radium and polonium.

Marie Curie's groundbreaking discoveries piqued the interest of physicists worldwide, including her daughter Irène Joliot-Curie. Irène and her husband Frédéric were the first persons to artificially synthesize a radioactive substance, when they bombarded boron with alpha particles to produce radioactive nitrogen. The Joliot-Curies shared the 1935 Nobel Prize in Chemistry for their landmark discovery. Since that time, more than a thousand radioactive isotopes have been artificially produced. [3]

From these rudimentary beginnings, physicists soon had quarks, muons, gluons, and many other subatomic particles dancing in their heads. Some were real; others imagined. With a healthy sense of humor, physicists described subatomic charm, flavor, and strangeness.

Fortunately, to understand nuclear waste you don't have to be a Nobel Prize-winning physicist, nor be able to distinguish a muon from a gluon. By simply grasping a few basics, one can have a working understanding of the seemingly forbidden world of nuclear energy and its wastes.

RADIOACTIVITY 101

All naturally occurring substances on Earth are comprised of one or more basic building blocks known as elements. Many of these elements are well known. Oxygen and nitrogen are in the air we breathe. Carbon is the essential element of life. The earth abounds with metals such as copper, iron, and zinc.

Elements are made up of atoms. *Atomos*, the Greek root of the word atom, means indivisible. Originally it was thought that the atom was the smallest indivisible particle. Twentieth-century physicists were to prove otherwise. The *nuc* in nuclear emphasizes the *nucleus* of atoms, home to its neutrons and protons.

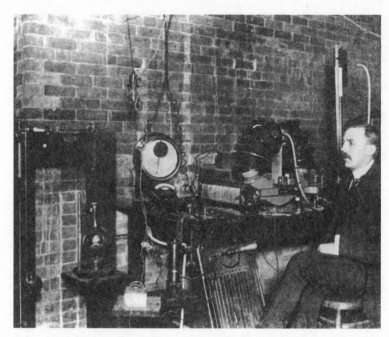

Figure 4.1 Ernest Rutherford at McGill University, Canada, in 1905.
Credit: McGill University, Rutherford Museum/Emilio Segrè Visual
Archives/American Institute of Physics/Science Photo Library.

Ernest Rutherford, who dominated early nuclear physics, gave us a
first glimpse of the properties of the nucleus. In 1909, under his direction
Hans Geiger (of Geiger counter fame) and Ernest Marsden shot a beam
of positively charged atomic particles at a thin metal foil. Rutherford
expected the high-energy particles to travel right through the foil with, at
most, very minor deflections in their paths. Most of the particles behaved
just as expected but, to Rutherford's great surprise, some were not only
deflected but actually ricocheted back. "It was almost as incredible as if
you fired a 15-inch shell at a piece of tissue paper and it came back and
hit you," he later remarked. [4]

In the same way that two positive ends of magnets repel, Rutherford correctly reasoned that the nucleus must have a tiny massive center
that carries a charge. The positively charged atomic particles in the beam
that had been deflected had made a "close encounter" with this dense
center of the atom. The particles that ricocheted back had made a direct
hit. In 1911, Rutherford announced his findings to the Manchester Literary and Philosophical Society.

Rutherford was the master of the simple, but elegant, experiment. In a very straightforward manner, his experiment revealed that the nucleus of the atom has some very impressive properties. The nucleus comprises more than 99.9 percent of the mass of an atom, yet it is extremely small. If the nucleus were the size of a ping-pong ball, the center of the nearest electron would be about one-third of a mile away. A ping-pong ball sized nucleus would have a mass of 2.5 billion tons. [5]

Nuclear reactions involve the protons and neutrons of the nucleus, so the number of protons and neutrons in an atom has great significance. Elements are uniquely defined by their number of *protons*. Hydrogen, the smallest and lightest element, always has one proton. Oxygen, the most abundant element in the Earth's crust, always has eight protons. Uranium, the heaviest element found in any significant way in nature, always has 92 protons. In contrast, atoms of the same element can have a different number of *neutrons*. These are referred to as *isotopes* of that element. All but about 20 of the naturally occurring elements have two or more isotopes.

Carbon, the building block of life, has six protons and three naturally occurring isotopes; in other words, three different types of atoms each with a different number of neutrons. The vast majority of carbon atoms have six neutrons. This common isotope is referred to as carbon-12 – with 12 denoting the sum of the protons and neutrons. A small percentage of carbon atoms have seven neutrons, also known as carbon-13. Even fewer carbon atoms have eight neutrons, making carbon-14; an isotope known for its use in dating archaeological finds. The numbers for different isotopes reflect their relative mass. Thus, an atom of uranium-238 is 17 times heavier than an atom of carbon-14.

Many naturally occurring isotopes are stable, meaning that their nucleus never changes. Others, referred to as *radionuclides*, are unstable; their nucleus changes by radioactive decay. Radioactive decay proceeds by emitting radiation until a stable form is reached. The rate of decay is predictable and commonly reported as a half-life, which is the time required for half of any given mass to decay to the next element or isotope in the series. Half-lives have an enormous range, from fractions of a second to billions of years. Consider uranium-238. It decays to thorium-234 with a half-life of 4.5 billion years (about the age of the Earth). Thorium-234 then decays to palladium-234 with a half-life of 24 days, which in turn decays to uranium-234 with a half-life of just under 7 hours. The process continues through 10 other radionuclides until stable (nonradioactive) lead-206 is formed.

One of the most important things to understand about radioactivity is that the half-life of a radionuclide is an inverse measure of the intensity of radiation it generates. The shorter the half-life, the more atoms that decay and emit radiation each second. Elements with shorter half-lives, like thorium-234 at 24 days, are more radioactive than those with longer half-lives, like uranium-238 at 4.5 billion years. On the other hand, half-lives also indicate how long the radiation from a given radionuclide will remain potentially dangerous. A rough rule of thumb is that the amount of a radionuclide remaining after 10 half-lives is so small that its radioactivity is no longer a serious threat. However, in the process of its decay it may create other radionuclides that extend the period of danger.

In the same way that the protons and neutrons in the nucleus give an element its nuclear properties, the surrounding electrons impart unique chemical properties. Chemistry is the science of how substances interact. Some are naturally attracted; others dislike each other intensely. Chemistry determines whether an element will form a compound that dissolves in water, attaches to a mineral surface, or precipitates out of solution. It thus controls how substances – in our case radioactive elements – behave in the environment.

THE PERIODIC TABLE: CHEMISTRY'S ROSETTA STONE

Walk into almost any room where chemistry is taught or practised, and a periodic table is likely to be hanging prominently on the wall. The periodic table was constructed to represent the patterns observed in the chemical properties of the elements. The present form of the table was conceived independently by the German chemist, Julius Lothar Meyer, and the Russian scientist, Dmitri Mendeleev, at about the time of the American Civil War. Mendeleev is given most of the credit because he emphasized how useful the table could be in predicting the existence and properties of still unknown elements. Many chemists have been involved in development of the periodic table over time. A simple form of the periodic table is shown in Figure 4.2.

The elements in the periodic table are listed across rows by their number of protons; also known as the atomic number. For example, the element with 84 protons (atomic number 84) is seen to be polonium. Polonium was discovered in 1898 by Marie and Pierre Curie in their search for the sources of radioactivity in uranium ore. Marie Curie named it after her beloved native country of Poland. This is the same element that attracted worldwide attention in the 2006 poisoning of Alexander

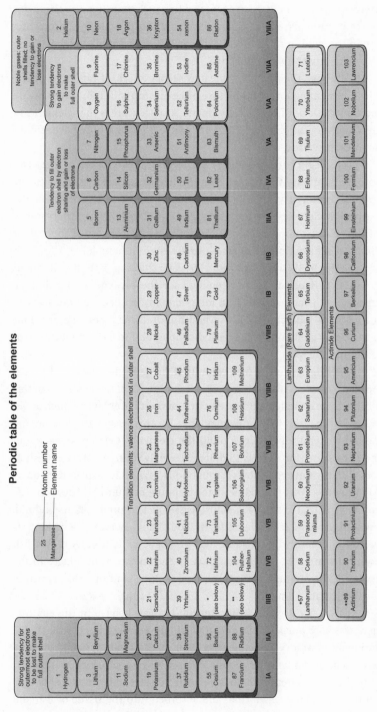

Figure 4.2 Periodic table.

Litvinenko, a former Russian agent residing in Britain. It is also the major source of radioactivity in cigarettes. Polonium provides an interesting line in the sand in the periodic table – all isotopes of all elements from polonium onward are radioactive.

Although the periodic table may seem esoteric, it can be very help-ful in developing a mental picture of key isotopes in radioactive waste and their behavior in the environment. A key feature is that elements in the same vertical column of the periodic table commonly have similar chemical properties.

Let's begin with cesium-137 and strontium-90. Together, these two isotopes dominate the hazard from high-level radioactive waste for its first few hundred years. About 99 percent of the radioactivity at the Hanford Nuclear Reservation, the largest site of high-level military waste, is from these two isotopes alone. Naturally, we are concerned about their chemical behavior once released into the environment. [6]

Cesium-137 and strontium-90 each have a half-life of about 30 years. This period is short enough for these isotopes to give off a super dose of radioactivity and long enough for them to emit danger-ous levels of radioactivity for several hundreds of years. The positions of these two elements in the periodic table reinforce the importance of containing them.

Cesium lies in the first column of the periodic table, along with the common elements sodium and potassium. All of the elements in this column have a single dangling electron in their outer layer, which is eas-ily lost in reactions with other elements. For this reason, they are among the most chemically reactive of all the elements. Cesium, like sodium and potassium, forms salts that are extremely soluble in water. Cesium goes with the flow. As it flows through soils, cesium is easily picked up by clay minerals. This is good news for trapping cesium-137. However, the bad news is that cesium-137 is also easily picked up by plants and ani-mals. From there, it biomagnifies in food chains, potentially making its way into humans who consume food grown in contaminated soil or fish from contaminated waterways. Once in the body, cesium lodges in the tissues of the stomach, intestines, liver, spleen, and muscles, emitting dangerous radiation.

The many ways in which cesium-137 can behave in the environ-ment was both purposefully and unwittingly demonstrated by Oak Ridge National Laboratory from the early 1940s to the 1960s. When the labora-tory discharged much of its liquid radioactive waste into local waterways, cesium-137 was later found to have settled on streambed sediments, con-taminated aquatic vegetation, and absorbed into the tissue of fish and other aquatic animals. In the 1960s, the laboratory scientists deliberately released cesium-137 into fields to study how the isotope would behave in the environment after a nuclear explosion. The end result was that cesium-137 contaminated groundwater and entered the nearby Clinch River through surface-water runoff and erosion. These and other early experiences provided hard-earned lessons on the importance of much more careful handling of cesium-137. [7]

Strontium-90 also potentially finds its way into living organisms once released into the environment. The most troubling feature of strontium-90 is that it is in the same column of the periodic table as calcium – one door down, in fact. Strontium and calcium are so chemically similar that our bodies don't know the difference. Strontium-90 that replaces calcium in our bones and teeth can remain inside the human body for years, continually emitting radiation and possibly causing cancer.

Concerns about strontium-90 go way back in the atomic age. Pick up any article about the dangers of atmospheric nuclear testing in the 1950s and 1960s, and you're sure to find mention of strontium-90. The fear was that strontium-90 in grass and hay from atmospheric fallout would become incorporated into cow's milk and from there be passed on to humans. They called it the "milk pathway." Even *Consumer Reports* conducted its own surveys of strontium-90 in milk, and accused the government of dragging its feet due to a huge conflict of interest. According to *Consumer Reports*, the Atomic Energy Commission was responsible for both "manufacturing" the fallout and having the final word on the "product's" safety. [8]

In another grass-roots effort, scientists and citizens in St. Louis began a *Baby Tooth Survey* in 1958 to look for strontium-90. Almost 300,000 toddler teeth were collected. The study concluded that St. Louis children born in 1964 had about 50 times more strontium-90 in their baby teeth than those born before the start of atomic testing. These efforts, combined with political campaigns, congressional hearings and continuing media coverage, ultimately led to a ban on atmospheric testing of nuclear weapons. [9–10]

Radium is another example of the link between the periodic table and behavior in the environment. Like strontium, radium lies in the second column of the periodic table and can substitute for calcium in our bones. Yet amazingly, radium was once touted for its curative powers. Spas featured radium-rich water. Household products, such as toothpaste and hair creams, advertised their healthful radium additives. Radium could make certain ordinary chemicals fluoresce, causing its use in self-luminous paints for watches, clocks, and instrument dials. So-called *radium girls* painted all those clock faces and dials. To get the necessary fine tip, they used their lips to shape the paintbrush. The "girls" even painted their faces with the radium paints to amuse their friends and husbands in the dark. The party ended when the radium girls began to develop astounding rates of anemia and bone cancer. The ensuing controversy ultimately had a significant impact on occupational labor law. Kurt Vonnegut later featured the story in his novel, *Jailbird*.

Radium-226 is the most notorious of the many radium isotopes, with a half-life of 1,622 years. It also has a pernicious radioactive decay product – radon-222. Like other elements in the last column of the periodic table, radon is a *noble gas*; as such it does not mix with other elements. Radon-222 has a half-life of only 4 days, but, if inhaled, its radioactive decay products can lodge in the lungs and radiate into body tissues for years. The greatest hazard from radon gas arises when it accumulates in a confined area, like a mine or basement. It is a primary isotope of concern in uranium mining and production of nuclear fuel. The ubiquitous occurrence of uranium in rocks also increases our exposure to radon. A typical person gets much of their annual exposure to ionizing radiation from radon emitted by rocks.

Public awareness of the dangers of radon in our everyday lives came from an unexpected occurrence at a nuclear power plant. In 1984, Stanley Watras, a construction engineer, kept setting off radiation alarms at the Limerick nuclear power plant northwest of Philadelphia. The problem was, he set off the alarms on his way *into* work. Upon investigation, authorities discovered exceedingly high levels of radon gas in Stanley's home. It turned out that this high radon level was the product of Stanley Watras living on a uranium-rich geologic formation known as the Reading Prong. State officials in Pennsylvania began sampling houses in the surrounding area and found many with high indoor radon levels. The problem was soon found in other States. Radon awareness grew rapidly and an entire industry for radon home testing and mitigation was soon born.

We conclude our discussion of radionuclides and the periodic table with the heaviest elements. Uranium, the principal element involved in nuclear fission, appears in a special row at the bottom of the periodic table. Prior to the discovery of nuclear fission, the periodic table ended with uranium. As such, it is the heaviest of the elements found in nature in more than trace amounts. Elements with larger atomic numbers are made artificially by a nuclear reactor or particle accelerator and are referred to as *transuranics*, meaning beyond uranium. Transuranics, such as neptunium, plutonium, and americium are of particular importance in waste from nuclear reactors. Many have extremely long half-lives. After several hundreds of years, when strontium-90 and cesium-137 die down, the transuranics become the dominant hazards from nuclear waste, along with a few other long-lived isotopes, like technetium-99 with a half-life of 211,100 years. Technetium (atomic number 43) was a textbook example of the use of the periodic table when, long before its discovery, Mendeleev predicted its chemical properties.

RADIATION AND ITS EFFECTS

Radioactive decay occurs in various ways, but three are important for understanding the decay of radioactive waste. The first two, *alpha* and *beta* particles, were discovered and named by Ernest Rutherford after the first two letters of the Greek alphabet. The third type was discovered by Paul Villard, a French physicist working in Paris contemporaneously with Marie Curie. Continuing the established pattern, it was named *gamma* after the third letter of the Greek alphabet.

We can thank Rutherford for making radiation nomenclature so simple. (He also named the *proton* and coined the term *half-life*.) As fate would have it, we can thank one other gentleman, too. Rutherford was born on a sheep farm in New Zealand. As a young man, he came second in a scholarship to attend Cambridge University. It was only when the winner decided to get married that Rutherford received the scholarship and a gateway to his unique career as the father of nuclear physics. [5]

An *alpha particle* consists of two protons and two neutrons. It is equivalent to the nucleus of a helium atom, the element with atomic number 2 in the periodic table. By losing two protons, the atom moves two steps back in the periodic table. For example, radium-226 decays to radon-222 through alpha decay. The drop is seen in the periodic table when radium (atomic number 88) drops to radon (atomic number 86). Alpha particle decay is very common for heavy radionuclides like uranium and the transuranics.

A *beta particle* is an electron ejected at very high speed. This "electron" is not one of those that surround the nucleus. Instead, by a marvel of physics, a neutron spontaneously changes into a proton and an electron. The proton remains in the nucleus, while the mutant electron is kicked out as a beta particle. The net effect of beta decay is that the atom moves one step *forward* in the periodic table. For example, cesium-137 (atomic number 55) decays by emitting a beta particle, taking one step forward to become barium-137 (atomic number 56).

Gamma rays are electromagnetic waves that may accompany, or follow closely, alpha and beta emission. They are basically the same as X-rays, but generally have more energy. The emission of gamma rays is one way the nucleus "works off" excess energy and "relaxes" to a lower energy state. However, a gamma ray changes neither the element nor the atomic mass of an atom. For example, barium-137 (resulting from the beta decay of cesium-137) is unstable with a half-life of a mere 3 minutes. Barium-137 relaxes and stabilizes itself by emitting a gamma ray. As a result, the decay of cesium-137 produces both beta particles and gamma rays.

The dangers differ greatly among the three types of radiation. Gamma rays are the most penetrating of the three kinds of radiation, requiring at least an inch of heavy metal, like lead, for effective shielding. Gamma rays are capable of penetrating deeply into the body. As a result, protection from gamma rays requires substantial shielding and handling of radioactive waste by remote control. Beta particles are absorbed by several feet of air, or by small thicknesses of metal or glass. If skin is unprotected, they can cause burns and penetrate human skin to a depth of a very small fraction of an inch. Alpha particles travel only two or three inches in air, and are unable to penetrate the outer layers of human skin.

The principal concern about radioactive waste is that when released into the environment it might be taken into the body through ingestion (drinking or eating) or by breathing, thus placing a source of radiation very close to vulnerable tissues. Although alpha and beta particles travel only a short distance and are easily shielded outside the body, they can cause serious harm within the body. By analogy to an athlete, alpha and beta particles are sprinters who pack a punch for the short run. On the other hand, gamma rays can "go the distance" by pacing themselves along the way. An isotope like cesium-137, that emits gamma rays as well as alpha or beta particles, is both a potential external hazard from the gamma rays and internal hazard from the alpha or beta particles – a double whammy.

Radioactivity is measured by the *curie* – named (of course) after Marie Curie. A curie is defined as a whopping 37 billion disintegrations (radioactive particles or rays) per second – equivalent to the decay rate of 1 gram of radium. Most uranium is about 0.0000003 (three-tenths of a millionth) curies per gram, making it only weakly radioactive. Radium is about 3 million times more radioactive than uranium. Strontium-90 comes in at 140 curies per gram and cesium-137 at about 90 curies per gram. Both are about 100 times more radioactive than radium and about 100 million times more radioactive than uranium. [11]

The *curie* is a useful unit for measuring the amount of radiation emitted. A few examples are listed below for comparison. These estimates are in units of millions of curies. [12–18]

U.S. defense wastes released into the environment (as of 1996)	3
Ocean dumping	4
Buried low-level waste	50
Chernobyl (1986)	100
Hanford releases to Columbia River (1944–71)	110
Tanks at Hanford, Savannah River, and Idaho (as of 2006)	800
Russian defense wastes released into the environment (as of 1996)	1,700
Uranium mine and mill tailings	3,000
U.S. commercial spent fuel (2010)	40,000

The danger of radiation is largely one of degree. In various controlled ways, radiation is used safely and productively in our everyday lives. For example, americium is one of the transuranics artificially created in a nuclear reactor. When Glenn Seaborg discovered this element, he named it for the Americas by analogy with the element directly above it in the periodic table, which had been named europium after Europe. Americium-241 is a major contributor to the long-term hazard of radioactive wastes. The same isotope is also used in tiny amounts as an ionization source in smoke detectors. While the smoke detector body shields us from the radiation, the smoke detector protects us from another hazard – fire. The typical home smoke detector emits 1 millionth of a curie.

NUCLEAR FISSION

Nuclear reactors produce energy through a process called *fission* – the splitting of the atom. Fission occurs when an atom of fissile material is struck by a neutron and becomes unstable and splits, producing fission fragments and high-energy neutrons. In a reactor, one of these supercharged neutrons strikes another fissile atom to maintain a steady chain reaction. Fission also releases heat, which is used to produce steam to spin an electrical generator. The only fissile material found in nature is uranium-235, which makes up less than one percent of natural uranium.

In 1938, Otto Hahn and Fritz Strassmann, two scientists working at the Kaiser Wilhelm Institute in Germany, discovered nuclear fission.

A third scientist, Lise Meitner, who collaborated with Otto Hahn at the Kaiser Wilhelm Institute, would most assuredly have shared in the discovery if she had not fled Nazi Germany earlier that year. A few other scientists, including Irène Joliot-Curie and Enrico Fermi, just missed the momentous discovery in their research.

Lise Meitner and her nephew Otto Frisch were the first to explain the Hahn and Strassmann results. Frisch coined the term "fission" by analogy with the process by which living cells divide. "Thereby the name for a multiplication of life became the name for a violent process of destruction," observes Richard Rhodes in his book, *The Making of the Atomic Bomb*. [19]

The discovery of fission first came to the public's attention in early 1939 after being discussed at a technical meeting at Columbia University and then being picked up by a *New York Times* reporter. The *Times* enthusiastically proclaimed a "Revolution in Physics" on their front page:

> Great news came out of the physical laboratories of Columbia University and the Kaiser Wilhelm Institute the other day. Slow neutrons were hurled at uranium. Out came two complete atoms, the one barium, the other still to be identified. In addition, the energy released, which is of the order of a hundred million volts, far exceeds that of the neutron that does the shattering... Romancers have a legitimate excuse for returning to Wellsian utopias where whole cities are illuminated by energy in a little matter. [20]

In spite of the enthusiasm of the *New York Times*, the possibility of harnessing the energy of the atom still seemed remote, if not fanciful. It was one thing to split a uranium atom by bombarding it with neutrons; it was quite another to create a self-sustaining chain reaction. But nuclear physics was on a roll. On December 2, 1942, Enrico Fermi and his colleagues pulled a neutron-absorbing rod out of a "pile" of graphite blocks plugged with uranium cylinders, constructed in a squash court beneath the University of Chicago sports stadium. Fermi's uranium pile soon went critical, creating the world's first controlled, self-sustaining nuclear reactor. The atomic age had begun, but only a few people took note. Of much greater concern to Americans that day was the start of wartime gasoline rationing.

At its peak, Fermi's reactor produced about half a watt of power, enough to dimly light a single Christmas tree light bulb. The reactor ran for less than five minutes. Nevertheless, the possibilities were enormous. Fermi demonstrated, in principle, that nuclear power could light up a city. Or destroy one. [19, 21]

Destruction would come first. In August 1945, less than three years after Fermi's reactor went critical, the United States dropped atomic bombs on Hiroshima and Nagasaki, Japan. It would take nine years after Fermi's successful reactor experiment to demonstrate the peaceful side of the atom.

5

The Cold War legacy

In the councils of government, we must guard against the acquisition of unwarranted influence, whether sought or unsought, by the military industrial complex.

President Dwight D. Eisenhower's last State of the Union Address,
January 17, 1961 [1]

The Manhattan Project was the largest, most complex project in human history. The endeavor took place at sites across the United States, Canada, and the United Kingdom, but most of the work occurred at three secret locations. Uranium was enriched at the Oak Ridge Reservation, nestled in the Appalachian valley and ridge province about six miles west of Knoxville, Tennessee. Plutonium was created and processed at the Hanford Reservation, along the Columbia River in southeastern Washington. Nuclear weapons were designed and manufactured at Los Alamos National Laboratory, location of a former boys' school on a remote mesa in New Mexico.

The monumental enterprise to produce nuclear weapons expanded during the Cold War. The Savannah River site, near Aiken, South Carolina, was established in 1950 to increase production of plutonium and tritium for use in nuclear weapons. Lawrence Livermore National Laboratory, built near California farming country, became the Nation's second weapons-design laboratory to hasten development of the thermonuclear bomb. Plutonium was machined into bomb components at Rocky Flats, Colorado. Nuclear weapons were assembled (and later disassembled) at the Pantex Plant, 17 miles (27 km) northeast of Amarillo, Texas. Altogether the Department of Energy (DOE) weapons complex, as it has come to be known, encompassed more than 100 sites distributed among 31 States and one territory. Individual sites range in size from a few acres

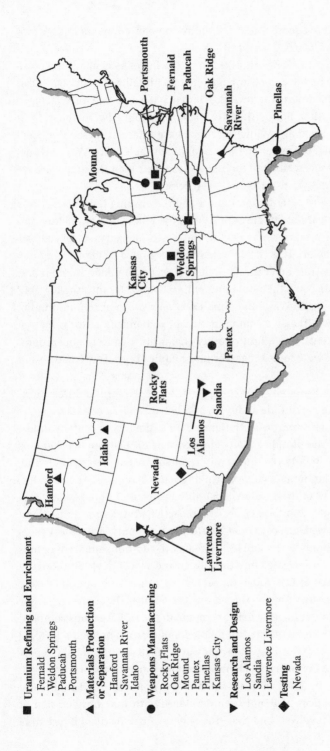

Figure 5.1 Major US nuclear weapons sites.
Source: Reference [5].

to hundreds of square miles. The entire complex covers more than two million acres (800,000 hectares). [2]

When Niels Bohr was asked in the spring of 1939 about the possibility of producing a nuclear bomb, he told his fellow physicists, "It can never be done unless you turn the United States into one huge factory." Years later, when Bohr visited Los Alamos, he told Edward Teller, "You have done just that!" While Bohr may have underestimated the possibilities of nuclear fission, he was right about the factory. What no one, including Bohr, seemed to foresee was that this monumental "factory" would produce waste unlike any seen before. [3]

During the Manhattan Project and the ensuing Cold War, efforts were focused almost single-mindedly on the production of nuclear warheads. Little attention was paid to the waste. Management of radioactive waste materials at the defense sites was guided by two general precepts: concentrate and contain high-level wastes in underground tanks; dilute and disperse the remaining radioactive wastes into the air, soil, and groundwater. Waste handling expenses were kept to a minimum. The result was massive contamination by radionuclides, toxic metals, organic solvents, and a host of other chemicals used in the operations. Often, these were mixed together in a complex brew. Few records were kept.

Soil and groundwater contamination at Hanford and Oak Ridge illustrate the magnitude and types of problems that have occurred. The soil and groundwater beneath Hanford are estimated to contain almost 2 million curies of radioactivity and between 100,000 and 300,000 tons of chemicals. Nearly 100 square miles (260 square kilometers) of groundwater are contaminated above drinking-water standards. At Oak Ridge, more than 40 million gallons (150 million liters) of liquid waste, containing about 1.2 million curies of radioactivity from strontium-90 and cesium-137, were disposed of in "waste burial grounds," where leaks contaminated groundwater and local streams. In addition, hundreds of tons of mercury were released into the ground and a local creek. The mercury had been used at Oak Ridge for separating lithium isotopes that were used to make tritium for thermonuclear bombs. These are but a few examples. Contamination exists throughout the DOE weapons complex, while Hanford remains the "mother lode" of cleanup problems. [4–6]

There were warnings along the way. In 1948, an internal AEC report (the Williams Commission) identified serious concerns about waste management at the nuclear weapons sites. The report stressed that waste disposal practices had "not been developed with full consideration of the hazards involved" and that "the degree of risk justified in wartime

Figure 5.2 Interior of a truck-mounted radiation counter taken to schools to monitor radiation levels in children who lived near Hanford. This was a common practice in the mid 1960s.
Source: US Department of Energy.

is no longer appropriate." The Williams Commission report was largely ignored, and waste practices essentially continued unchanged. [5, 7]

For decades, all information about atomic energy, including the wastes, was "born classified." As a result, releases of radioactivity and chemicals into the environment at the Nation's atomic weapons plants were shrouded in secrecy. Over time, the veil of secrecy was gradually removed and revelations about contamination emerged from the weapons facilities. In the 1980s, US government claims that the weapons sites were exempt from State and federal pollution control laws began to be overturned in court. As the contamination legacy of the DOE weapons complex was unfolding, the era of large-scale nuclear weapons production was coming to an end. The thaw in Cold War relations culminated in the breakup of the Soviet Union in 1991. [8]

Today, a large part of the DOE weapons complex has been shut down or placed on standby. Thousands of people arrive at the sites each work day, not to develop nuclear weapons, but to participate in the largest environmental cleanup in history. The cleanup bill for the DOE weapons complex will run into the hundreds of billions of dollars, likely exceeding the cost of developing the nuclear weapons in the first place. Cleanup is expected to take until at least 2070. Even then, contaminants will remain, requiring long-term management and possibly further actions to prevent spreading. [9]

High-level radioactive wastes are stored at three sites in the DOE weapons complex – Hanford, the Savannah River site, and Idaho National Laboratory. In 1997, about 100 million gallons (380 million liters) of liquid high-level waste from reprocessing for nuclear weapons were stored in underground tanks at Hanford and Savannah River, enough to fill about 10,000 tanker trucks. In addition, there are about 1,900 capsules (20-inch-long cylinders) filled with cesium and strontium concentrated from the tanks at Hanford. The Idaho National Laboratory holds reprocessing wastes converted to dry granules, as well as liquid wastes. In 2006, the radioactivity of the reprocessing wastes at Hanford, Savannah River, and Idaho were estimated to be about 320 million, 426 million, and 41 million curies, respectively. [10]

The three sites also store spent fuel. More than 2,000 metric tons of defense reactor spent fuel is stored at Hanford. The fuel, much of which was highly corroded, has been dried and placed in storage canisters. The Savannah River site and Idaho National Laboratory store spent fuel returned from university, government, and foreign research reactors – the latter from fuel supplied by the United States to foreign governments under the Atoms for Peace Program. The Idaho National Laboratory also manages spent fuel from nuclear submarines, aircraft carriers, and training reactors belonging to the US Navy. The status of the spent fuel and high-level wastes varies considerably among the three sites.

HANFORD

Manhattan Project scientists believed that an atomic weapon could be constructed using either uranium-235 or plutonium-239. Given the cost and difficulty of concentrating uranium-235 in the highly enriched form needed for bombs, plutonium seemed to offer the greatest chance for success. Moreover, plutonium-239 is more explosive than uranium-235. A pilot plant for plutonium production was built at Oak Ridge, but large-scale production at the Tennessee site was rejected. A serious accident could contaminate large numbers of people in nearby Knoxville, eliminate project secrecy, and result in what General Groves feared would be the "congressional investigation to end all congressional investigations." A more remote location was needed. [5, 11]

In December 1942, Lt. Col. Franklin Matthais from the US Army Corps of Engineers and two engineers from DuPont were dispatched to search for potential sites for large-scale plutonium production in the western USA. Among the sites visited, Hanford stood head and shoulders above the rest. The site had electricity from the newly constructed Grand Coulee Dam, abundant cold river water for cooling the reactors, a railroad line, a mild and dry climate, good construction materials, and a small population. In addition, the Columbia River furnished a "convenient dispersal mechanism" for contaminated reactor-cooling waters. The site was officially selected in January 1943. In March 1943, the families of about 1,300 people received letters telling them they were being evacuated. No one was told why. Many were given less than a month to pack up and move out. [5]

Nine reactors for plutonium production were built on the Hanford Reservation, widely spaced along one of the last free-flowing reaches of the Columbia River. Irradiated fuel from these reactors was transported by specially shielded rail cars to a plateau in the middle of the Hanford site where chemical reprocessing plants recovered plutonium and uranium. Five enormous chemical reprocessing facilities operated over the history of the site. The long, windowless concrete buildings are known as *canyons*, because their interiors resemble a gorge in a deep valley between steeply vertical cliffs. The canyons were comparable in size to the Empire State Building lying on its side. The first two reprocessing plants were nicknamed the Queen Marys, after the famous ocean liner. In 1950, the Atomic Energy Commission boasted to Congress that "if all of the radioactive materials that have ever been accumulated outside of Hanford were to be added to a single batch of material going through the [Hanford] plutonium plants, the increase in radioactivity would be scarcely noticeable. Hanford handles radio materials in tons, not grams." [12]

All of the Hanford reactors have now shut down. The last reprocessing facility ceased operations in 1990. But the waste remains. During their operation, the Hanford reprocessing plants generated more than 500 million gallons (2 billion liters) of high-level waste that was piped into 177 large underground tanks. The tanks range in size from 55,000 to 1 million gallons; most are 500,000 gallons (2 million liters) or larger. A typical tank is 75 feet (23 m) in diameter and 30 to 40 feet tall – comparable in size to the US Capitol Dome, but buried about 10 feet (3 m) underground. [5]

During the 1940s, Hanford workers could not install enough tanks to hold all the waste from the reprocessing operations. Consequently,

Figure 5.3 Inside the B reprocessing plant at Hanford. Workers are monitoring radiation levels atop one of the heavily shielded and remotely operated reprocessing cells used to dissolve uranium fuel and extract plutonium.
Source: US Department of Energy.

in these early years some high-level waste was discharged directly into the ground. Later, millions of gallons of liquids were treated to separate out the cesium and then poured into the soil. Waste volume was further reduced by letting the waste boil off from the heat and the use of special evaporators. Today, about 53 million gallons of high-level waste remain – enough to fill about 2,600 railroad cars forming a train 26 miles (40 km) long. [5]

The Hanford tanks were considered a "cheap" temporary method of disposal. The assumption was that either a more permanent way of dealing with the wastes would be implemented before tank failure or, if the tanks failed, new tanks would be constructed and the wastes pumped into them. This confidence proved sorely misplaced. Management of the Hanford tanks has been marked by hastily contrived expedients with long lasting and often unanticipated consequences for future waste retrieval.

Stainless steel was in short supply during wartime, so the first tanks were constructed using a single layer of carbon steel as a liner. Even after the shortage of stainless steel ended, use of the cheaper carbon steel continued. Because the highly acidic wastes would eat right through carbon steel, sodium hydroxide was added to neutralize the wastes. Neutralization doubled the waste volume and caused metals and most radionuclides to form an insoluble sludge that settled to the bottom of the tanks. The continued use of carbon steel proved to be one of the more costly short-term expedients.

The first tank leaked in 1956, after less than ten years of service. About half of the 149 single-shell tanks are known, or suspected, to have leaked a total of about one million gallons of high-level waste directly into the ground. The most serious tank leak is estimated at more than 100,000 gallons (380,000 liters). Six weeks passed before Hanford supervisors realized that the tank was emptying out in excess of 2,000 gallons per day. The single-shell tanks were not designed with systems to detect leaks, so no one knows for sure exactly how many tanks have leaked or by how much. [5, 13]

In spite of the leaks, the AEC continued using single-shell tanks until 1968 when double-shell tanks began to be constructed. These improved tanks enclose the waste in a double layer of carbon steel separated by a space between them called an annulus – a kind of cup-and-saucer arrangement with the annulus serving as the saucer to collect and monitor leaks. All of the liquid wastes in the single-shell tanks were drained into the 28 double-shell tanks, leaving the sludge behind. The double-shell tanks have proven more resistant to leaks, but they too have a limited life. The last tank was built in 1986.

The waste is a complex mixture of radioactive and nonradioactive chemicals. Each tank holds a unique blend, resulting from different reprocessing techniques, different chemicals used in efforts to remove selected radionuclides, and transfers of waste among tanks.

The addition of chemicals to the tanks led to new problems. When organic compounds were added to extract strontium, breakdown of the compounds produced flammable gases that created the possibility of a chemical explosion. Cement and diatomaceous earth were added to soak up liquids, resulting in the formation of hard, crystalline layers.

Figure 5.4 Hanford double-shell tanks under construction in 1978.
Source: US Department of Energy.

Approximately 150 tons of ferrocyanide were added to about 20 tanks to precipitate out cesium before disposing of residual liquids into the ground. Later, it was discovered that the ferrocyanide could react with nitrates to cause excessive heat buildup, and possibly an explosion. The crisis was resolved as a result of chemical breakdown of the ferrocyanide and more shuffling of wastes among the tanks, but not until after considerable expense and controversy had ensued.

The top safety concern for nearly all of the 1990s was the build-up of hydrogen gas trapped beneath a hardened crust in one of the double-shell tanks. Every few months, the hydrogen lifted the tank's waste as much as a foot, until it was released as giant "burps" of flammable gas. During these burping episodes, the slurry resembled a thick brew of oatmeal at a slow, rolling simmer. In 1993, a seven-storey-tall mixer pump was installed, acting as a huge eggbeater that allowed for a steadier gas release. The extra safety measures for this one tank cost $30 million. [5, 14]

Studies of how to deal with the highly radioactive tank waste began over 50 years ago. Progress has been slow. Constantly shifting priorities have thwarted cleanup efforts. Several major false starts have resulted in vestiges of aborted projects. The $286 million foundation of a huge plant to vitrify high-level waste came to naught. The plans to mix low-level radioactive waste generated by tank cleanup with cement grout were halted due to the public's concern about the long-term ability of grout to isolate the waste – but not before about $200 million had been spent on a plant and storage facilities. [15–16]

Removing the waste is proving as challenging as what it took to create it. The highly radioactive waste comprises multiple forms – liquids, slurry, sludge, and a *saltcake* that precipitated as the waste evaporated. Some of the waste has crusted on internal tank surfaces and resists removal. There is also untold residual waste in underground pipelines, pumps, and valves. Access to tank interiors is limited to a few portals that extend up to the surface. Tools to pump, dislodge, or mix the waste must be inserted through these openings. A daunting challenge is figuring out how to retrieve waste from corroded carbon steel tanks that leak.

Back in the days when all this highly radioactive waste was being generated, recordkeeping was not a priority. As a result, no one really knows what is in any given tank. Waste composition is important for several reasons – to avoid reactions that might clog pipes or build up hazardous gases; to assure that the ingredients are acceptable for making glass logs; and to meet the waste acceptance requirements at a geologic repository.

To top it all off, characterizing tank waste is extraordinarily expensive. Costs vary from thousands of dollars for limited chemical analyses of a liquid grab sample collected using the "bottle-on-a-string" method, to hundreds of thousands of dollars for core samples of sludge. The extreme heterogeneity of waste composition in each tank, and in *each part* of the tank, requires multiple samples. [17]

The Hanford Federal Facility Agreement and Consent Order, commonly known as the Tri-Party Agreement, requires DOE to meet federal and State cleanup regulations. In spite of deep mistrust on the part of Washington State, the Tri-Party Agreement was a landmark agreement that enabled DOE and the State to begin working together. The Agreement spells out specific activities, timelines, and a hefty fine should DOE not meet their obligations. In early 2011, "new, realistic but aggressive" deadlines were set for treating all tank waste by 2047 and closing all the tank farms by 2052 – one more testament to the complexity of the problem. [5, 18–19]

Beginning in the late 1960s, cesium and strontium were chemically removed from some of the single-shell tanks and encapsulated in metal cylinders (capsules), each about 20 inches (50 cm) long. About 1,900 capsules are stored in pools at Hanford, comprising approximately

40 percent of the radioactivity from reprocessed waste at the site. The cesium and strontium capsules, along with more than 2,000 metric tons of spent fuel that was never reprocessed, sit at Hanford awaiting final disposal. [5, 10]

Delays in retrieving and treating the tank wastes almost match the delays in opening a geologic repository. After decades of debate, the world's largest radioactive-waste treatment plant is finally under construction to treat the tank waste. Projected to cost more than $12 billion, startup is planned for 2019. Fortunately, cleanup is further along at the other major plutonium-producing facility – the Savannah River site.

SAVANNAH RIVER SITE

During the 1950s, the Savannah River site began to produce plutonium and tritium for nuclear weapons. Five reactors, two reprocessing plants, and support facilities were built on 310 square miles (806 square kilometers) of former farmland and swamps bordering the Savannah River. The river provided good quality water for the reactors, while the southern climate allowed a long construction season to get the site quickly up and running.

Reprocessing wastes were stored in 51 underground tanks similar to the double-shell tanks at Hanford. Most of the tanks have a carbon steel inner wall and outer concrete wall with an annulus between them. Although only one tank is believed to have leaked waste to the environment, several have a history of cracks or leakage into the saucer-like annulus. Most tanks have a dense "forest" of pipes that circulate cooling water. These cooling coils remove heat produced from the radioactive decay, but impede waste removal. [10]

The Savannah River tanks have a slightly higher share of radioactivity than those at Hanford, but the waste is chemically simpler. As at Hanford, sodium hydroxide was added to neutralize the wastes, once again causing radionuclides to settle as sludge. But the Savannah River tanks were spared the other chemical additions that have plagued Hanford. As a result, tank cleanup is simpler and much further along than at Hanford.

Various innovative approaches have been explored for cleaning up the tanks at the site. Robotic arms have been designed that can cut, dig, and lift the wastes, yet that are small enough to pass through the tank openings and flexible enough to reach the edges of the tanks. Tethered, remote-controlled vehicles, such as a mini-bulldozer, have been inserted into the tanks. In March 2009, a remote-controlled wall-crawler, complete with brush and nozzle assembly, began operations to clean waste that had leaked into the annulus between the walls of two emptied tanks. The magnetic crawler beamed closed-circuit television shots of the wall as it moved along, controlled remotely by workers. When workers saw a salt deposit, the crawler pressure-sprayed water on the area and brushed clean the leak site. [10, 20]

On March 12, 1996, the Nation's first factory to solidify (vitrify) high-level waste from nuclear weapons production began operation at the Savannah River site. Energy Secretary Hazel R. O'Leary presided at the opening ceremony with much fanfare and news coverage. With a price tag of $2.4 billion, the factory was a marvel of automation. An operator, Clint Oglesby, sat at a control panel with ten joysticks for manipulating cranes and cameras. Only being able to see 9 inches (23 cm) at a time, it was "a bit like building a ship in a bottle," Oglesby said. At a rate of one to two quarts per minute, officials estimated that it would take 25 years to solidify the site's 36 million gallons (144 million liters) of high-level waste. [21]

High-pressure water jets and mixing pumps are used to loosen the sludge and pump it into pipelines to the factory, where the waste is mixed with molten glass at extremely high temperatures (over 2,000 degrees Fahrenheit) in a special device called a *melter*. The mixture is poured into stainless steel canisters, two feet in diameter, ten feet long, and weighing more than two tons. After cooling and hardening, the containers are sealed with a huge jolt of electricity. The cost of producing each *glass log* exceeds $1 million. By March 2010, about 2,900 glass logs had been produced. The operation is expected to finish by 2030. Once again, the logs await, indefinitely, final disposition to a geologic repository. The Savannah River site also stores a wide variety of spent nuclear fuel assemblies from domestic and foreign research reactors. [17, 22]

IDAHO NATIONAL LABORATORY

After Japan crippled the US Pacific fleet in December 1941, the Navy quickly expanded its West Coast bases and looked inland for support facilities. On April 1, 1942, the Naval Ordnance plant at Pocatello, Idaho, was established. Access was good. Pocatello was along a transcontinental highway, as well as being home to one of the largest railroad terminals in the country. Lying east of the coastal mountain ranges, it was also reasonably secure. During World War II, guns from warships were sent on a regular basis to the Idaho facility for refurbishment. The flat, treeless terrain provided ample open space to test-fire the guns. [23]

Designation of the site as the National Reactor Testing Station followed in 1949. Later named the Idaho National Laboratory (INL), this windswept, high desert overlies ancient basalt lava flows of the eastern Snake River Plain. On a clear day, the peaks of the Grand Tetons are visible more than 100 miles to the east. The site hosts the world's largest collection of nuclear reactors, with more than 50 built and operated over the years. Only a few are now in operation. Clusters of reactors and laboratories are scattered miles apart across an area almost the size of Rhode Island.

The design of the first nuclear-powered submarine, the *USS Nautilus*, took place at INL under the direction of Admiral Hyman Rickover. A highly disciplined man of immense energy, Rickover believed that the nation that was first to develop a "nuclear engine" would rule the oceans of the world. "Our enemies are working on such engines; we *must* be first," he steadfastly maintained. Rickover insisted on perfection and personally oversaw every detail of the project. To make sure everything would fit properly, the reactor was cocooned in a full-sized replica of two *Nautilus* hull sections. The hull section containing the reactor rested in a "sea tank" of water 40 feet deep and 50 feet in diameter. [23]

Operations at INL have recovered uranium-235 from many types of fuel – from on-site reactors, university and test reactors, commercial power plants, and Navy ships. Each had unique cladding and chemistry, challenging the chemists and engineers to develop new formulas in small-scale pilot plants. Damaged fuel and core debris from the Three Mile Island nuclear power plant also made its way to INL.

Unlike Hanford and Savannah River, chemists at INL developed an alternative to pouring liquid high-level wastes from reprocessing into almost endless underground tanks. The approach, known as *calcination*, removed water from the waste and reduced it to a solid. Calcination takes its name from its most common application to make cement from calcium carbonate (limestone). The process is carried out in furnaces or kilns of various designs. The scientists at INL used a *fluidized bed*. In this approach, hot air flowed into a bed of sand-like material suspending the grains like popcorn being air-popped in a movie theater lobby. The hot "fluidized" grains were sprayed with liquid high-level waste. While the heat vaporized the water, the remaining part of the waste adhered to the solids. [23]

At the end of 1963, calcination was started up at full scale. Within a year, half a million gallons of liquid high-level wastes had

Figure 5.5 Damaged reactor core materials being shipped from Three Mile Island to Idaho National Laboratory.
Source: US Department of Energy.

been transformed into 7,500 cubic feet of solid – a better than 9 to 1 reduction in volume. The end product, known as calcine, has the consistency of laundry detergent. The calcine was placed in giant bins which were grouped together in a "bin set" inside a thick reinforced concrete vault, and surrounded by earth and gravel shielding. By May 2000, when the operation was shut down, nearly all of the liquid waste from reprocessing of spent fuel at INL had been calcined.

The design life of the calcine storage bins is asserted to be 500 years, compared to a few decades for the tanks at Hanford and Savannah River. Although the 500-year design life has been challenged, the situation is much less urgent than the liquid wastes stored in tanks at Hanford and Savannah River. In 2010, the Energy Department decided to use hot isostatic pressing to convert the calcine into "ceramic-like" waste forms. The remaining liquid waste at INL (about a million gallons) was left in its original state in stainless steel tanks. The sludge and saltcake that greatly complicates tank cleanup at Hanford and Savannah River does not exist at INL. Unfortunately, the practices at INL were never transferred to the other sites. [10, 24]

Nevertheless, the waste operations at INL have their own problems. The bins contain numerous internal obstructions, and the first bins were built without openings for the calcine's eventual removal. The INL site also has its share of soil and groundwater contamination. Prior to 1984, treated wastes from reprocessing spent fuel were discharged directly into the basalt aquifer beneath the site.

The citizens of Idaho have long feared that INL will become a permanent dump for nuclear waste. For many years, Idaho sought a commitment from the federal government that the spent fuel and reprocessing wastes would be removed from the State. As a result of this unflagging pressure, the Department of Energy and US Navy signed the Idaho Settlement Agreement, in October 1995. This binding agreement outlines what wastes can enter the State and what wastes must leave, and by when. The high-level reprocessing waste must be treated and ready for final disposal, and the spent fuel must be completely removed, by 2035. Failure to meet this deadline carries a penalty of $60,000 a day. In addition, if DOE fails to meet any of the agreement milestones, the State may ask the federal court to halt further spent fuel shipments to INL. Just as for spent fuel at the Nation's commercial power plants, the federal government is legally committed to a deadline for removal of high-level waste from INL – with the American taxpayer on the hook as a backup plan. [23]

TANK CLEANUP – IN THE EYE OF THE BEHOLDER

How clean is clean? This seemingly simple question has confronted industry and communities since Congress began passing a series of environmental laws in the late 1960s. Tank cleanup is no exception. At what point can the tanks be closed? Must they be removed altogether? What waste can be buried on site? What waste must be sent to a geologic repository?

A key underlying question in these debates is what constitutes "high-level waste?" The 1982 Nuclear Waste Policy Act defined high-level waste in vague terms based on the waste *source*, not its radioactivity. The Act defines high-level waste as the highly radioactive part of reprocessing waste and "other highly radioactive material" that the Nuclear Regulatory Commission (NRC) determines needs permanent isolation. The NRC decided to include spent nuclear fuel in the high-level waste category.

Treatment, packaging, and shipment of 100 million gallons (380 million liters) of tank waste from Hanford, the Savannah River site, and INL to a geologic repository would be extraordinarily expensive. Moreover, radionuclides comprise much less than one percent of the mass of the tank waste. To reduce costs and the space requirements for a geologic repository, the plan is to chemically separate the waste into two "streams," a high-activity stream and a low-activity stream. The high-activity portion, which contains most of the radioactivity, theoretically goes to a geologic repository. The low-activity portion will be disposed of on-site. At the Savannah River site, where they are beyond the planning stage, the low-activity waste is being immobilized in a cement waste form for shallow onsite disposal. At Hanford, planning for on-site disposal is still underway. The low-activity waste is somewhat benignly referred to as "waste incidental to reprocessing." [16, 24]

Processing waste into two separate waste streams inevitably leaves residual amounts of cesium, strontium, and transuranics in the low-activity waste. Current practices for handling low-activity waste are far better than the Cold War approach of simply dumping the waste into the local soil and groundwater. Nevertheless, separating out low-activity waste for on-site disposal has been a source of heated debate. In the words of opponents, such a practice would create "national sacrifice areas" within the defense sites.

In 2002, the Natural Resources Defense Council, Snake River Alliance, and Confederated Tribes and Bands of the Yakima Nation sued DOE over the practice of on-site disposal of the low-activity waste. To DOE's dismay, the federal district court in Idaho ruled in favor of the plaintiffs. This ruling was subsequently reversed by the US Court of Appeals, on the grounds that the case was not yet "ripe" for judicial determination. Although given a temporary reprieve, the Energy Department had no assurance that their plans for waste separation and tank closure would eventually hold up in court. [10]

Left in legal limbo, DOE obtained a remedy from Congress. Senator Lindsey Graham (R-SC) attached an amendment to the Ronald Reagan National Defense Authorization Act of 2005, to allow some waste from spent fuel reprocessing to be legally defined as *not* high-level waste, thereby qualifying it for on-site disposal. Environmental groups immediately cried foul – a deal had been cut behind closed doors with no debate or public input. Worse yet, this amendment had been attached to a bill to support the troops, making it difficult to vote against the idea. For jurisdictional reasons, Graham's amendment applies only to INL and the Savannah River site, not to Hanford. [25]

In addition to the debates over what wastes must go to a repository, there's another cleanup problem – it's impossible to remove all the waste from the tanks. Over time, this residual waste has hardened and chemically bonded to the tank structure, making it almost part of the tank. Because the tanks cannot be completely cleaned, some people argue that they must be removed. The counter argument is that removing the tanks would result in additional hazards to workers for negligible, if any, benefits far in the future.

Removing the tanks would also be a massive undertaking. The removal of all single-shell tanks at Hanford would add up to enough steel to build 14,000 cars or 47 sports arenas, such as Seattle's King Dome; enough concrete for the foundations of 30,000 small houses; and 130,000 cubic yards of contaminated soil. As such, plans are to remove as much of the radioactivity as possible (a goal of 99 percent or more) from the tanks and then to fill them with grout or other stable material. [16]

By 2006, only two of the 246 tanks and bins at the three sites had been cleaned and closed. Cleanup of the tank waste remains a long-term and extremely expensive legacy of the Cold War era. And yet, the tank wastes are not the greatest danger from the Cold War operations. Separated plutonium – the product of the grand enterprise – presents a much greater danger to society. [10]

AWASH WITH PLUTONIUM

The Atomic Energy Act of 1946 established that the President is the ultimate authority for nuclear weapons production. At least once each year, the President would determine how many weapons should be manufactured. In April 1947, David Lilienthal, soon to be confirmed as the first chairman of the AEC, gave President Truman his first briefing on the existing supply of atomic bombs and nuclear materials. A stunned Truman was informed that the Nation had exactly zero nuclear weapons ready for use. [26]

That was 1947, in the incipient stages of the Cold War. Forty-four years later in 1991, the United States and Russia agreed to end the Cold War by dismantling some 30,000 nuclear warheads. In doing so, the world was suddenly bequeathed more than 100 tons of pure, weapons-grade plutonium. The US and Russia began dismantling thousands of warheads each year at the Pantex facility near Amarillo, Texas and in Siberia, producing stockpiles of plutonium "pits." Each pit is about the size of a grapefruit. Many bombs worth of separated plutonium also remained at Hanford, the Savannah River site, Los Alamos, and elsewhere in the DOE weapons complex. The world was awash with plutonium. [27]

Aside from being a tempting target for theft, long-term storage of this plutonium from dismantled warheads on US or Russian soil would hardly convince the world that either country was serious about disarmament. The plutonium had to be destroyed or disposed of in a manner that assured that any later retrieval and reassembly into bombs would be out of the question. Industry and government churned out thousands of pages of documents on options, ranging from detonation of underground nuclear explosions to launching the plutonium into space. The group holding the most sway in these discussions was a plutonium study group formed by the National Academy of Sciences. Wolfgang Panofsky, former director of the Stanford Linear Accelerator Center, chaired the group. He was a superb choice.

At the end of World War II, Wolfgang "Pief" Panofsky was considered one of the most brilliant and promising young physicists in the United States. His father, Erwin, had been a world-famous art historian and a professor in Germany. When Hitler came to power in 1933, Jews teaching at German universities were dismissed from their positions. Soon thereafter, Erwin was offered a permanent position at Princeton's new Institute for Advanced Studies. [28]

At the age of 15, Panofsky entered Princeton. Four years later, he graduated with highest honors and began graduate work in physics at Caltech. After completing his Ph.D. he stayed on at Caltech teaching evening classes and conducting war-classified research. His work soon came to the attention of Berkeley physicist Luis Alvarez, who brought Panofsky into the Manhattan Project at Los Alamos. His job entailed developing shock-wave calibrators that would be used to measure the yields of the atomic bombs dropped on Hiroshima and Nagasaki.

When the war ended, Panofsky joined Alvarez at Berkeley in the Nation's premier physics department. Twenty-six-year-old Pief had landed in the physicist's *Big Apple*. With the field of particle physics still in its infancy, he demonstrated a unique brilliance for machine building. Spending his days working in Ernest Lawrence's Radiation Laboratory (dubbed the Rad Lab), he played a key role in developing the earliest tools of particle physics – the cyclotrons, the synchrotrons, and the linacs. Using war-surplus radar gear, Panofsky and Alvarez built the first linear proton accelerator. Regarded as a precious resource in the physics department, Panofsky envisioned a long and productive career at Berkeley. Then unforeseen circumstances intervened.

In the spring of 1949, the California State legislature proposed adding an anticommunist and antisubversion clause to the State's oath of allegiance, applicable to all State employees. Berkeley president Robert Sproul went one step further by proposing that all university employees be required to sign an additional loyalty oath declaring that they were not members of the Communist Party. The regents agreed. Many faculty members were outraged, arguing that their academic freedom and constitutional rights were being violated. For Panofsky and the other physicists involved in classified research in the Rad Lab, refusing to sign the oath could have cost them their positions.

When prestigious eastern universities learned of the turmoil at Berkeley, they saw this as an opportunity to build up their own physics departments. Panofsky received offers from Harvard and Columbia, yet with apparently little thought he turned them down. Berkeley was his natural element. In addition, the young physicist and his family liked California. With no wish to leave, Panofsky signed the loyalty oath.

On June 23, 1950 the board of regents fired 31 faculty members, including two professors in the physics department. Panofsky had hoped to weather the storm, but after the firings he informed Ernest Lawrence and Luis Alvarez that he had decided to leave Berkeley. Deeply dismayed at the prospect of losing his young star, Ernest Lawrence devoted all of his considerable influence and connections to work something out.

While Lawrence was setting up clandestine meetings between Panofsky and one of the regents, a flurry of letters made their way from Princeton to Berkeley. Pief's parents – who had lost home, work, and country because of fascism – were deeply alarmed that their son might be bribed into staying. In one of his more forceful letters, Erwin wrote: "But what is now, I feel, imperative is that you do not, under any account, accept a continuance of your appointment at Berkeley. If you did so, you would be considered as one who had allowed himself to be bribed by about the worst enemy of academic freedom in the whole United States." Erwin signed off with a dramatic flourish, telling Pief that he would share his last piece of bread with him and his family, but he didn't know how he could face his friends if his son allowed himself to be bribed into staying at Berkeley. His parent's missives were probably unnecessary. Once the decision was made, he never appeared to change his mind. In his autobiography, Panofsky gives only the briefest account of this difficult period in his life.

More prestigious job offers arrived and, once again, Panofsky turned them all down. In an audacious move, he decided to go across the bay to Stanford University. In the early 1950s, this now world-class university was not known for high academic standards. In many ways, Stanford was just one more private university catering to rich kids who enjoyed the excellent climate and proximity to San Francisco. However, after World War II, the university embarked on a serious plan to expand its science and engineering departments and staff them with world-class faculty. When Panofsky informed his friend and mentor, Luis Alvarez, of his decision to go to Stanford, Alvarez pleaded: "Oh Pief you'll fade

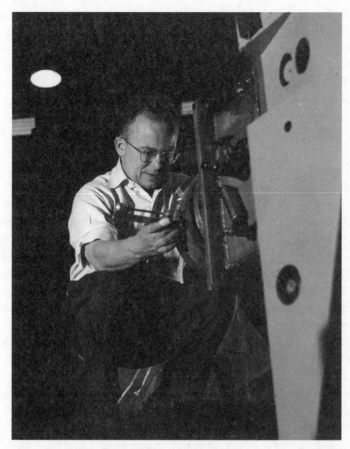

Figure 5.6 Wolfgang Panofsky adjusting an analyzing magnet used in meson experiments performed with the 1-Bev linear electron accelerator at the High Energy Physics Lab, Stanford, CA, 1959.
Source: Emilio Segrè Visual Archives. Copyright Stanford University, David Rhinelander / Stanford News Service; Stanford News Service Number NS5142.

away at Stanford, nothing goes on there, you'll never be able to do any significant research."

Luis was wrong, and Berkeley's loss became Stanford's gain. Panofsky became one of the giants of twentieth-century experimental physics. Initially, he worked on improving the Mark III linear accelerator, which led to a series of landmark experiments. The need for a significantly larger energy machine had been discussed in the Stanford physics department for almost a decade. Through Panofsky's pioneering vision, he showed the foresight and courage to embark on a five-year-long battle to construct the Stanford Linear Accelerator Center (SLAC). Eventually funded at enormous cost, this two-mile (3 km)-long "laboratory" would rank among the most productive research facilities ever constructed. For over two decades, Wolfgang Panofsky served as the first Director of SLAC, overseeing research that led to many new discoveries, and three Nobel Prizes, in the fundamental nature of matter.

Stepping down as Director of SLAC in 1984, Panofsky turned his attention to science-related international issues. His role in the Manhattan Project had profoundly influenced his thinking on the ethical responsibilities of scientists. For eight years, he chaired the Committee on International Security and Arms Control of the National Academy of Sciences. Assuming the role of a high-level government advisor on vital issues of arms control and international security, Panofsky influenced presidents from Eisenhower to Carter. He helped to secure the Atmospheric Test Ban Treaty during the Kennedy Administration, and the Anti-Ballistic Missile Treaty during the Nixon Administration. In his later years, he helped found Stanford's Center for International Security and Arms Control.

Among his many awards and accomplishments, Panofsky also had a life-long commitment to open scientific exchange across international boundaries. At a time when the world had become ideologically polarized, he cemented SLAC's ties to laboratories in the USSR and the People's Republic of China, helping to bring these nations into the worldwide scientific community.

Panofsky's stellar scientific background, combined with his science-without-borders work, made him a natural choice for chairing the National Academy of Sciences plutonium study group. He and his committee considered more than 30 different options for ridding the world of its excess plutonium. In the end, the committee concluded that only two options deserved serious consideration. One was mixing the plutonium with uranium and burning it in commercial reactors. The second was mixing it with highly radioactive waste. Neither solution was entirely satisfactory, but either could be made to work. For both options, the ultimate destination was a geologic repository and, importantly, both would lead to what the committee called the "spent fuel standard." [29]

The spent fuel standard required that the plutonium be made as inaccessible for use in weapons, as the plutonium in the spent fuel from civilian power reactors. In effect, it must be made too hot to touch. Ironically, this high radioactivity that makes high-level waste so dangerous was the desired end game for plutonium. As one official said, the goal is to turn plutonium "from the nastiest thing on the planet to the second nastiest thing on the planet." [27]

The first option proposed by Panofsky's committee was to mix plutonium with uranium in a mixed oxide (MOX) fuel for use in commercial reactors. Although most of the plutonium would remain with the spent fuel, once it passed through a reactor core a terrorist would have great difficulty getting near it. Even 50 years later, the radiation level a meter away would be enough to assure a lethal dose in less than an hour. In addition, spent fuel assemblies are much too heavy to cart off, making them easy to keep track of. Proponents of the MOX option also argued that it made sense to use the plutonium for generating electricity, even though MOX is far more expensive than uranium as a nuclear fuel. [30]

Opponents of the MOX option argued that it ran counter to long-standing US nonproliferation policy, which bans reprocessing and the use of plutonium in civilian reactors. Panofsky argued that it was "simply wrong" to link the MOX option to the acceptance of reprocessing for commercial reactors. The intent of the MOX option was to destroy plutonium we wish we didn't have, while the intent of reprocessing is to generate new plutonium. [31]

The second option proposed by Panofsky's committee involved mixing the plutonium with liquid high-level waste from Hanford or the Savannah River site, vitrifying it into giant glass (borosilicate) logs, and disposing of the logs in a repository. The committee noted that vitrification of tank wastes was scheduled to begin soon at the Savannah River site and might be modified to include plutonium. Vitrification raised fewer security risks in handling than the MOX option, because the process of mixing plutonium with high-level waste would be easier to safeguard than the more complex process of fabricating MOX fuel. However, research was needed to address potential problems of criticality or recovery for weapons use. It was not entirely clear that the vitrification option would meet the "spent fuel standard."

Although a number of groups condemned the MOX option as sending the wrong signal to the world, it had one major advantage – it was the only approach acceptable to the Russians. The Russians view their plutonium stocks as a national treasure come by at enormous sacrifice, and are unwilling to adopt a "throwaway" option such as vitrification into glass logs. They were also skeptical that plutonium would be sufficiently irretrievable from the glass logs to meet the spent fuel standard. Given the Russian government's precarious control over the plutonium extracted from its dismantled warheads, it was critically important to get Russia onboard with a plan of action.

The NAS committee proposed that either approach was acceptable, and left the political decision to others. Because some so-called "scrap" plutonium is too contaminated with impurities to be used in MOX fuel, both approaches may be needed. Panofsky reminded everyone involved that the surplus plutonium was "a clear and present danger" to national and international security. "The name of the game is security, not economics," he emphasized. To make his point, Panofsky cited the "value" of one ton of plutonium. To a government budget officer, it was minus $25 million. To an energy conservationist, it was about 1,000 megawatts of electricity for one year. To the Russians, it was "sunk cost" to be recovered, corresponding to about 2,000 man-years of past socialist labor. To Saddam Hussein, it was 250 nuclear bombs. [27, 32]

On December 9, 1996, Energy Secretary Hazel R. O'Leary announced the Department's intention to pursue a dual track, using both approaches proposed by Panofsky's committee. "The arms race is over," she said. "Our struggle now is to get rid of this sea of plutonium." In 2000, the United States and Russia signed the Plutonium Management and Disposition Agreement, in which each country committed to dispose of 34 metric tons of surplus weapons-grade plutonium. This was considered a first step, with more to come later. [33]

In 2002, the Bush Administration announced its preference for the MOX option. Plans for the vitrification option were canceled. In 2007, construction of a facility to fabricate the MOX fuel began at the Savannah River site. The goal was to begin operations by 2016 and continue for approximately 14 years, providing MOX fuel equivalent to about one year's worth of electricity to 50 million homes. Construction of a second facility at the Savannah River site to process the wastes began in 2009. A third facility to disassemble pits from nuclear weapons and prepare the plutonium for the fabrication plant is still in the design phase. [34–35]

The complexity of the effort, changing plans, and reliance on a single option have caused considerable anxiety in the State of South Carolina, which fears that if the plans fall apart, the State is left holding the bag for indefinite long-term plutonium storage. The government has yet to sign up a single customer for the MOX fuel. The most likely customer, the Tennessee Valley Authority, delayed the decision after the Fukushima Dai-ichi disaster. One of the damaged Fukushima reactors had six percent of its core made from MOX, causing added concern about the disaster. [36]

While some progress is being made in dealing with the stocks of military plutonium from dismantling bombs, no comparable policy exists for the larger and growing worldwide plutonium stocks that come from civilian reprocessing of spent fuel. This sea of plutonium is piling up in the UK, France, Russia, Japan, and a few other countries. It is more difficult to assemble a plutonium bomb from this "reactor-grade" plutonium (due to greater accumulation of plutonium isotopes other than plutonium-239 in commercial reactors), but a bomb can still be made. Equally stringent protection policies are essential for both reactor-grade and weapons-grade plutonium.

Experience has shown that tight security systems, safeguarding a wide range of valuable materials, can occasionally be breached by determined and clever groups. Finding an acceptable long-term solution for high-level radioactive waste is not only critical for dealing with nuclear energy, but would remove one more obstacle in efforts to make the world safer from nuclear bombs.

6

The peaceful atom and its wastes

Nuclear energy was conceived in secrecy, born in war, and first revealed to the world in horror. No matter how much proponents try to separate the peaceful from weapons atom, the connection is firmly embedded in the minds of the public.

K.R. Smith [1]

In June 1948, Robert Oppenheimer delivered a stinging indictment to the AEC Commissioners for dragging their heels on what he considered the most crucial application for atomic energy – nuclear power. While harboring serious misgivings about the weapons application of atomic energy, Oppenheimer fervently believed in the peaceful uses of the atom. "We despair of progress in the reactor program," the Father of the Atomic Bomb told the Commission. As scientific director of the Manhattan Project, Oppenheimer had steered the project to its successful and horrific completion in an astonishing two-and-a-half years. Now, almost three years since the war ended, the AEC had made next to no progress with nuclear power. [2]

Oppenheimer was not alone in his frustration. Enrico Fermi considered reactor development to have lost its zest. In the early days of atomic energy, a period that Oppenheimer cheerfully referred to as one of "general deviltry," a small group of scientists was allowed to work with almost complete independence. Such autonomy had been replaced with mounting lists of rules and regulations. From the perspective of these atomic pioneers, the AEC was increasingly bogged down and handicapped by hesitancy. The days of vision and daring appeared to be in the past. [2]

By the end of the year, the Commission finally made some progress on nuclear power by deciding which prototype reactors to build. Yet, the program remained stymied until a testing site was selected. In the late 1940s, most reactor R&D was being conducted at Argonne National Laboratory near Chicago. To Fermi and Oppenheimer, Argonne seemed like the logical place to build a series of large experimental reactors. Others disagreed. Edward Teller argued that the "calculated risks" taken by the Manhattan Project could not be justified in peace time, and certainly not in the suburbs of the Nation's second largest city. Teller favored a remote place far away from Argonne. As head of the AEC Reactor Safeguard Committee, Teller prevailed. [3]

Candidates for the reactor testing site had to meet several criteria. Fewer than 10,000 people and no national defense sites could be in the vicinity. Water and electrical power should be plentiful. Finally, the AEC must have complete control of the property. The search narrowed from 20 candidates to two finalists: Fort Peck along the Missouri River in Montana and the Naval Proving Ground in southeastern Idaho. In February 1949, the Idaho site was selected and construction was soon underway on 400,000 acres of high, windswept desert. First called the National Reactor Testing Station, the fledgling site would eventually grow into the Idaho National Laboratory. [4]

Six months later, the Soviet Union simultaneously detonated its first atomic bomb and kicked the Cold War into high gear. Shortly after the Soviet explosion the theoretical physicist, Eugene P. Wigner, gave a speech at Oak Ridge National Laboratory. Having conceived many of the early principles for reactor development, Wigner had often questioned why the high hopes for nuclear power had not come to fruition. Part of the answer was that weapons had been a higher priority. Yet, to Wigner's mind, winning the war also eliminated the crucial element of competition – in this case, with Germany. He believed that the lack of competition had profoundly affected atomic advancement for both war and peace. With the Soviets entering the atomic age, the competition was back. "We will stop glorifying our past," Wigner told the scientists. [2]

A major impediment to nuclear energy development was the perceived shortage of uranium. As a result, the AEC hoarded uranium for military use. To help address the uranium shortage, Walter H. Zinn, director of Argonne Lab and a key scientist in reactor design and development, proposed building a "breeder" reactor that would generate electricity and manufacture plutonium at the same time. The AEC finally agreed to provide a small amount of uranium for the experiment, and scientists from Argonne began to assemble the Experimental Breeder Reactor-1 (EBR-I) at the new National Reactor Testing Station. Not far away, Admiral Rickover was overseeing development of the prototype for the first nuclear-powered submarine.

(a)

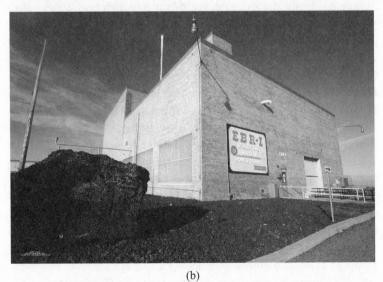

(b)

Figure 6.1 Experimental Breeder Reactor-I: (a) First electricity production
by nuclear energy, illuminating four light bulbs, and (b) National Historic
Landmark.
Source: US Department of Energy.

In May 1951, Zinn flew in from Chicago for the first attempt to bring EBR-I to criticality. More than eight years earlier, Zinn had been one of Fermi's assistants in charge of constructing the first successful nuclear reactor at Stagg Field. This time, however, the test was a flop – the amount of uranium-235 needed to bring EBR-I to criticality had been underestimated. Zinn requested more of the precious uranium-235 fuel from the AEC and tried again three months later. This time EBR-I went critical, but the fuel rods still needed to be reconfigured. Over the fall, the scientists ran low power tests until Zinn arrived on the scene a few days before Christmas. [2, 4]

On December 20, 1951 the historic experiment finally began. EBR-I was started up and the team gradually increased the power. At 1:23 p.m. power began to flow from the football-sized reactor core through the turbine generator, lighting up a string of four light bulbs. For the first time, a usable amount of electricity was produced from nuclear power. Walter Zinn brought out a bottle of champagne. After a second experiment the next day, Zinn chalked his name on the wall beside the generator and invited those present to do the same, recording this historic achievement with the simple proclamation: "Electricity was first generated here from Atomic Energy on Dec. 20, 1951. On Dec. 21, 1951 – all of the electrical power in this building was supplied from Atomic Energy." Afterwards, one of the engineers sketched a devilish figure exhaling smoke to add a little flair to the wall display. [4]

EBR-I is the Kitty Hawk of nuclear power. Housed in its original small and Spartan cinderblock building, EBR-I is now a registered National Historic Landmark. It sits in the company of Valley Forge, the Hoover Dam, and Chicago Pile-1, where Fermi's reactor first sustained a nuclear reaction. President Lyndon Johnson came to Idaho on August 26, 1966 to dedicate the site. Lady Bird Johnson, AEC Chairman Glenn Seaborg, and Admiral Rickover were among the dignitaries present. [4]

In addition to producing the first nuclear-powered electricity, EBR-I also suffered the first meltdown in American reactor history. In November 1955, Walter Zinn was pushing the reactor to high temperatures as part of experiments when an operator misunderstood an instruction and reinserted a control rod too slowly. Alarms went off and everyone quickly evacuated the building. The event produced no sound, no smoke, no explosion, yet melted about half the reactor core. No one was hurt, so the AEC decided not to inform the public. Several months later the news leaked out in the national press, causing a stir. The meltdown was a premonition of the vulnerability of nuclear reactors to human error – a vulnerability that would be demonstrated years later by the Three Mile Island and Chernobyl accidents. [4]

SPENT NUCLEAR FUEL

From this humble beginning, the "front-end" of the modern-day nuclear fuel cycle has developed into a sophisticated industrial undertaking. It begins with the mining and chemical extraction of uranium. The uranium is then enriched in uranium-235 and fabricated into pellets the size of pencil erasers. These pellets are then loaded into 12-foot (4 m)-long fuel rods, usually made of a zirconium alloy. The metal fuel rods are bundled together in groups of 100 to 300 to form fuel assemblies. After about three to six years in a reactor, the efficiency of the fission reaction is significantly reduced by the buildup of fission products, and the fuel rods are removed. It is now time for the "back-end" of the nuclear fuel cycle. Here, progress has been stymied.

At the time of its removal, spent nuclear fuel is about *one million times* more radioactive than when the uranium fuel rods were placed into the reactor. If the fuel rods were left in the open air, the metal surrounding the nuclear material would melt or self-ignite. As a result, the spent fuel is immediately immersed in a pool of water to cool it and to block the radiation. The heat output drops by 99 percent in the first year and by another factor of 5 by the time the spent fuel is 5 years out. Even then, it is extraordinarily hot. [5]

Fission products – the "ashes" of the fission fire – make up about three to four percent of the spent fuel, but cause most of its radioactivity. Although hundreds of fission products are formed in a reactor, most have very short half-lives and decay within days to weeks of their formation. A decade later, the radioactivity is dominated by cesium-137 and strontium-90. After several hundred years, when the cesium and strontium have died down, the fission product of most concern is technetium-99, with a half-life of about 211,000 years. There are no significant fission products with half-lives between 30 years for cesium-137 and 211,000 years for technetium-99 – a remarkable gap. Technetium-99 and the even longer-lived fission product, iodine-129 with a half-life of 16 million years, are very soluble and mobile in groundwater. These two radionuclides are of considerable long-term concern.

A particularly troubling part of the waste is the *transuranic* component – all those elements heavier than uranium. These long-lived radioactive isotopes differ from the fission products in that the neutrons do not split the atom, but rather are captured by it. The transuranic component is mainly a blend of several plutonium isotopes with lesser amounts of americium, curium, and neptunium. Although the transuranic elements make up only about one percent of the spent fuel, they constitute the main source of the long-term nuclear waste problem. Nearly all of their isotopes are hazardous, long-lived alpha emitters. Many of these isotopes have half-lives from tens of thousands to millions of years.

Plutonium-239 is the transuranic isotope of most critical importance because of its high concentrations in the spent fuel and association with nuclear weapons. Plutonium-239 is created from the interaction of neutrons with uranium-238, the dominant uranium isotope. In a conventional reactor, uranium-238 rarely splits when hit by a neutron. It more likely absorbs the neutron to become the slightly heavier uranium-239. This sets off a quick sequence of beta decay transformations that convert the uranium-239 to plutonium-239. The artificially created plutonium-239 is even more fissile than the naturally occurring uranium-235. When spent fuel is removed from a reactor, plutonium-239 is typically contributing more than half the power being generated.

While the spent nuclear fuel is extremely hazardous, the volume of waste is small compared with most energy sources. If an American got all of his or her electricity from nuclear energy, that person's share of the spent fuel over a typical lifespan would weigh two pounds and fit into a single Coke can. This low volume of high-level waste convinced many nuclear pioneers that the waste problem would be trivial. But the Coke can analogy is not the full story. Large volumes of lower-level radioactive wastes are produced during uranium extraction and enrichment. Uranium enrichment also provides a troubling link between the peaceful and non-peaceful uses of nuclear energy. [6]

URANIUM ENRICHMENT

Uranium-235 is unique in being the only isotope to occur naturally (beyond trace amounts) that has the coveted property of being *fissile* – in other words, is capable of sustaining a nuclear chain reaction. Uranium-235 is rare, making up only 0.7 percent of the natural uranium. More than 99 percent of uranium atoms are uranium-238. To serve as fuel for most nuclear power plants, the uranium-235 content must be increased (enriched) to between three and five percent. A uranium-based atomic bomb requires enrichment to at least 20 percent, and more "ideally," greater than 90 percent uranium-235.

Enrichment is challenging, as the two uranium isotopes have virtually identical chemical properties. Only the weights differ, and these almost imperceptibly. It is this small difference in weight that is exploited in the enrichment process. Enrichment processes use a large number of identical stages to produce successively higher concentrations of uranium-235.

Gaseous diffusion was the first enrichment technique used in large-scale production. In this technique, solid uranium is converted into a gas, uranium hexafluoride (called *hex* for short). The hex gas is circulated over and over through fine filters with tiny openings. The lighter uranium-235 atoms pass through the filters slightly more easily than the heavier uranium-238 atoms. Step by step, the hex gas is enriched in uranium-235. After sufficient enrichment, the gas is converted back to solid form for fabrication into a bomb or nuclear fuel.

Gaseous diffusion plants cover large areas and consume huge amounts of electricity. The first gaseous diffusion plant was built at Oak Ridge, Tennessee, as part of the Manhattan Project. The U-shaped plant was four storeys high with two half-mile-long sections. It covered an area the size of 35 football fields. Largely as a result of its three uranium enrichment plants, Oak Ridge used one-seventh of the Nation's electricity during the peak of the Manhattan Project – about the same electricity usage as all of New York City. Ironically, much of this electricity was generated by strip-mined coal.

Over time, a much more efficient approach for enrichment was developed by using gas centrifuges. A gas centrifuge works much like a classic centrifuge, with a hollow cylinder tube that is spun at very high speeds about its axis. The rotation creates a strong centrifugal force so that the heavier gas molecules containing uranium-238 move toward the outside of the cylinder and the lighter gas molecules, rich in uranium-235, move inward toward the axis of rotation. Gas centrifuges require much less energy to achieve the same separation as the older gaseous diffusion process.

Modern centrifuges can be quickly converted from peaceful to non-peaceful uses to obtain bomb-grade uranium. Unlike the gargantuan gaseous diffusion plants, centrifuges are easy to hide and produce little detectable signal. For these reasons, uranium enrichment facilities in the hands of countries like Iran and North Korea are of grave concern. While purporting to be enriching uranium for nuclear power, they could be (on the sly) bringing it up to bomb grade. [7]

Decades of uranium enrichment for military and civilian uses have generated enormous quantities of remnant uranium. The Department of Energy stores more than 700,000 metric tons of this *depleted* uranium in metal cylinders at its former enrichment plants in Kentucky and Ohio.

Although considered a waste product and an environmental liability, depleted uranium is one part of the nuclear fuel cycle that can be put to use. Depleted uranium could serve as a future fuel for a fast reactor or, if uranium prices are sufficiently high, re-enriched and used in a light-water reactor. Depleted uranium is also potentially useful for nuclear waste containers, serving to dilute fissionable uranium to inhibit criticality. A few other markets exist. Civilian applications, resulting from its high density, include use as counterweights for airplane rudders and elevators. The military has made use of depleted uranium by shaping it into projectiles that penetrate enemy armored cars and tanks with ease. It also makes good shielding to protect *our* armored cars and tanks. These military uses are controversial because of uranium's chemical toxicity. [8]

Niels Bohr was the first to recognize that the rare isotope uranium-235 was responsible for nuclear fission. Bohr's epiphany came a few months after the discovery of fission, as he walked across a snowy Princeton University campus. Bohr continued to his office borrowed from Albert Einstein, where, without a word and to the amazement of fellow physicists, he sketched the entire theory on a blackboard. Princeton physicist, John A. Wheeler collaborated with Bohr to work out the fine points. On September 1, 1939, the same day Germany invaded Poland and launched World War II, Bohr and Wheeler published the first detailed explanation of nuclear fission and the role of uranium-235. This groundbreaking work helped make possible the brave new world of harnessing the atom for both war and for peace. [9, 10]

While the name John Wheeler doesn't have the same level of public recognition as Niels Bohr, he was one of the foremost theoretical physicists of the twentieth century. Wheeler coined the term *moderator* for the material that slows neutrons in a nuclear reactor, as well as christening and popularizing the term *black hole*. Wheeler was a virtual hero of the first nuclear reactor at Hanford. He had insisted on "over-designing" the reactor due to a theoretical physicist's version of a nagging feeling. If it were not for John Wheeler, the reactor would have failed ("poisoned" by the fission product xenon), critically delaying plutonium production. Wheeler's talent and contributions also saved him, and his reputation, in one of the more bizarre incidents of the Cold War. [11]

During the first week of January 1953, Wheeler received a highly classified document detailing the design and operating principles of the first hydrogen bomb. Pressure of work did not give him time to read it, so he brought it with him on a trip to Washington, DC. Wheeler took special precautions to keep this and other highly classified documents in his possession during the overnight train ride, yet the next morning they were gone. After a frantic search, he found the packet with all the documents intact, but for one – the hydrogen bomb document was missing. When Wheeler arrived at the Capital, he reported the loss to the congressional Joint Committee on Atomic Energy, who immediately ordered Pullman officials to impound the sleeping car and all the train's laundry and trash. No document. They proceeded to partially dismantle the Pullman car. The document was gone. [12]

It was hard to imagine how anyone could have lost a more sensitive document. It was also unfathomable that information of this sensitivity could have been transported without an armed escort. The loss was promptly reported to the newly elected President Eisenhower. Appalled by such an incredible security lapse, Eisenhower summoned the five AEC Commissioners to demand an explanation. After lining them up like five errant schoolboys, Eisenhower gave vent to an extraordinary display of anger. *In the Army*, Eisenhower informed them, such a security lapse would be dealt with *swiftly and severely*.

Eisenhower became convinced the theft was a Soviet "inside" job. On the night of January 6, when Wheeler made his ill-fated trip, many Rosenberg sympathizers were traveling to the Capital to demonstrate for clemency for the two Russian spies. Eisenhower not only denied clemency, but the Rosenbergs were dealt with swiftly and severely. In comparison with the document Wheeler had lost, the atomic secrets the Rosenbergs gave the Soviet Union were relatively inconsequential. Unlike the Rosenbergs, however, no one questioned Wheeler's integrity and loyalty to the United States. The chairman of the AEC defended Wheeler to the President-elect as "a scientist of exceptional abilities, a man so gifted the nation could not afford to lose his services." Wheeler received a reprimand and, for him, the matter ended.

7

Recycling

Perpetual care is neither difficult nor costly, chiefly because the inherent volume of nuclear waste is small.

Chauncey Starr and R.P. Hammond (1972) [1]

Recycling as a means to reduce waste and conserve resources is viewed as good for the environment. In a society that routinely recycles aluminum cans, plastic containers and cardboard, the idea of recycling nuclear waste seems self-evident. So why aren't we doing it?

In the nuclear world, recycling refers to reuse of nuclear spent fuel after its recovery through reprocessing, although the two terms *recycling* and *reprocessing* tend to be used interchangeably. Reprocessing was well developed by the atomic weapons program long before commercial reactors went on-line. The world's first industrial-scale reprocessing plant began operation at Hanford in December 1944. The plant produced the plutonium that went into the bombs exploded at the Trinity test site in New Mexico and dropped on Nagasaki, Japan.

In the early days, it was taken for granted that reprocessing would be incorporated into the commercial application of nuclear energy. Advocates of reprocessing pointed out that it would close the nuclear fuel cycle. Since spent fuel contains 95 percent of the original uranium plus some plutonium, conceivably there is lots of potential nuclear energy remaining that would be "thrown away" in the once-through cycle. The only problem is that reprocessing of spent nuclear fuel turned out to be a technologically complex, expensive, accident-prone, and very messy business – *and* only marginally addressed the waste problem.

The standard worldwide approach to reprocessing is known as PUREX for plutonium-uranium extraction. The basic process doesn't sound like much of a big deal. When spent reactor fuel rods enter a reprocessing plant, they are first chopped up into small sausage-sized pieces. Then nitric acid is added to dissolve the fuel, leaving behind the metal cladding as waste. At this point, the uranium, the fission products, and the transuranics are all in one solution. The next step involves mixing this radioactive soup with an organic solvent, causing the uranium and plutonium ions to migrate into the solvent. The fission products and transuranics other than plutonium stay behind. Finally, the uranium is separated out and *voilà*, pure plutonium is born. In a typical reprocessing operation, more than 99 percent of the uranium and plutonium is recovered from the spent fuel, while the liquid high-level waste retains most of the other radionuclides.

Curiously, the AEC began to encourage private industry to enter the fuel reprocessing business in the mid 1950s, before there were civilian nuclear-power plants. No one stepped forward. Within a few years, however, the Commission's interests coincided with the State of New York's ambitions to become a leader in futuristic technologies, such as atomic power. In 1961, New York purchased land at West Valley, a sparsely settled hamlet amidst rolling wooded hills and dairy farms about 35 miles (55 km) south of Buffalo. The newly designated Western New York Nuclear Service Center seemed a natural fit for a reprocessing plant. After prolonged negotiations, a construction permit was granted for the Nation's first commercial nuclear fuel reprocessing plant. The plant was built and operated by a new corporate entity, Nuclear Fuel Services. In its eagerness to get into this exciting new technology, the State rashly agreed to assume responsibility for the waste. Nuclear Fuel Services, for its part, would contribute to a "perpetual care fund" for the waste. [2–3]

Operating on the principle "build it and they will come," Nuclear Fuel Services opened its doors for business in 1966. During the five years of the plant's operation, there were still very few commercial power reactors to supply fuel for reprocessing. The AEC made up the difference by contributing spent fuel from its Hanford defense operations – adding up to more than 60 percent of the fuel reprocessed at West Valley.

The AEC insisted that Nuclear Fuel Services set the price for reprocessing artificially low. Trying to do reprocessing on the cheap had major consequences. First, the "perpetual care fund" was completely inadequate for dealing with the waste. Second, reprocessing requires special care to protect workers from excessive radiation exposure. While no expense had been spared on remote-controlled operation and maintenance for the military reprocessing plants at Hanford and the Savannah River site, these innovations could not be afforded at West Valley.

As a result, equipment failures were time consuming and expensive to repair. "Transient" workers were hired to help keep full-time workers within acceptable limits of radiation exposure. The West Valley plant manager described how he once "used six guys to get one nut off." Each of these temporary workers might work for three minutes, but be paid for four hours. Reprocessing was, as one AEC official noted, "the dirty end of the nuclear business." [4]

The plant closed in 1972, ostensibly for upgrading and expansion. It never reopened. Within a few years, Nuclear Fuel Services announced it would be withdrawing from the reprocessing business. The company cited the cost of meeting new seismic requirements, but the plant required many expensive improvements. Furthermore, it was evident that the reprocessing plant would not be able to compete with newer planned facilities. When the lease expired at the end of 1980, the State of New York suddenly found itself the owner of over half-a-million gallons of orphaned high-level radioactive waste.

From 1966 to 1971, the West Valley plant reprocessed more than 600 metric tons of fuel, and generated 600,000 gallons (2.3 million liters) of liquid high-level waste. As another consequence of the low revenue, Nuclear Fuel Services used a carbon steel tank to store most of the waste. Using carbon steel instead of the more expensive, corrosion-resistant stainless steel made it necessary to neutralize the waste. As at Hanford, this short-term expediency had long-term ramifications. Neutralization increased the waste volume, caused formation of an insoluble sludge, and made waste retrieval more hazardous and costly. About 85 percent of the radioactivity became trapped in the sludge at the bottom of the tank. The sludge could not be re-dissolved in acid without also dissolving the tank. Furthermore, steelwork protruding from the tank floor, described as a "giant waffle iron," interfered with attempts to remove the sludge. Another design problem was that access to the tank was limited to a small hole through the roof. The tank had an expected useful life of 40 to 50 years. Something had to be done. [5]

In 1980, after several years of haggling between the State and federal government over who should be responsible, Congress enacted the West Valley Demonstration Project Act. The Act designated West Valley as a testing ground for vitrification – the process of solidifying liquid high-level waste into glass logs enclosed in stainless steel containers. Vitrification of the tank waste finally began in 1996 and was completed in 2002, resulting in 275 canisters indefinitely stored on-site. These highly radioactive glass logs are considered part of the DOE high-level defense wastes, and have been designated first-in-line among defense

Figure 7.1 Inside the vitrification plant at West Valley, New York.
Source: US Department of Energy.

wastes should a geological repository for high-level waste ever open for business. [6]

Eventually, the Nuclear Regulatory Commission must license the decontamination and decommissioning of the entire West Valley site, including the reprocessing plant and all the waste facilities. A small amount of irretrievable high-level waste remains in the tank. The total cleanup is expected to take 40 years and could cost over $5 billion in 2006 dollars. New York is responsible for 10 percent of the cleanup costs, making it the only State that has been forced to pay for cleanup of high-level radioactive waste. [7]

Two more attempts at commercial reprocessing followed after West Valley. Both were costly and neither reprocessed any wastes. In 1967, a small reprocessing plant was constructed by General Electric at Morris, Illinois. The plant was within a stone's throw of the Dresden nuclear power station and was intended to serve a cluster of reactors within a short shipping distance. General Electric developed its own chemical process, but never tested it beyond the laboratory scale. Once the plant was built, test runs with dummy fuel caused repeated clogging of pipes and valves. The $64 million Morris plant simply did not work and was scrapped. In 1971, a large reprocessing plant was built in Barnwell, South Carolina, adjacent to the Savannah River weapons plant. Facing uncertainties about the future of reprocessing, major regulatory changes, and rising costs, the Barnwell plant never opened. After investments of more than $250 million, the plant officially closed at the end of 1983. [8]

The failed efforts at West Valley, Morris, and Barnwell illustrate how the technological and economic pitfalls of reprocessing were greatly underestimated. Concurrently, the benefits of reprocessing to reduce waste had been overestimated. Although the total radioactivity of the waste is reduced as a result of removing plutonium and uranium, the waste still contains the fission products, small amounts of residual plutonium, and transuranics such as americium and curium. Significantly, there is little reduction in the heat generated by the waste, which is the main limiting factor on repository capacity.

Reprocessing also leads to *additional* waste. Cutting up the fuel cladding causes release of gaseous radionuclides, including carbon-14 and iodine-129. These must be captured or vented to the atmosphere. Considerable transuranic waste is also created, including the chopped fuel cladding and plutonium-contaminated filters, protective clothing, used equipment, and laboratory wastes. For these reasons, in comparing reprocessing with the once-through cycle, the American Physical Society concluded in 1978, "on balance the two waste disposal situations are comparable." The bottom line is that recycling high-level nuclear waste does not have the same clear advantages as recycling aluminum cans. [9]

In the early years of nuclear power, the principal driver behind reprocessing was the scarcity of uranium. It was believed that easily recovered, and thus low cost, uranium reserves would be exhausted within a few decades by a rapidly expanding nuclear industry. Breeder reactors would be required to sustain the industry into the twenty-first century, with reprocessing an essential component for the breeder reactors to keep operating. As time went on, it was discovered that uranium was more plentiful than originally believed. In addition, world nuclear capacity plateaued at levels of less than one-tenth of what had been projected for the year 2000. Finally, reprocessing without breeder reactors yields only modest uranium savings, on the order of 15 to 20 percent. [10]

According to the Nuclear Energy Agency, there are 5.5 million metric tons of identified uranium resources, with an estimated additional 10.5 million metric tons that remain undiscovered. This amounts to roughly a 230-year supply for today's worldwide fleet of 440 or so nuclear reactors. Like any mineral resource, uranium reserve estimates depend in large part on price. As the price goes up, it becomes economical to mine lower grade and less accessible ores, while increasing exploration for new reserves. If the price of uranium goes high enough, it could even become economical to separate uranium from the Earth's abundant phosphates, yielding an additional 22 million metric tons. The 4.5 *billion* metric tons of uranium in seawater could provide up to a 60,000-year supply at present consumption rates, if it could somehow be extracted economically. [11]

During the boom years after World War II, the western United States became the world's leader in producing uranium. Today, 60 percent of world uranium production comes from Canada, Australia, and Kazakhstan, with only four percent from the United States. Australia has the largest known recoverable resources, with about one-quarter of the world total. Unlike petroleum, a considerable amount (though certainly not all) of today's economically recoverable uranium is in stable countries friendly to the United States. In 2010, a study by nuclear experts sponsored by the MIT Energy Initiative concluded: "There is no shortage of uranium resources that might constrain future commitments to build new nuclear plants for much of this century at least." [12]

In recent years, uranium fuel for nuclear power plants has come from an unexpected source – dismantled US and Soviet bombs. Under a program known as "Megatons to Megawatts," Russia agreed in 1993 to eliminate about 500 metric tons of highly enriched uranium recovered from surplus Soviet warheads. This highly enriched uranium, containing 90 percent or more of the highly fissionable uranium-235, is down-blended to five percent or less and fabricated into commercial nuclear fuel. By 2009, almost half of the United States' electricity provided by nuclear power came from Russian warheads. Looking at it another way, an average of one out of every 10 light bulbs in the United States is powered by uranium from former Soviet nuclear warheads, many of which were once targeted at US cities. The 20-year Megatons to Megawatts pact expires in 2013. As this is a relatively cheap source of uranium fuel, the nuclear industry is lobbying for continuing the arrangement. [13]

Uranium is not the only element that can be used to generate nuclear power. Thorium, named after the Norse God of thunder, is another. Thorium is not fissile but when its major isotope, thorium-232, captures a neutron it morphs into the fissile isotope uranium-233. The process is similar to the way capture of a neutron by uranium-238 yields plutonium-239. The use of thorium for nuclear power was studied as early as 1946 at Oak Ridge National Laboratory, but the most advanced development has taken place in India – a country with relatively little uranium but large thorium deposits. Although more difficult to employ for nuclear power, thorium is about three times more abundant in the Earth's crust. In addition, thorium-based spent fuels contain fewer long-lived radioactive elements. [14–15]

As concerns about the supply of uranium abated, the danger of plutonium as a potential source of nuclear bombs took center stage in the reprocessing debates. All three fissile isotopes (uranium-233, uranium-235, and plutonium-239) can be used to make nuclear bombs, but the fissile uranium isotopes can be diluted with natural uranium to attain nearly harmless status. This is not possible with plutonium because there is no source of "safe" plutonium for dilution. As a result, plutonium is the radionuclide of most concern with reprocessing.

PLUTONIUM RECYCLING AND NUCLEAR PROLIFERATION

Plutonium is warm but, unlike other forms of high-level nuclear waste, not too hot to touch. In addition, as an alpha-emitter it is virtually harmless outside the body. The first plutonium produced as part of the Manhattan Project was placed in an inconspicuous wooden box, driven by two military officials from Hanford, Washington to Portland, Oregon and from there traveled by regular train to Los Alamos. [16]

In 1950, a sensational story hit the press when a 28-year-old research scientist, Sanford Lawrence Simons, was jailed for possession of a small vial of plutonium. Mr. Simons claimed it was just a "souvenir" from Los Alamos, where he had worked more than four years earlier. He had kept the plutonium under his house to hide it from his children. Simons was reported to be little perturbed by his plight and talked freely with the press. His "souvenir" story seemed to check out, as he had no known underworld or communist connections. Two days after Simons' arrest, Senator Brien McMahon (D-CT), Chairman of the Joint Committee on Atomic Energy, assured the public that under the present-day safeguards such a theft was impossible without the loss being detected. The story disappeared from the news. [17–18]

In spite of Senator McMahon's assurances about the accurate accounting practices, maintaining an exact inventory of plutonium is all but impossible. From 1956 to 1964, some 944 kilograms of plutonium were unaccounted for at Hanford – enough for more than 100 Nagasaki bombs. Experiences at the Savannah River plutonium weapons plants were similar, except that the amount of plutonium at the end of the inventory period was *greater* than the amount supposedly present before reprocessing. Neither case generated much concern. Security at the weapons plants was extremely tight and there were plenty of benign explanations for the discrepancies – plutonium held up in piping, sampling mistakes, uncertainty in the amount of plutonium produced in the reactor, etc. [5]

The story was quite different in the commercial reprocessing industry, where security clearances were much more relaxed. In referring to commercial reprocessing, AEC Commissioner Clarence E. Larsen quoted an "unavoidable" loss rate of one to two percent – while discrepancies in plutonium inventories at West Valley were on the order of two to four percent. If such discrepancies actually represented stolen material, such losses could supply the plutonium for numerous bombs. Without the same tight security used in reprocessing at the weapons plants, one could never be absolutely sure that some of the unaccounted for plutonium had not been stolen. [5, 19]

At the time of Sanford Simons' reported theft, the possibility that terrorists or rogue nations might acquire plutonium and build a nuclear bomb had not yet emerged as an issue. This changed dramatically in December 1973, when the *New Yorker* magazine published a three-part profile of Theodore B. Taylor, a former designer of atomic bombs at Los Alamos. Taylor pointed out that information about how to make an atom bomb was readily available from unclassified sources. Obtaining plutonium was the hardest part. A clever graduate student could figure out the rest. Taylor's continued warnings about the possibility of a do-it-yourself atom bomb received widespread attention in the scientific community and in the popular press. In the early 1970s, these fears were compounded by growing terrorist activities worldwide. [20]

In spite of these concerns, the AEC continued full speed ahead with its development of the breeder reactor. With plutonium as the fuel *and* the product, the breeder was the most proliferation-prone approach to nuclear power conceivable. By 1973, the breeder reactor consumed nearly half the total US energy R&D budget. Its most fervent advocate, Glenn T. Seaborg, served as AEC Chairman from 1961 to 1971. As a co-discoverer of plutonium, Seaborg saw the whole world revolving around plutonium with himself at center stage. According to Richard Hewlett, official historian for the AEC, it was of paramount importance to Seaborg "that he had discovered a new element that would be the salvation of mankind." In a speech given in October 1970, Seaborg sketched a picture of the USA depending on plutonium for 70 percent of its electricity by the year 2000. The AEC projected that more than 500 breeder reactors would be in operation by the end of the century, with possibly 2,000 such reactors by the year 2020. [19, 21–23]

In this new "plutonium economy," plutonium would be separated from the irradiated fuel by reprocessing, transported to a fuel fabrication facility, and incorporated into fresh fuel assemblies. The whole scenario would depend on large amounts of plutonium, perhaps hundreds of shipments per week, being bustled hither and yon to reprocessing plants, fuel fabrication facilities, and then to a breeder reactor. Tens of thousands of kilograms of separated plutonium would be placed in open commerce annually. By comparison, less than 10 kilograms of this "reactor-grade" plutonium is needed for a Nagasaki-like bomb. While solidly behind breeder-reactor development, even the AEC Commissioners admitted the likelihood that such a vast traffic of plutonium would spring a leak onto a worldwide black market. [19]

India's detonation of a nuclear bomb in May 1974 provided a stark reminder of these dangers. The bomb was made from reprocessed plutonium from a civilian reactor provided by Canada with US technical support. The Indian scientists announced their "peaceful nuclear explosion" by proclaiming in a coded message to Prime Minister Indira Gandhi, "The Buddha is smiling." Meanwhile, the rest of the world was frowning. Soon thereafter, an ill-timed proposal by the Nixon Administration to sell reactors to Egypt and Israel exacerbated the plutonium controversy and cast doubts about the government's prudence in commercializing plutonium. [24–25]

In spite of these growing concerns, in the summer of 1974, the AEC decided the time was ripe to begin licensing reprocessing facilities for commercial reactors. As required under the National Environmental Policy Act, the AEC released a four-volume environmental impact statement on their plans. The report, which assumed that the details and costs of an adequate safeguard program would be worked out later, was roundly criticized. By the following May, the newly formed Nuclear Regulatory Commission (NRC) announced that questions about the feasibility of safeguarding plutonium from theft needed further study. Environmental groups praised the new-found independence of the four-month-old agency, while the Atomic Industrial Forum, an industry group, termed the move "deplorable." [24]

During the 1976 presidential campaign the rise, seemingly from nowhere, of a peanut farmer (turned Governor) from Georgia as the Democratic presidential candidate raised the debate about reprocessing several notches higher. Having trained as a nuclear engineer in Admiral

Rickover's nuclear navy, Jimmy Carter knew a thing or two about nuclear energy. During the campaign, he called for a reassessment of US domestic nuclear policy and expressed deep concern about the international spread of nuclear technology and materials.

Under growing campaign pressure, President Ford initiated an intensive White House review of nuclear policy. Five days before the election, in a statement that shook the nuclear establishment worldwide, Ford temporarily banned nuclear fuel reprocessing in the United States. While supporting increased use of nuclear energy, Ford said, nonproliferation objectives must take precedence over economic and energy benefits if a choice must be made. A few months after taking office, President Carter took it one step further and deferred commercial reprocessing indefinitely.

Just prior to Carter's announcement, a panel of 21 influential scientists, economists, and political scientists sponsored by the Ford Foundation had come to a similar conclusion. This was not an antinuclear crowd. The Ford Foundation panel included Secretary of Defense Harold Brown, former director of the Princeton Institute of Advanced Study Carl Kaysen, and Stanford physicist Wolfgang Panofsky. The panel favored once-through light-water reactors and coal to meet the Nation's energy needs. Although they acknowledged the greenhouse effect from burning coal, this seemed less imminent as the Earth's climate was presumed to be in a cooling trend. [26]

The Reagan Administration reversed Carter's policy of "indefinitely deferring" reprocessing of commercial nuclear fuel. However, Reagan insisted that a reprocessing plant must be run by the private sector. It soon became obvious that private industry was not interested. No company was willing to step forward and accept the risk of a potentially large money-losing proposition with considerable attendant political uncertainty.

Reagan's new Secretary of Energy, a former dentist turned Governor of South Carolina, James B. Edwards, soon added to the reprocessing controversies. With a perceived need for plutonium for bombs and spent fuel piling up in pools at nuclear power plants, Edwards had an idea. He suggested that the waste problem and military need for plutonium could be solved simultaneously simply by reprocessing the spent fuel to obtain plutonium for bombs. Edwards' proposal to divert civilian reactor plutonium to weapons would have undermined two decades of US nonproliferation policy and caused the nuclear power industry in the USA untold bad press, at home and abroad. Peter Bradford, an NRC Commissioner, bluntly summed up the problem: "The average nuclear utility realizes that it does not need the controversy and that most of its customers do not want the feeling that when they turn on their lights, they are also turning on the local atomic bomb factory." The Reagan Administration soon backed away from the idea. In 1993, the Clinton Administration reinstated US opposition to reprocessing. [5, 25]

Safeguarding plutonium in the USA was one issue, assuring control in other countries was quite another. With his announcement of indefinite deferral of reprocessing, President Carter was, in effect, saying "we don't reprocess and you don't need to either." This message was marginally successful. Contracts for reprocessing facilities in Brazil, Pakistan, and South Korea were dropped, yet little else changed. The time had long since passed when the United States could wield a father-knows-best approach to the spread of nuclear technology, particularly among countries without the coal and uranium resources of the United States. [27]

France, India, Japan, Russia, the UK, and China (since 2010) continue to reprocess nuclear fuel. Reprocessing in these countries has not always gone smoothly. As an example, a leak of nuclear fuel dissolved in nitric acid forced the closure of the UK reprocessing plant at Sellafield in 2005. The leaked solution was enough to half fill an Olympic-size swimming pool, and contained 20 metric tons of uranium and 350 pounds (160 kg) of plutonium. The leak went undetected for more than eight months. Fortunately, the highly radioactive solution drained into a holding tank. The operator was fined £500,000 ($1 million) and the plant remained closed for two years. [28]

The French mix plutonium from reprocessing spent fuel with uranium in a mixed oxide (MOX) fuel for reuse, which gives them 15 to 20 percent more energy from the original uranium. Once the MOX fuel is used, it is difficult to reprocess again, and is shipped back to the reprocessing facility for indefinite storage. Like the initial spent fuel, the MOX spent fuel has fission products that make it difficult to access and make into a bomb. France has been unable to burn MOX fuel at the rate at which it comes from reprocessing plants, and their stockpiles of separated plutonium have been growing. Until recently, the UK also produced MOX fuel, but at a very low rate, resulting in the world's largest stockpile of separated civilian plutonium. Russia stores large amounts of separated plutonium from spent fuel reprocessing, with a long-term goal of using it in fast-breeder reactors. Japan and Germany also have civilian stockpiles, most of it stored in France and the UK. As of 2010, the global stockpile of separated civilian plutonium had grown to 255 metric tons – enough to make more than 30,000 nuclear weapons. [29]

Until recently, France and the UK generated substantial revenue by reprocessing the spent fuel of other countries (particularly Japan and Germany). This arrangement conveniently provided political cover that the waste issue was being addressed. This solution turned out to be only temporary, however, as the contracts with both France and the UK require that high-level waste from reprocessing (or its equivalent) be returned to the country of origin. As the reprocessing waste began to return home, these countries recognized the pitfalls of sending their nuclear waste abroad. It was cheaper to store the waste at home and save the reprocessing fee. As a result, France and the UK have lost virtually all their foreign customers. Without these subsidies, reprocessing is a money-losing proposition. [30]

Nowhere has reprocessing been more controversial than in Germany. The Gorleben site was originally to be a full-service nuclear waste management center that would store, reprocess, fabricate MOX fuel, and dispose of high-level waste. In 1979, a contentious international hearing on Gorleben led to tens of thousands of protesters in the streets of Hannover, the capital of Lower Saxony. Afterwards, Prime Minister Ernst Albrecht of Lower Saxony, who had originally proposed the site, concluded that reprocessing was politically infeasible. The concept of an integrated nuclear waste management center was subsequently abandoned and replaced by a plan to use the site as a high-level waste repository with attached interim storage.

With a reprocessing capability nowhere in sight, Germany turned to France and the UK to reprocess its spent fuel. This became the next focal point of the boisterous German antinuclear movement. Each major transport of nuclear waste through Germany was accompanied by massive demonstrations, often mixed with disruptive and sometimes violent forms of protest.

Under duress, the German government agreed to ban shipments to reprocessing plants after 2005, but the country continues to receive back its reprocessed waste – accompanied by huge protests. In November 2011, France's twelfth and final shipment of reprocessing waste to the Gorleben facility was true to form. Along the German portion of the route, 20,000 police were deployed to try to maintain order as protestors staged sit-ins, chained themselves to the railroad tracks, and sabotaged the railway. Additional shipments from the UK are scheduled to begin in 2014. [31]

Russia stands in stark contrast to Germany as a fervent believer in reprocessing. The country has long been interested in providing reprocessing and nuclear waste disposal services for the world at-large to help finance the cleanup of its massive nuclear waste and contaminated sites. Russia has long had a program to "take back" spent fuel of Russian origin for reprocessing from commercial and research reactors in Central and Eastern Europe. Because of their limited reprocessing capabilities, however, most of this spent fuel sits in pool storage. Nonetheless, Russia hopes that other countries with high-level waste will someday find it more economical and convenient to ship their waste to Siberia rather than dispose of it themselves. In 2001, Russian President Vladimir Putin signed into law a bill permitting import of foreign spent fuel for indefinite storage on Russian territory. The idea of an international repository, with or without reprocessing, is not new. In the early 1980s, several Western countries negotiated with China for the construction of deep geological repositories in the Gobi Desert and other remote areas. [32–34]

Research continues in the United States, France (the world leader in reprocessing technology), and other countries to develop alternatives to the plutonium-generating PUREX for reprocessing. The PUREX process was developed specifically to separate plutonium for bombs, so virtually any other reprocessing approach would be more proliferation-resistant. Yet, serious questions remain about the alternatives. Just how proliferation-*resistant* (compared to the Holy Grail of proliferation-*proof*) are they? What new wastes do they create? And the ever-present bottom line, how much would it cost?

Most research in the United States has focused on two types of reprocessing. One of these, UREX+ (for uranium extraction), is chemically similar to PUREX but leaves the plutonium mixed with other radioactive metals. The other, pyrochemical reprocessing, is based on high-temperature electroplating of plutonium plus other transuranics onto giant electrodes that have been inserted into a chemical bath of waste. Both of these reprocessing methods result in a radioactive mix that, although not completely fail-safe, is safer than having pure plutonium lying around. Research on these reprocessing techniques is often tied into proposals for development of fast reactors. [35]

FAST REACTORS

The term *fast* in fast reactor refers to the speed of neutrons within the reactor core. Neutrons produced by fission move at about 30 million feet per second (10 million meters per second). If a neutron is too fast – and 30 million feet per second qualifies as too fast – it will shoot right through a uranium-235 atom. For this reason, most reactors worldwide and all commercial power reactors in the United States use water or some other "moderator" to slow down the neutrons to optimum levels for causing fission. *Slow* versus *fast* are relative terms here – the slowed down neutrons still travel at speeds of around 7,000 feet per second. Because commercial reactors in the United States use ordinary water for the coolant, engineers call them *light-water reactors.*

In contrast, a *fast reactor* requires liquid sodium or some other special coolant that allows neutrons to maintain their high velocity. Fast neutrons are not as good at fissioning the scarce uranium-235, but are readily snapped up by the abundant uranium-238 to transform it into fissile plutonium-239. The breeder reactor, a type of fast reactor, uses this principle to convert uranium-238 atoms into plutonium at a faster rate than plutonium atoms are fissioned and destroyed.

A fast reactor can also be designed to make it *burn* more plutonium than is *bred*, serving as a kind of nuclear incinerator that converts plutonium and other transuranics into shorter-lived radionuclides. This waste still needs to be contained in a geologic repository, but only for a few centuries instead of for hundreds of thousands of years. Although still largely in the R&D stage, these fast "burner" reactors have a strong appeal.

Unfortunately, in spite of 50 years of research and billions of dollars of investment since the first fast reactor (Zinn's EBR-I breeder reactor), no fast reactors have been successfully deployed on a commercial scale. Admiral Rickover, the father of the US nuclear navy, experimented with a fast reactor for *Seawolf*, the second US nuclear submarine. He aptly summarized the problems: these reactors are "expensive to build, complex to operate, susceptible to prolonged shutdown as a result of even minor malfunctions, and difficult and time-consuming to repair." [36]

The Enrico Fermi Nuclear Generating Station in Michigan was the first American commercial fast-breeder reactor, but operated only from 1963 until 1972 before a partial meltdown and other problems led to a failed license renewal and subsequent decommissioning. The only other commercial fast-breeder reactor in the United States was the Clinch River Breeder Reactor in Tennessee. The project was halted in 1983 when Congress cut funding after years of spiraling cost overruns and growing concerns about the dangers of nuclear proliferation, should the plutonium generated by breeder reactors get into the wrong hands.

China, India, and Russia continue with fast-breeder reactor programs. Japan's $6 billion Monju fast-breeder reactor came on-line in 1994, after almost three decades on the books. The reactor, named after the Buddhist sage who symbolizes wisdom, closed just over a year later. The reactor faced considerable political opposition and operational difficulties, including a fire caused by a leak of its liquid sodium coolant. While controversy has swirled for years around plans to reopen the facility, the Fukushima disaster may have put the decisive nail in the coffin. Probably the most creative failure was the Kalkar fast reactor in Germany. In another no-go, this $4 billion reactor was auctioned off in 1995 for 2.5 million euros and converted into an amusement park. First called Kernwasser Wunderland (Nuclear Water Wonderland), for obvious reasons it was later renamed Wunderland Kalkar.

NUCLEAR ALCHEMY

A fast reactor is a form of artificial transmutation, or modern-day alchemy. Transmutation – the conversion of one element to another – harks back to the medieval alchemists who tried to change lead and other base metals into gold. Alchemists believed this quest could be accomplished through chemical reactions, in particular by using a legendary substance known as the *philosopher's stone*. As the modern theories of chemistry evolved, the idea of transmutation came to be associated with magic and superstition.

Then came the surprise. In 1901, Frederick Soddy and Ernest Rutherford discovered that transmutation occurs naturally through the process of radioactivity. When Soddy realized that radioactive thorium was changing into radium, he reportedly shouted, "Rutherford, this is transmutation!" As the story goes, Rutherford snapped back, "Soddy, don't call it transmutation. They'll have our heads off as alchemists!" Soddy later joined the ranks of 11 Nobel Prize winners trained by Ernest Rutherford, an unsurpassed record. Among his accomplishments, Soddy proved the existence of *isotopes* – a name he adopted from the Greek for "at the same place," in reference to all isotopes of an element having the same position in the periodic table. [37–38]

Since Rutherford's time, physicists have artificially transmuted many elements into others. Even into gold. In 1941, scientists at Harvard University used fast neutrons to transmute mercury into gold. However, this accomplishment was not quite what the medieval alchemists had in mind. Aside from the expense of the operation, the *gold* was radioactive with isotopes having half-lives of only a few days. [39]

Artificial transmutation of long-lived radionuclides has long been explored as a possible way to reduce radioactive waste. After years on the back burner, the concept seemed to come of age with the election of George W. Bush in 2000. With concerns about climate change and declining fossil fuel resources spurring renewed interest, nuclear power was suddenly back in the limelight. By 2002, the Bush Administration was requesting millions of dollars for research on reprocessing and fast reactors through the new Advanced Fuel Cycle Initiative. After Congress passed the Energy Policy Act of 2005, additional support for the nuclear industry came in the form of tax incentives, loan guarantees, and stream-lined procedures for licensing new nuclear power plants.

The anticipated nuclear renaissance was not just in the USA. Many nuclear nations, particularly in Asia, wanted to expand their existing nuclear programs. In addition, more than two dozen newcomers were exploring nuclear energy. Among these were Algeria, Egypt, Indonesia, Iran, Jordan, Libya, Nigeria, Thailand, Turkey, Vietnam, and Yemen. While some of these countries have stable forms of government, others are far from stable. And some are not so friendly with the United States. These countries would need fuel to power their new reactors, leading to concerns that they would develop their own uranium enrichment and reprocessing facilities. Whether initially intended or not, such facilities could lead to the ingredients for nuclear bombs. [40–41]

GNEP

Smitten by the talk of a worldwide nuclear renaissance, while also concerned about nuclear proliferation, in 2006 George W. Bush announced an ambitious new initiative – the Global Nuclear Energy Partnership (GNEP). The GNEP initiative was going to pair one of the new reprocessing methods that did not create separated plutonium (UREX+ or pyrochemical reprocessing) with fast reactors to burn the fuel from the reprocessing plant.

According to the GNEP vision, new technologies for reprocessing and fast reactors would be set up in the United States and a few other select nations. These *fuel-supplier* nations would provide aspiring nuclear nations with conventional reactors and nuclear fuel. In exchange, the recipient nations would agree to return their spent fuel to the nation of origin and pledge not to develop uranium-enrichment or spent-fuel reprocessing capabilities. The USA and the other fuel-supplier countries would reprocess the spent fuel for them using the advanced reprocessing and fast-reactor technologies.

The basic idea of GNEP was to provide reliable access to nuclear fuel at reasonable cost, along with spent-fuel services, as incentives for countries not to develop nuclear technologies that also might be used for weapons production. To accomplish this goal, the United States would build the largest reprocessing plant in the world – large enough to serve the equivalent of all 103 commercial reactors in the USA.

The reality didn't fit the goal. GNEP turned out to be a hastily conceived, half-baked proposal, hyping nuclear technologies that are still in the R&D stage. Large-scale implementation of either UREX+ or pyrochemical reprocessing is at least 20 to 25 years away. Even under the best circumstances, it would be decades before GNEP would be up and running. Should that day arrive, keeping up with the output from the proposed US reprocessing plant would involve the considerable feat of actually bringing on line more than 30 of the troublesome fast reactors – each in the 1,000 megawatt range. To top off this fantastical scenario, DOE proposed to separate the primary heat-generating fission products (cesium-137 and strontium-90) from the high-level waste for separate storage and decay over several hundred years. If there's such a thing as a nuclear waste wish list, GNEP would qualify. [42]

A key element of the GNEP initiative called for the return of spent nuclear fuel to the fuel-supplier nations. Such an arrangement would circumvent the fuel-recipient countries from reprocessing their own waste. Left unsaid, however, was that no country in the world had yet licensed a high-level waste repository, even for its own waste. The idea that it would be politically acceptable to take waste from other countries in such an open-ended way was another case of wishful thinking.

Contrary to their billing, fast reactors are not a silver bullet for the waste problem. Only a fraction of the recycled elements are transmuted while the fuel is in a fast reactor. As a result, numerous recycles would be required to rid the waste of its long-lived transuranics. After each cycle through a fast reactor, the fuel must be reprocessed, the remaining uranium and transuranics fabricated into new fuel, and the fuel returned for yet another pass through the fast reactor – and each "pass through" creates more waste. Assuming an ideal case that a fast reactor destroys 25 percent of the transuranics with recycle intervals of six years, it would take 16 cycles and 96 years to achieve a 100-fold mass reduction. And that's just for *one* batch of fuel. Furthermore, the technology for fabricating fuel that has undergone multiple cycles in a fast reactor has yet to be developed. Finally, although the transuranics would be greatly reduced, long-lived fission products like technetium-99 (half-life of 211,000 years) and iodine-129 (half-life of 16 million years) would still remain. These radionuclides are highly mobile in groundwater and may contribute more to the long-term radiation dose than many transuranics. [43–44]

By removing the cesium-137 and strontium-90 (as well as the transuranics), the long-term temperature increase of the repository would be decreased by about 20-fold, greatly reducing the repository footprint. This was a major driver of the GNEP concept, as a true renaissance in nuclear power had been significantly stymied by the waste problem. The Bush Administration claimed that, with GNEP in place, the proposed Yucca Mountain repository would suffice for the rest of this century. Meanwhile, the reprocessing facility would become a de facto long-term storage site for the separated cesium and strontium, as well as the transuranics that await further processing in a fast reactor. Of course, if the fast reactors did not come on-line as planned, the high-level waste would pile up at the reprocessing facilities. [27]

The price tag to get GNEP up and running could be as much as $100 billion. The program's success would require sustained support and full funding through several successive presidential administrations, thereby requiring a demonstration of political will all but lacking in the nuclear waste arena. What most worried many of its critics, however, was the emphasis on an accelerated schedule to resolve the program's many R&D challenges. The Energy Department's eagerness to implement unproven technology was a prescription for yet another schedule-driven failure. [40]

For all the cost and effort, serious questions remained about how much the program would actually contribute to nonproliferation. Proliferation resistance is only marginally achieved by keeping other transuranics with the plutonium because the radioactivity of this mix is not much higher than that of plutonium itself.

The GNEP initiative ran into almost immediate resistance from other nations, from Congress, and from proponents and opponents of nuclear energy alike. Even many advocates of the long-term research conducted through the Advanced Fuel Cycle Initiative considered GNEP premature. Non-nuclear countries saw the GNEP restrictions as a challenge to their inalienable right to develop nuclear technology for "peaceful purposes" – as spelled out in the Nuclear Non-Proliferation Treaty. In addition, many countries wanted to be on the "A" team to supply the new nuclear fuel services; few wanted to be on the "B" team with restrictions on their activities. India, with three small reprocessing plants, was seriously affronted when asked to join GNEP, not as a provider of reprocessing power, but as a client. The requirement for recipient nations to forgo uranium enrichment or reprocessing was later dropped, thereby changing the GNEP program to a "carrot" rather than "carrot and stick" approach – seriously undermining the whole nonproliferation point. [40, 45]

When asked to review the GNEP initiative, a National Academy of Sciences review committee was unanimous that "the GNEP program should not go forward." For a group composed of scientists from many different fields and perspectives, rarely has the Academy come out with such a unanimous finding. The US Congress agreed with the Academy and, except for some R&D, gutted funding for the program. [44]

In March 2008, near the end of the Bush Administration, the National Academy of Sciences held a two-day *Summit on America's Energy Future*. John Holdren, soon to become President Obama's science advisor, summarized the prevailing view. Reprocessing spent fuel makes nuclear energy "more complicated, more expensive, more proliferation-prone, and more controversial." Holdren summarized by saying, "If you want nuclear energy to be rapidly expandable, and to take a bite out of the climate change problem, you want to make it as cheap as possible, as simple as possible, as proliferation-resistant as possible, and as non-controversial as possible, and that means you don't want to reprocess any time soon." [46]

This is not to suggest that long-term research on alternative nuclear fuel cycles should not be pursued. A proliferation-resistant reprocessing technology that is economically viable and helps to address the nuclear waste problem is a worthy goal, but is by no means ready for large-scale commercial application.

8

Dry cask storage

Life is what happens to you while you're busy making other plans
Attributed to Allen Saunders, Betty Talmadge, John Lennon, and
others [1]

The Maine Yankee was situated on Bailey Point, along the scenic
Maine coast near the town of Wiscasset – according to locals, the State's
prettiest village. From its startup in 1972, this 900 megawatt pressurized water reactor produced much of Maine's electric power until it was
decommissioned in 1997. The Maine Yankee was one of the largest commercial nuclear power plants of its time. Originally licensed for 40 years,
cracks were discovered in the steam generator tubes in 1995, and the
plant was shut down for almost a year for repairs. Investigations by the
Nuclear Regulatory Commission identified so many problems that
the expense of repairing them became too costly. After only 25 years,
the Maine Yankee was turned out to pasture.

What had not gone smoothly during the Yankee's last years of operating life became a second-to-none decommissioning. Working together,
stakeholders broke new ground in a number of areas that won international acclaim for innovation and excellence. The Maine Yankee was one
of the first large commercial power reactors to complete its eight-year
decommissioning safely and within budget. *Firsts* included the first ever
use of explosives to safely demolish the reactor containment building.
All plant structures were removed to three feet below grade. The site
was then cleaned up radiologically to a significantly higher level than
required by the Nuclear Regulatory Commission. Decommissioning also
included the largest single campaign to move spent nuclear fuel from
wet to dry storage. The Maine Yankee holds the record for the largest dry
cask storage project for a decommissioned plant in the United States. [2]

Figure 8.1 Dry cask storage at decommissioned Maine Yankee site. Courtesy Maine Yankee.

Today, Bailey Point is once again a pristine meadow graced with wildflowers and fringed with woods. A coastal estuary sparkles through the trees. The State of Maine plans to convert this idyllic 800-acre (320 hectares) meadow into some form of commercial or municipal use. But there's a problem. On approximately 12 of these acres, 64 concrete casks stand like forgotten sentinels from another world. The casks are fenced-in by a double chain-link fence topped with razor wire. Standing almost two storeys tall, they look like Midwestern grain silos lined up in rows seven deep on concrete pads – a definite eyesore.

Yet, this is the least of the problem. Housed inside these silos are more than 1,400 spent nuclear fuel rods. For now, and into the indefinite future, the spent nuclear fuel from the fully decommissioned and long gone Maine Yankee is staying right where it is. There is nowhere else for the waste to go.

US nuclear power plants generally have an operational cycle of 18 to 24 months. At this time, the reactor is shut down and one-quarter to one-third of its highly radioactive fuel assemblies are removed. For a minimum of five years, these fuel rods are stored in a deep pool where the water blocks radiation and cools the fuel. It was originally assumed that, after this cooling-down period, the fuel would be reprocessed or shipped to a geologic repository. After reprocessing of commercial nuclear waste was banned during the Carter Administration, and as geologic disposal continued to be mired in technical and political controversy, the spent fuel in these storage pools began to reach its own version of critical mass.

For decades, the strategy was to "re-rack" the spent fuel and consolidate the individual fuel rods, to get as much waste as possible into the pool. The hope was that there would be some place for all this ingeniously packed waste to go by the time the pool filled up. Alas, no such luck.

After the terrorist attacks of September 11, 2001, the newly formed Department of Homeland Security and the Nuclear Regulatory Commission (NRC) sponsored a study by the National Academy of Sciences to investigate the possibility of terrorists breaching the security system of nuclear storage pools and unleashing radioactive waste into the environment. With certain qualifications, the task force reported back in the affirmative: If terrorists have sufficient knowledge, they could get into one of these compounds and partially, or completely, drain the pool. In such an event, the thermally hot fuel rods could self-ignite and release large quantities of radioactivity into the environment. [3]

This disaster scenario was not necessarily a given. The study concluded that these consequences could be avoided if power plant operators took prompt action to reduce the loss of pool coolant. It also recommended certain preventative measures to be undertaken. The fuel should be rearranged in the pools so that the hottest fuel assemblies were surrounded by relatively cooler assemblies. Water-spray systems should be installed that would be able to cool the nuclear assemblies even if the pool drained. As a result of these findings, the NRC mandated further safeguards at all operating reactor sites. [4]

In 2011, the nuclear crisis at the Fukushima plant in Japan brought worldwide attention to the dangers of storing spent fuel in pools. In the aftermath, experts reminded people that about three-quarters of US spent fuel is submerged in pools. Many of these pools are full, with some containing four times the amount of spent fuel that they were designed to handle. [5]

Compared with leaving the spent nuclear fuel in pools beyond the standard five-year stay, dry casks offer some distinct security advantages. Dry cask storage is a passive system that relies on natural air circulation for cooling, and it divides the inventory of spent fuel among a large number of discrete, robust containers. These factors make it more difficult to attack a large amount of spent fuel at one time and also reduce the consequences of such attacks. To move all the spent fuel older than five years from pools to dry casks would cost an estimated $3.5 to $7 billion. While this cost seems exorbitant, proponents argue that the price tag is small compared with the potential dangers presented by prolonged storage in numerous tightly packed pools. [6]

Dry cask storage has another distinct advantage – it's dry. As of 2009, leaks from pools had been reported at five reactor sites. The NRC assures us that the near-term public health impacts have been negligible. Nonetheless, contaminated groundwater can be extraordinarily expensive, if not impossible, to clean up. [4]

Dry cask storage systems are known as Independent Spent Fuel Storage Installations (ISFSIs). The method was developed in Canada in the 1970s, when the Canadians were still in a quandary over whether to reprocess spent fuel or write it off as waste. A holding pattern ensued and the resource/waste began to be transferred to highly reinforced steel and concrete containers. American nuclear experts saw no reason why the concept of dry cask storage could not be used here, with certain modifications. US nuclear power plants use light-water fuels, which stays in the reactor longer and has about three times as much decay heat per ton as Canadian nuclear plants. In addition, our fuel assemblies are longer, so they would require a taller silo than the ones the Canadians were building. These problems were hardly insurmountable. As a US official commented, "it is so simple – no pumps, no fans, no filters." [7]

NRC-certified dry storage casks are typically made of an inner canister of reinforced stainless steel, surrounded by a thick concrete overpack (outer cask). The inner steel canister resembles a giant metallic thermos, 14 feet (4 m) long and 3 feet (1 m) wide. The canisters are loaded underwater in the storage pool, so that workers are shielded from radiation. After the rods are inserted into a canister, the water is pumped out and multiple lids are bolted or welded on. The canister is then filled with helium, or some other inert gas, that conducts heat away from the waste and prevents oxidation of the fuel rods. Finally, the canister is transported to an outdoor concrete storage pad where it is loaded into the concrete overpack. Fully loaded, each cask weighs 100 tons or more. Natural convection through vents in the concrete provides passive cooling of the inner metal canister, that can reach temperatures of 400 °F (200 °C) or higher from the ongoing radioactive decay. Each cask costs about $1 million and holds half to a full year's worth of a reactor's spent fuel.

The ISFSIs are no-fuss facilities. No building protects the casks from rain, sleet, or snow. A chain-link fence topped with razor wire encloses the compound, with floodlights, motion detectors, and cameras strategically installed. Workers regularly emerge from a nearby security and operations building. Their main job is to walk the site and make sure the cask's air vents remain free of snow or debris. There's probably a gun somewhere. And that's pretty much your standard Independent Spent Fuel Storage Installation.

The first ISFSI in the United States was licensed in 1986 for a nuclear power plant in Surry, Virginia – across the river from historic Williamsburg. Fifteen years later, 20 nuclear power plants were using ISFSIs for storing spent fuel. By 2009, there were more than 1,000 dry casks at 44 sites in 31 States. Almost 8,000 dry casks are anticipated by 2040. Business is good for dry cask manufacturers. [8–9]

As the agency responsible for licensing and overseeing ISFSIs, the Nuclear Regulatory Commission officially views dry cask storage as *temporary*. Dry casks were originally licensed for a period of 20 years. As the opening of a geologic repository has been pushed further and further into the future, so has the renewal licensing of dry casks – now going well beyond their stated design life of 50 years.

With the future of nuclear energy inextricably entangled with the imperative for safe storage of spent fuel, the NRC has responded with periodic *Waste Confidence* rulemaking. The roots of this rulemaking go back to 1976, when California instituted a moratorium on building any new nuclear power plants until a federally approved method for permanent disposal of spent fuel was assured. Taking a cue from the successful moratorium, the Natural Resources Defense Council petitioned the NRC to conduct a waste confidence hearing to basically determine whether California was right. The NRC denied the request, asserting that as a matter of policy "it would not continue to license reactors if it did not have reasonable confidence that the waste can and will in due course be disposed of safely." [10]

A few years later, the NRC was again pressured to hold a waste confidence hearing – this time in response to lawsuits over licensing additional on-site storage at the Vermont Yankee and Prairie Island (Minnesota) nuclear reactors. In 1984, eight years after the original request for a ruling, the NRC issued its first Waste Confidence Decision. The Commission concluded that there was "reasonable assurance" that a repository would be available by 2007–2009 when the licenses for the Vermont Yankee and Prairie Island expired and additional storage was needed. This NRC *confidence* was largely based on the recent passage of the Nuclear Waste Policy Act, which stipulated that a geologic repository would be open for business by the end of the century. The NRC also concluded that storage in pools or dry casks will be available and safe for at least 30 years past the reactor's license expiration. With these rulings in place, the NRC promised to revisit the issue from time to time. [11]

The first revisit came in 1990. Amendments to the Nuclear Waste Policy Act had been passed a few years earlier, and Yucca Mountain was the only site that would be studied for a repository. But progress was much slower than anyone had foreseen. The availability of a geologic repository by 2007–2009 no longer seemed like such a sure bet. With a more somber outlook, the NRC was reasonably sure that a repository would open "within the first quarter of the twenty-first century." [12]

The NRC reviewed their findings once again in 1999. Remaining steady in their *confidence*, this time the Commission concluded that "experience and developments since 1990 had confirmed the findings and made a comprehensive reevaluation of the findings unnecessary." [13]

The first years of the twenty-first century saw renewed interest in nuclear power. The partial-meltdown crisis at Three Mile Island had receded into an almost forgotten past, while global warming appeared to loom in the almost immediate future. In October 2008, the NRC decided to undertake a review of the Waste Confidence findings "as part of an effort to enhance the efficiency of combined operating license proceedings for applications for nuclear power plants anticipated in the near future." Cutting through this bureaucratic lingo, applications to build new nuclear power plants were piling up, and the licensing process needed to pick up the pace. [4]

During the course of the Waste Confidence review, the Obama Administration announced that Yucca Mountain would no longer be considered for a repository. With dry cask storage suddenly the only game in town, two of the NRC Commissioners balked at declaring continued confidence in an eventual repository. After one of the opposing Commissioner's terms ended, the NRC announced its updated Waste Confidence Decision, in December 2010. With a geologic repository an ever-receding mirage, the NRC now concluded that sufficient repository capacity will be available "when necessary." Aside from what some may call "irrational exuberance," this fence-straddling took the NRC out of the messy business of repository prediction. Simultaneously, the NRC extended their confidence that storage (in dry casks plus pools) was safe, with no significant environmental impact, for as long as 60 years beyond the licensed life of any reactor, for a possible total of 120 years. The Commission explained that the previous 30-year safety lifespan was not a *technical limitation* on storage, but merely reflected the expectation for when a repository might begin receiving spent fuel. The NRC further directed its staff to conduct an in-depth review of existing spent fuel storage, as well as to draft new rules that would allow storage for more than 120 years – or perhaps even 300 years *or more*. Increasingly, the approach to waste confidence looks like shooting an arrow at a wall, drawing a bulls-eye around it, and proclaiming yourself an excellent marksman. In June 2012, the US Court of Appeals agreed with this perspective, ruling that the NRC has no basis for its waste confidence position. [14–15]

Since the first ISFSI was licensed 25 years ago, the NRC has staunchly affirmed, and reaffirmed, the safety of dry cask storage. Indeed, dry casks generally have performed well. There have also been some problems. At one point, operators at the Surry plant had to open up several casks because of faulty seals. At a Wisconsin reactor site, a welding torch ignited pent up hydrogen gas with sufficient force to dislodge the 4,000-pound (1,800 kg) lid and tilt it ajar on top of the cask. Such isolated problems are, perhaps, inevitable. Yet, the larger question remains: how did the NRC come up with their 60-year confidence in on-site storage with the possibility of extending this for several centuries?

It turns out that all this confidence rests on the passive and robust nature of dry casks, and *one* study that comprehensively looked at *one* dry cask. The study, entitled *The Dry Cask Storage Characterization Project*, was conducted on a 15-year-old cask by the Idaho National Laboratory. The cask had been loaded with spent fuel assemblies from the Surry Nuclear Power Plant, in 1985. When opened in 1999, the cask and stored fuel assemblies were examined for signs of deterioration and aging. As noted by the Electric Power Research Institute (EPRI), a co-sponsor of the study, this is "the only test to-date that has examined fuel that has been in dry storage for a significant time (15 years), at conditions realistically representative of those expected for the storage of LWR [light-water reactor] fuel." [16–17]

The Idaho National Laboratory team first inspected the concrete pad that supports the 100-ton casks. Overall, the pad had maintained its structural integrity, although it did exhibit a network of fine cracks over the entire surface. In the freezing and thawing climate that includes most of the USA, hairline cracks in horizontal surfaces widen and deepen with time. Nonetheless, at the time of inspection this network of cracks posed no structural problem and so the pad passed inspection.

Next, the concrete cask exterior was inspected. Having been exposed to all weather conditions for the past 15 years, they found small corroded areas on the cask cooling fins. The bolts on the top lid appeared in good condition, but on the cask bottom cover they found heavy corrosion on one of the bolts. After removing this corrosion, they discovered bolt head pitting and surface irregularities. The cask exterior passed inspection.

The primary top lid has both metal and elastomer O-rings, serving the dual purpose of keeping the inert gas *in* and oxygen *out*. Oxygen getting into the canister would, over time, corrode the inner steel canister containing the fuel rods. Overall the O-rings and seals were in excellent condition, though they discovered a few "pinpoint indentations" and a partially open splice joint in the elastomer O-ring. The lid and O-rings passed inspection.

The interior cask and the steel basket holding the fuel rods could only be very partially inspected by remote camera. Out of the 14-foot length of the inner cask, they could see only about the top foot of the sidewall. By positioning the camera at the end of 21 narrow fuel tubes, they were able to see limited portions of the steel cask bottom. This inspection revealed some minor marks and blemishes, but the nickel coating they could see was still intact. The inspectors concluded that the inner cask was performing well.

Then the fuel assemblies were remotely lifted and examined. They were free of physical damage and, with some unexplained changes in coloration, seemed okay. Selected fuel rods were removed from one fuel assembly, visually examined, and then shipped to Argonne National Laboratory for thorough examination. The fuel assemblies passed inspection.

The tests continued, with temperature readings of the cask exterior, a radiation survey, checks for helium gas leaks, and so forth. When they finished, Idaho National Laboratory concluded there was *no* evidence of degradation of the dry cask *important to safety* from the loading of the cask in 1985 up to the time of testing in 1999. In a follow-up document, the Electric Power Research Institute provided a qualified endorsement to the longer-term applicability of the study, noting that "while the tests tell little about the defect mechanisms and the potential for failures in long-term storage, it is important to remember that no gross breaches occurred, and fuel assemblies/canisters were able to be pulled out of the casks with no adverse effects, signs that longer-term storage is feasible." [16–17]

The Idaho National Laboratory study demonstrated that, overall, the cask performed well over a 15-year period. The inspections also illustrated a number of things that can go wrong. While the concrete outer cask protects the inner canister from weather, it also makes it difficult to monitor conditions inside – where the action is.

There are significant limitations with generalizing from this study to current dry cask storage. The cladding and fuel in the test cask were in mint condition. This is not always the case at current sites, which sometimes store damaged fuel rods. Today's nuclear fuel is also much hotter and more radioactive than the fuel in the test, and the materials used for fuel cladding and other components have changed.

In addition, many dry casks are in corrosive marine environments, not represented by the Idaho climate. The most obvious limitation, however, is that the test covered only a 15-year period out of a possible century or longer. A lot of wear and tear can take place during an additional 85 to 285 years.

Dry cask proponents argue that the cask functioned well during the most severe conditions of radioactivity and heat. The temperature of the fuel in the test cask dropped by more than half, from about 660 °F (350 °C) in 1985 to about 300 °F (150 °C) in 1999. Proponents also argue that spent fuel does not have to stay in the original cask for the full 100-year period. At some point, it could be put into a new cask via either a wet or dry transfer. Which raises another question – since it's never been done – of how easily the casks can be changed out after long-term aging. Likewise, the ability to handle and transport the spent fuel after perhaps a century or more of degradation is unknown.

Once a particular brand of dry cask is approved by the NRC, it can be used at any reactor site without further environmental analysis. No site-specific study is required, and no separate environmental impact statement. This *generic licensing process* has been a source of controversy. An example is the Palisades Nuclear Plant in Michigan, which was the first to receive the go-ahead for dry casks under the reactor's general operating license. At the Palisades plant, the reactor is anchored to bedrock, while the dry casks stand on a concrete pad built on a sand dune 150 feet (45 m) uphill from Lake Michigan. [18]

There is also a stark contrast between the scientific rigor required for dry cask storage and that required for licensing a geologic repository. A thoughtful, substantive analysis comparing the risks of extended on-site storage over a period of 60 to 100 years, or more, with the risks of a geologic repository, has never been done.

In 2007, the NRC conducted a pilot study to develop a methodology for risk assessment of dry cask storage. This was a study of *how to study* risk assessment. A specific cask system and site were evaluated, along with selected hypothetical risks that included dropping the cask during handling, earthquakes, floods, high winds, accidental aircraft crashes, and nearby pipeline explosions. These risks could be reasonably estimated. The pilot study did not include other more difficult to quantify risks – such as unloading of the cask, fabrication errors, misloading of spent nuclear fuel, and effects of aging. Interestingly, the results of the Idaho *Dry Cask Storage Characterization Project* were not considered transferable to this study, as it was unclear whether the two casks would experience similar conditions. The final report estimated very low risk for the limited situations, but carefully avoided drawing any regulatory implications. [19]

Meanwhile, neither the cask manufacturers nor the utilities are comfortable with keeping spent fuel on-site in dry casks for anywhere near a century. At a 2009 meeting, representatives of the nuclear power industry minced no words on this subject: [20–21]

> Leaving the spent fuel onsite for extended periods of time was never intended and is not responsible. (Dry cask manufacturer)

> We really don't want the fuel on site for that long. Our sites really weren't characterized for that type of storage. (Utility executive)

> It is not ethical, basically, to plan for long-term storage without pursuing a well defined repository program. (DOE employee speaking as a citizen)

Not everyone agrees. Nevada politicians have fought long and hard to keep a geologic repository from being built in their State. Senate Majority Leader Harry Reid (D-NV) contends, "if you produce nuclear power, you just leave it where you produce the energy." [22]

Few people would be comfortable with following Senator Reid's advice. Open-ended reliance on dry casks is not a responsible strategy, yet on-site storage using dry casks could play an important role as one component of a responsible nuclear waste disposal program. After an initial cooling period, it is safer than pool storage and less likely to contaminate nearby water resources. The heat and radioactivity of spent fuel diminish rapidly within the first few decades. Lower heat and radioactivity make the waste easier to handle and transport, allow for denser packing of the waste in a geologic repository, and increase confidence in predictions of the long-term behavior of the waste in a repository.

During the past few years, both industry and regulators have caught on to the limited technical basis for very long-term dry cask storage. In 2009, the Electric Power Research Institute formed an Extended Storage Collaboration Program to evaluate long-term storage in dry casks and their future transportability. Meanwhile, storage pools at reactors are filling up. With nowhere else for the spent fuel to go except into on-site dry casks, the *only* option is now in danger of being touted as the *best* option. [23]

9

Interim storage

One day he is at war and the next day at peace. He has many colors.

Cochise [1]

Having failed to meet a January 1998 deadline for beginning to accept spent nuclear fuel, the Department of Energy (DOE) is now besieged by lawsuits for breach of contract. By 2010, the government had paid about $760 million in settlements, and DOE estimates $13 billion in potential liability costs if the government does not start accepting nuclear waste by 2020. [2]

With a geologic repository nowhere in sight, and the government in a legal and financial bind, the idea of a centralized *interim* site is often talked up. This site would serve as a way station between the nuclear power plants and final transfer to a geologic repository. On paper, this looks like a win–win. The federal government could take charge of the waste and stop hemorrhaging from lawsuits. The utilities would get the waste off their lots. People who live near nuclear power plants would get the waste out of their neighborhood. Proponents of nuclear energy could claim progress with at least some aspect of nuclear waste disposal. But is interim surface storage any more possible, or palatable, than geologic underground disposal? And how *interim* is interim?

Interim surface storage is not a new idea. In the early 1970s the Atomic Energy Commission (AEC), caught off guard by the sudden rejection of the first candidate repository site near Lyons, Kansas, needed a backup plan. They proposed constructing a surface storage facility at one or more existing nuclear sites to temporarily store high-level waste while other options were pursued. The AEC was soon forced to back down when the Environmental Protection Agency and environmental groups argued that the storage facility could easily become a de facto permanent disposal solution.

In 1982, the Nuclear Waste Policy Act became the first law to address interim storage of spent fuel. The Act provided for development of a long-term "monitored retrievable storage" (MRS) facility in conjunction with geologic disposal. The three words comprising MRS were carefully chosen. The site would be carefully *monitored*, and the Act used the term *retrievable storage* to be perfectly clear that this was not a permanent solution.

Initially, DOE identified three potential sites for an MRS facility. All were near Oak Ridge, Tennessee, but no one bothered to consult Tennessee in advance. Governor Lamar Alexander learned of the State's "good fortune" only after the decision was made. This was not the first time a Tennessee Governor was left out of the nuclear loop. When Oak Ridge was constructed in the 1940s as part of the secret Manhattan Project, the Governor was not informed. In 1974, congressional aides changed the name of Oak Ridge National Laboratory to Holifield National Laboratory in honor of retiring California Congressman Chet Holifield, a prominent member of the Joint Committee on Atomic Energy. Once again, no one bothered to consult the Governor. The name change was short-lived. [3]

The local Oak Ridge community endorsed an MRS facility, with some conditions. These included assurance it would not become a de facto repository, well-defined payments, and a fixed schedule for cleaning up Oak Ridge's extensive environmental problems. In spite of local support, a public opinion poll showed that approximately 90 percent of Tennesseans were opposed. Governor Alexander, a strong advocate of nuclear energy, knew how to read the tea leaves – an MRS facility was not welcome in his State. "This may be the first time Tennessee has objected to having federal dollars frittered away inside its border," observed science writer Eliot Marshall. Presumably, Marshall was thinking of the Clinch River breeder reactor and the Tennessee Valley Authority, both well-connected to the nuclear industry. [4–5]

The Tennessee Valley Authority (TVA) was FDR's most ambitious New Deal project. Intended in part as a flood control project, the TVA also fulfilled FDR's dream to reduce the control of private companies over electrification and to bring cheap electricity to the poverty stricken Tennessee Valley. The project involved a massive infusion of federal dollars, resulting in federal control of the region's electric power business. Ronald Reagan, who was President during the MRS-Oak Ridge controversy, was well aware of the federal government's influence wielded through the TVA. Years earlier, during his acting career, Reagan referred to the TVA as one of the problems of big government. He was promptly fired as host of General Electric Theater.

The TVA provided the immense amounts of electricity needed by Oak Ridge to enrich uranium for the Manhattan Project. David Lilienthal, known as *Mr. TVA*, became the first chairman of the Atomic Energy Commission. In the 1970s, TVA embarked on an ambitious program for nuclear reactor construction. Seventeen nuclear reactors were planned, though only six were completed. Among TVA's nuclear power plants is the Browns Ferry Nuclear Plant, the largest in the world when it began operation in 1974. Barely a year after opening, the Browns Ferry plant became infamous when a worker, using a candle to search for air leaks, accidentally set fire to some cables in one of its three operating units. They now use feathers to detect leaks.

By the 1980s, Tennesseans had apparently had enough, and the MRS project caved-in to political pressure. Soon after, Congress nullified the selection of Oak Ridge but still authorized DOE to construct an MRS facility – just not in Tennessee. The 1987 Amendments to the Nuclear Waste Policy Act also came with conditions for any future MRS work. First, an MRS site could not be selected until a geologic repository site was recommended. This finally occurred 15 years later with the presidential site recommendation for Yucca Mountain in 2002. Second, construction of the MRS facility could not begin until DOE receives the license to construct a repository. This condition is a long way from being realized. In effect, central interim storage is firmly anchored to a geologic repository in the current law.

A new chapter in the search for an interim storage site began with the creation of the Office of the Nuclear Waste Negotiator, as part of the 1987 Amendments to the Nuclear Waste Policy Act. The idea was to have a nuclear waste ambassador to conduct a nationwide search for a volunteer community to host an MRS facility.

David Leroy, the first negotiator and former Lieutenant Governor of Idaho, made every effort to do it right. Leroy insisted that the proposed site must be truly voluntary. All dialogues were terminable at any time by the prospective host. Communities were free to enter and exit the process at will. A community would be considered as a candidate only if the Governor clearly endorsed the request. As a sweetener, Congress authorized a "signing bonus" of $5 million dollars per year for the host community prior to shipment of the waste and $10 million per year during the operation of the MRS facility. Behind the scenes, Leroy was authorized to negotiate a benefits package well in excess of these figures.

Leroy stressed the importance of dealing fully with safety concerns before discussing dollars. Study grants were offered to allow communities to investigate the risks and benefits of hosting the facility, without making a commitment. Leroy emphasized that all aspects were negotiable, including the choice of technology, oversight, size, timeframe, and ownership. Leroy contended that he was more committed to maintaining a credible and open process than finding a volunteer. He viewed himself "as the guardian of the process, rather than the guarantor of the result." This approach was the polar opposite of the unilateral selection of Yucca Mountain in the same Act that created Leroy's office. [6]

Regardless of his high sense of fair play, Leroy's overtures were met with a resounding silence by the Nation's Governors. Hired to be a negotiator, he was treated like a pariah. No Governor could afford the political heat of talking with him. A handful of counties expressed interest, only to be blocked by their respective Governors. The county commissioners of Grant County, North Dakota, applied for a study grant and were promptly recalled by an angry electorate. The county kept the $100,000 grant. [6, 7]

With the States out of the picture, Native Americans were the only remaining recourse. As sovereign nations, the Native American tribes could not be vetoed by any State Governor. Twenty-four tribes applied for study grants. Some were soon withdrawn or never moved past the first phase. In the end, two tribes persisted – the Mescalero Apache in New Mexico and the Skull Valley Band of the Goshute Nation.

The interest by these tribes is remarkable, given the history of Native Americans with the US government and its "trail of broken treaties." Native American communities already shouldered a disproportionate burden of radiation contamination from uranium mining and atomic bomb testing in Nevada. The historical record did not inspire a lot of confidence or trust. Would the government move the nuclear waste onto tribal lands and then forget about it?

The States where these tribes are located were none too happy about the prospect of nuclear waste being stored within their borders. The possibility of an MRS facility was very unpopular in New Mexico, a State that had witnessed the first atomic explosion and now hosted the Waste Isolation Pilot Plant for military transuranic waste. Senator Jeff Bingaman (D-NM) sponsored legislation that would require interested tribes to gain the cooperation of State and local officials before receiving study grants. Congress went even further by canceling the entire study grant program in October 1993. In December 1994, authorization for the negotiator expired. Congress failed to reauthorize the office – not because the voluntary siting process was deemed a failure but, rather, because it looked like a volunteer actually might be found. [6]

Although the negotiator's office was terminated, the issue refused to die. The Mescalero Apache began working outside the negotiator's purview with a private consortium of utilities headed by Northern States Power Company of Minnesota. Northern States was apparently motivated to work with the Mescalero when an appellate judge ruled against the utility's plan for dry cask storage at its Prairie Island station, on the grounds that such storage might become permanent.

Descendants of the warriors Geronimo and Cochise, the Mescalero Apache once ranged over much of the southwestern United States and parts of Mexico. The Mescalero had a record of entrepreneurship and successfully dealing with the modern world. The tribe's long-time chairman, Wendell Chino, was fond of saying, "The Navajos make rugs, the Pueblos make pottery, the Mescaleros make money." Laser-guided saws at their lumber company turned out 16 million board-feet a year. They also ran a successful casino and ski resort complex. On a clear day, the Mescalero's ski chairlifts are visible from the Trinity site, where the first atomic bomb was tested, near Alamogordo, New Mexico in 1945. [8, 9]

The Mescalero Apache hired technical experts, studied nuclear waste, and visited nuclear waste storage facilities. The tribal government converted an old laundromat into a nuclear information center, complete with Geiger counter, sample radioactive materials, and dummy nuclear fuel assembly. Chairman Wendell Chino argued that because the Mescalero Apache have a cultural tradition of harmony with nature and protecting the earth, the tribe was a better guardian of nuclear waste than mainstream America. He spoke of radiation as "ghost bullets." These bullets were deadly if mishandled. Yet, if treated with respect, they could earn millions of dollars for schools and social services, while also providing high-tech jobs to lure technically trained Mescalero Apaches back to the reservation. [9]

Not everyone in the tribe agreed. A significant minority had strong reservations about the proposed MRS facility. The debate pitted descendants of Geronimo and Cochise on opposite sides. Joseph Geronimo, grandson of the legendary warrior, was adamantly opposed. Silas Cochise, great-grandson of the famous chief, was a proponent and manager of the waste project. Rufina Laws, a tribal member who had recently lost an election against Wendell Chino for Tribal Chairman, conducted a door-to-door campaign against the proposal and spoke widely of her vision of "glowing liquid flowing down the slopes of the sacred Sierra Blanca." Grace Thorpe, daughter of the great Olympic athlete Jim Thorpe and leader of the National Environmental Coalition of Native Americans, charged, "Yesterday, they gave us smallpox-infested blankets, and today they give us the MRS." [6, 10]

The tribal council sponsored a referendum in January 1995 to gain support for the project. Initially, the measure was defeated, but was then passed a month later after a successful second-vote petition by Wendell Chino. A good part of the reversal appeared to be negative reaction to non-Native American environmental activists, who implied that only white people know how to handle such complex matters.

The tribe stipulated that any MRS facility would be for temporary storage only and would accept only a fraction of the spent fuel from utilities. Title to the nuclear waste must remain with the generating utility, thereby ensuring that the liability would remain with the utility in accordance with the Price–Anderson Act. Ultimately, the negotiations broke down over the financial terms.

With the Mescaleros out of the picture, the utility consortium – now consisting of eight utilities and called Private Fuel Storage – turned to the tiny Skull Valley Band of the Goshute Nation whose reservation is located 45 miles (72 km) southwest of Salt Lake City. Like the Mescalero Apaches, the Goshutes had also fought long battles with the white men, when the Mormons poured into Utah in the 1800s. By 1912, the tribe had been decimated and pushed onto 18,000 acres of parched land covered by cheat grass, bitterbrush, and greasewood in Skull Valley.

The north–south trending valley is flanked by the Stansbury Mountains that rise abruptly to the east and the juniper-pinion-laced ridges of the Cedar Mountains to the west. The Wild West had bypassed Skull Valley. The Pony Express route, often attacked by the Goshutes, ran to the south. The trail blazed by the Donner Party, in their ill-fated attempt to find a shortcut to California, lies to the north.

The Goshutes were poor and almost completely forgotten. Only a few of the tribe's 120 or so members lived on the reservation. Tribal businesses consisted of a mini-mart along the dusty highway and land leased to the Hercules Corporation for testing satellite launch rockets. The MRS site was a potential bonanza for the tribe. The proposal was to store 40,000 metric tons of spent nuclear fuel in 4,000 dry casks set on a huge concrete slab – almost two-thirds of the spent fuel proposed for Yucca Mountain. If everything went according to plan, it was to be a huge open-air nuclear waste parking lot.

The Goshutes own a forbidding piece of real estate, in part because of the harsh environment, but even more so on account of the neighborhood. The Department of Defense tests biological and chemical weapons at nearby Dugway Proving Grounds, a restricted military zone about the size of Rhode Island. In 1968, an accidental release of VX nerve gas killed thousands of sheep in Skull Valley, along with countless jackrabbits, antelope, and other native wildlife. The Deseret Chemical Depot, lying to the southeast, once stored more than 40 percent of the Nation's chemical weapons – rockets, missiles, and mortars packed with sarin, mustard gas, and VX. The Depot houses the Nation's first chemical weapons incinerator, which began to destroy these deadly agents in 1996. A giant magnesium plant to the north, along the shore of the Great Salt Lake, had the distinction of being listed by the EPA as the Nation's worst air polluter for several years. The Utah Test and Training Range to the west contains the largest overland airspace for training and weapons testing by the Department of Defense. The only missing ingredient in the area's toxic mix was high-level nuclear waste. [11]

The Goshute tribe and utility consortium applied for a 50-year license from the Nuclear Regulatory Commission (NRC). The State government and environmental groups were adamantly opposed, as were some Goshutes. Utah Governor Michael Leavitt, who grew up downwind of the Nevada atomic bomb tests, said he would do whatever it takes to stop the plan. Since Utah could not veto the plan, they had to resort to other means. [12]

On the paved road from Interstate 80 to the reservation, the State erected a sign proclaiming "High Level Nuclear Waste Prohibited." This largely symbolic gesture was followed by the seizing of two county dirt roads that ran near the reservation, effectively blocking a railroad spur to the site. In 2002, this effort to stop development was denied by a federal judge. The State then appealed to the NRC to deny the tribe's request, citing dangers posed by earthquakes and from the potential crash of an F-16 fighter jet from the nearby Air Force training range. In March 2003, an NRC board was persuaded by the latter argument and ruled against the tribe and utility consortium. The board said it might reconsider its decision if the Air Force would reduce the number of flights over the reservation, or the consortium could demonstrate that the casks could withstand an F-16 crash. In May 2005, a panel of administrative law judges ruled that the casks, which were designed to withstand high-speed collisions with locomotives and 2,000 °F (1,100 °C) jet fuel infernos, could withstand such a direct hit. Four months after this ruling, the NRC authorized a license. [13, 14]

Efforts to block the transportation routes to the reservation continued on other fronts. In January 2006, the Cedar Mountains to the west suddenly became the first new federally designated wilderness area in

Utah in more than two decades. All mechanized equipment is banned in wilderness areas, including trains that would run on the proposed rail line to the Goshute reservation. The successful push for a wilderness designation came from an alliance of environmentalists and Utah's Republican politicians, for once facing a common enemy in the Goshute plans. The critical rail route to the reservation was effectively closed down.

The Utah congressional delegation also turned to Interior Secretary Gale Norton for help. The MRS proposal required permits from two agencies in her department: the Bureau of Indian Affairs and the Bureau of Land Management. In September 2006, the Department of the Interior blocked the interim storage site. As trustee of Indian lands, Interior claimed it could "derive no confidence from the public record" that there would be someplace for the fuel ultimately to go. "It's like the broken treaties," lamented Garth Jerry Bear, a proponent and son of the tribal chairman who originally supported the idea. [15]

Proponents of an MRS facility persist – as do many questions. Would it erode impetus and political support for repository development and compete for the same resources? What is the risk of it becoming a de facto permanent repository? Where do you locate an MRS facility for geographic equity? Who pays? What about the added risks and costs of transportation and handling? Is it really worth moving the waste twice?

The real drivers for an MRS facility are political. The ongoing delay of a geologic repository has left DOE with an obligation to take spent fuel, but with no place to put it. The nuclear industry is also in a difficult position. How can they propose expanding nuclear power without having even an *interim* solution to the waste problem? Finally, States are pressuring their representatives in Congress to get the used fuel off their decommissioned sites.

In 1996, the Nuclear Waste Technical Review Board, a presidentially appointed oversight group, concluded that there were no compelling technical reasons for moving commercial spent fuel to a centralized storage facility. According to the Board, the risks were essentially the same for at-reactor and centralized storage. But the pressures from the impasse in licensing a geologic repository continued to build. [16]

Responding to this increasing pressure, in 2008, Congress asked DOE to develop a plan to take custody of spent nuclear fuel that is currently stored at decommissioned reactor sites. At this time, 14 commercial nuclear reactors in the United States were permanently shut down. Ten of these, like the Maine Yankee, are storing waste at sites with no nuclear operations. Nonetheless, Congress was asking DOE to undertake a task that Congress had expressly forbidden them from doing. Clearly stated in the Nuclear Waste Policy Act, the Department of Energy has no authority to take spent nuclear fuel for interim storage until a geologic repository is licensed for construction. [17]

In July 2010, the Goshutes' proposal for interim storage was suddenly back in play when a US District Court judge rejected the Department of the Interior actions blocking the project. The judge ruled that the Interior Department must defer to Indian landowners "to the maximum extent possible." Utah's congressional delegation immediately issued a joint statement condemning the decision. "The plain simple fact is that we will never allow this facility to be built," stated Senator Orrin Hatch. [18]

Since passage of the Nuclear Waste Policy Act in 1982, the search for a willing community to host an interim storage facility has twice been successful. In the 1980s, the local community of Oak Ridge, Tennessee endorsed, along with some very reasonable conditions, an MRS facility. In spite of this willingness, most Tennesseans were opposed and Congress shut it down. In the 1990s, the Goshutes' willingness to host an interim storage facility was systematically trounced by the State of Utah. In both cases, NIMS (not in my State) superseded the in-my-backyard willingness of local community and tribe.

10

A can of worms

Newspapers are unable, seemingly, to discriminate between a bicycle accident and the collapse of civilization

Attributed to George Bernard Shaw [1]

On the evening of July 6, 1994, Dave Prudic arrived in the town of Needles, California. True to form, the temperature hovered around 100 degrees, yet it felt almost cool compared to the afternoon high of 108 degrees. Among meteorologists, Needles was the record-breaker – often registering the highest summertime high in the Nation, and occasionally for the world, as well as the highest *low*. It didn't cool down much in Needles during the summer months.

The 1994 population was around 4,500. Many worked for the BNSF railroad, as the town was a crew change point. Aside from the train coming in, not much was happening in Needles these days. Then again, not much had ever happened in this small western town sitting on the California side of the Colorado River and getting its name from a group of pointed rocks on the Arizona side.

The town has a couple of claims to fame. Charles Schultz, creator of the *Peanuts* cartoon, had lived in Needles as a child. Schultz had gone on to immortalize the town with Snoopy's brother, Spike, who lived in the desert outside Needles and spent his days lying on rocks talking to cacti. Needles also made news in 2008 when it threatened to leave California and become part of Nevada or Arizona. The town shakers and movers weren't particularly choosy about which State would take them. The problem was, Needles was on the wrong side of the river.

Dave grabbed a bite and a room, then settled down to review his presentation. Had he known how tomorrow's meeting would affect him personally, and the United States Geological Survey generally, he might not have slept well. In fact, he might not have slept at all. But Dave had no such sixth sense. He was just a down-to-earth, hard-working scientist giving one more presentation of his study on the hydrology of desert sediments. Dry stuff – the kind of thing that can put you to sleep. Dave turned off the light and slept like a baby.

The next morning, the dining hall of the local Elk's Lodge was bustling with scientists, representatives and consultants from the California Department of Health Services and US Ecology, a professor from San Diego State University, and any number of concerned citizens from the State of California. The purpose of the three-day meeting was to examine scientific concerns about the proposed Ward Valley low-level radioactive waste-disposal site, located about 20 miles (30 km) west of town. For the State of California, and the Nation at large, low-level waste was viewed as a fairly straightforward and manageable problem to solve – compared to the ongoing headache of what to do with high-level waste.

Largely defined by what it's not, low-level radioactive waste (LLRW) is not spent fuel from nuclear power plants, high-level reprocessed waste, transuranic waste, or uranium mill tailings. What's left is more or less classified as low-level waste, then further divided into Class A, B, or C. Class A and B waste becomes relatively harmless in 100 years or less. Class C waste may take somewhere in the neighborhood of 500 years. There is also "greater than Class C" LLRW waste that is to be disposed of separately.

For several decades, there has been ongoing debate about what qualifies as low-level waste. To be classified as such, the waste must not exceed 100 nanocuries of transuranic elements (like plutonium) per gram of waste. A nanocurie is one billionth of a curie. The allowance for even the tiniest amount of plutonium in low-level waste has been a lightning rod for controversy, no matter how small the actual risk.

Low-level radioactive waste comes from a variety of sources. In the overall stockpile, medical labs and hospitals contribute a small portion. Unbeknownst to most of us, medicine has gone radioactive. Perhaps half of all patients who now enter US hospitals are touched, one way or another, by radionuclides used in diagnosis, treatment, and equipment sterilization. Universities and research institutions also generate LLRW. The current understanding of DNA and the genetic code has relied heavily on insights from radionuclide detective work. By far and away, however, the most LLRW comes from nuclear power plants, in the form of used pipes and plumbing parts, protective clothing, rags, and anything else that happens to get contaminated. [2]

Figure 10.1 Cardboard waste drums randomly dumped into trenches.
Note water in foreground.
Source: US Department of Energy.

Initially under supervision by the AEC, the early days of LLRW disposal were pretty much *dump now, ask questions later* enterprises. These were strictly low-tech operations. The first LLRW disposal facilities resembled huge landfills, with burial trenches sometimes 1,000 feet long. Just like at good old-fashioned dumps, trucks would back-up and roll the cardboard or steel transport canisters into the trench. Some of the canisters were damaged from this unceremonious unloading, others were leaking when they arrived at the facility, yet these containers were buried along with the rest. They also didn't keep records. As a result, the number of leaking canisters will never be known. Regulation banning disposal of liquid waste was an unheard-of safeguard in those days. Compounding this problem, the burial trenches were left open to rain and snow until the trench was full. Helped along by Mother Nature, it wasn't long before the liquid waste from leaking canisters was spilling out of the trenches and working its way down to the water table.

By 1971, commercial sites were operating in six States – Illinois, Kentucky, New York, South Carolina, Washington, and Nevada. By the end of the decade the sites in Illinois, Kentucky, and New York were closed, largely because of liquid radioactive waste leaking beyond the burial trenches. In 1979, the year of the Three Mile Island partial core meltdown, two of the three remaining sites – Hanford, Washington and Beatty, Nevada – were temporarily shut down as a result of leaking transport containers arriving at their facilities. There was also a PR nightmare when one of the trucks carrying radioactive waste caught fire at the Beatty site. That same year, South Carolina decided it wasn't going to be the national dumping ground for low-level waste and decided to limit the amount of waste they were willing to accept, while also threatening to close their facility. [3]

Although soon rescinded, these actions provoked a crisis. Some medical facilities with too much waste piling up reported that they were within two weeks of having to shut down. Columbia, Harvard, and Duke threatened to close their medical research labs. The following year, residents of Washington State overwhelmingly passed a law prohibiting disposal of most nuclear waste from out-of-State. The US Supreme Court struck the law down as unconstitutional, stirring up a maelstrom of anger. Something had to be done. [3–4]

Notorious for being quick to react and slow to act, Congress broke some kind of record. In December 1980, the Low-Level Radioactive Waste Policy Act was enacted into law with little debate. The Act required States to take responsibility for their own commercial LLRW. It also left some wiggle-room for any State that had an aversion to dumping this waste within its own borders. A State could go it alone, if they so desired, or they could form compacts with other States. The basic idea was simple. One of the States would build and operate the facility, and the other States in the compact would pay to dump. Amendments to the Act in 1985 permitted any regional compact to exclude wastes from nonmember States, motivating everyone to get with the program.

The ensuing compacts resulted in some strange and sometimes fickle bedfellows. Vermont (and Maine for a while) joined Texas to become the Texas Compact. The Dakotas joined California and Arizona to form the Southwest Compact. Delaware and Maryland, originally in the Northeast Compact, later switched to the Appalachian Compact. And so on. Some States, like Massachusetts, decided to go it alone.

It took a while to work out the details of who was going to join which compact, and what was in it for them. It took a while to decide which State within a compact would develop the waste facility. And it took even longer to meet all the safety rules and regulations pertaining to site selection and licensing, before an LLRW facility could open for business.

By 1993, in just 13 years – which in the world of nuclear waste disposal might qualify as a major miracle – California was in the home-stretch of opening the first federally approved and licensed low-level radioactive waste facility in the Nation.

No real surprises here. California basks in its reputation as a national trend-setter. California also lays claim to the Mojave Desert, one of the driest places on the face of the planet. Producing nearly nine percent of the Nation's LLRW, the State sorely needed a safe place to put it. After studying any number of possible sites, Ward Valley at the eastern end of the Mojave Desert was chosen for the Southwest Compact's LLRW disposal facility. In this remarkable situation, where neither NIMS nor NIMBY had become a show-stopper, there remained one hitch – the proposed site was on federal land and required a land transfer from the Department of the Interior to the State of California. [5]

The Mojave is desert reduced to bare essentials, stripped clean of nearly everything but baking rock, sparse and singed chaparral, and the occasional dry wash. To most, a virtual moonscape. Yet, the discriminating viewer sees a dominance of creosote bush peppered by Mormon tea, wolfberries, Fremont dalea, and succulents such as beaver tail and Mojave yucca. Environmentalists see critical habitat for the endangered Mojave Desert tortoise, while Native American tribes in the area see sacred homeland.

The geologic features of Ward Valley bear colorful names that reflect, in part, the Native American heritage. The 50-mile (80 km)-long valley is bounded by Piute, Little Piute, Old Woman, Stepladder, and Turtle Mountains. Homer Wash, an ephemeral stream that only contains water during wet periods, runs south through the center of the valley and discharges into Danby Dry Lake. The name *Dry Lake* is not an oxymoron but a general indication of the aridity of this desert environment.

As in the case of high-level radioactive waste, water getting into low-level waste poses the main problem with long-term containment. At Ward Valley, water didn't look like it was going to be a problem. Gregg Larson, representing the Midwest Compact, said of the Ward Valley site, "If we can't get a facility built at that site, you sort of wonder how you can get a site built anywhere." [6]

THE NEEDLES MEETING

On the morning of July 7, 1994, the attendees sipping coffee and recovering from heat stroke in the Elk's Lodge in Needles knew exactly why the meeting was being held in this outpost of civilization. A field trip to the Ward Valley site was planned for the third day of the meeting. Being the only town in the vicinity, Needles was the logical place to meet. From all over the country, scientists flew to major hubs like Las Vegas, rented cars, and settled in for the long drive across the baking desert. Never before had such a concentration of scientific muscle descended on Needles. One by one, they rolled into town and dove for the shower.

The bigger question of *why* this meeting was even being held, confused and irked certain attendees. Representatives from US Ecology, who had been awarded the lucrative contract to build and run the Ward Valley LLRW facility, were definitely irked. So were employees and consultants from California's Department of Health Services, the State agency responsible for licensing and regulating the facility. Having adhered to all the hundred-plus safety regulations, as far as they were concerned, this three-day meeting was a time-consuming, nit-picking formality. After all, they were in the home stretch – the California Department of Health had already issued a license for site operation.

The reason for the meeting began the previous year, when Howard Wilshire and two other US Geological Survey (USGS) geologists sent a memorandum to Bruce Babbitt, Secretary of the Interior, stating they had a number of concerns about the safety of the Ward Valley site. Would the Secretary be interested in reviewing these concerns? Babbitt's office, of course, said yes. Send them in.

The USGS is in the Department of the Interior, yet bypassing their own agency and going straight to Interior was a breach of protocol. The first order of business would have been for Wilshire and his colleagues to discuss their concerns with other Survey scientists, a number of whom were experts on southwestern hydrology and related LLRW disposal issues. The soon-to-be-dubbed *Wilshire Group* were geologists and their most serious concerns involved water-related issues outside their fields of expertise. Howard Wilshire would later defend his actions while responding to a stinging criticism from Philip H. Abelson, editor of *Science* magazine: "In response to a written request from staff of the Secretary of the Interior, I and two other career USGS geologists forwarded a two-page internal memorandum." [7–8]

Another problem soon became apparent. The Wilshire Group based their concerns on an early draft of the environmental impact statement (EIS) for the Ward Valley site. US Ecology contended that, had they read the final EIS, they would have discovered their issues had been addressed. US Ecology sent copies of this letter to a number of powerful individuals, including California's Governor Pete Wilson, an ardent advocate of Ward

Valley. Senator Barbara Boxer of California, an outspoken opponent of nuclear power and Ward Valley, soon jumped into the fray in support of the Wilshire Group. The issue quickly catapulted its way to the US Congress. [9]

Meanwhile, the Wilshire Group steamed ahead. In December 1993, they sent a detailed report to Secretary Babbitt discussing seven technical concerns, including certain modifications after now having read the final EIS. By far, the most alarming issue they raised was that the waste could work its way to the Colorado River. Such a possibility, raised by three USGS scientists, offered new hope to the opponents of Ward Valley. Two months later, the Los Angeles-based antinuclear group, the Committee to Bridge the Gap (as in *the gap* between nuclear dangers and the survival of the human race), warned, "Tens of thousands of people could come down with cancer if this occurs." [6, 9]

The USGS suddenly had a serious problem on their hands. Three of their geologists had raised issues outside their field of training and expertise, and they were not backing down when presented with evidence to the contrary. In addition, the report they sent to Secretary Babbitt had not gone through the peer-review process that precedes release of any USGS scientific document. The Survey has a tradition of scientific freedom, as long as the work of its employees is subject to the checks and balances of *peer review*. USGS headquarters finally decided that the scientists could raise these concerns as individuals, but not in an official capacity as USGS employees. This subtle distinction was soon lost on the media and general public. [10–12]

Opponents to Ward Valley lobbied for the Wilshire Group's views to be given consideration because they were qualified earth scientists. In an attempt to settle the matter, Secretary Babbitt turned to the National Academy of Sciences (NAS). The Secretary requested that the NAS convene a committee of experts with the training, research, and field experience to evaluate the seven issues raised by the Wilshire Group.

The meeting in Needles marked the start of the NAS investigation. At ten o'clock sharp, the attendees dumped their coffee or got a refill and filed into the conference room. The room was soon packed with attendees and concerned citizens, who would be given the opportunity to state their concerns during a session at the end of each day.

Facing the room sat Dr. George Thompson, chair of the committee. As an internationally recognized earth scientist, Dr. Thompson had a long list of impressive awards, honors, and accomplishments to his credit. He had taught at Stanford University for over four decades and was one of two on the committee who were members of the prestigious National

Academy of Sciences. At 74, Thompson was alert and focused. After a lifetime of moving among the world's scientific elite, he still had the unaffected down-to-earth smile that had become a personal trademark. No one had a complaint with Dr. Thompson chairing this investigation. He was renowned for his ability to remain objective.

Another scientific star on the NAS committee was Dr. G. Brent Dalrymple, who looked more like a western cattle rancher than a renowned scientist. Dalrymple was one of the visionaries who helped establish the theory of plate tectonics. Besides being a member of the National Academy of Sciences, he received the National Medal of Honor in 2005. Equivalent to the Congressional Medal of Honor, this is the Nation's highest award for scientific achievement. Even in the presence of such illustrious company, the other 15 members of the committee were all highly accomplished and respected experts in their fields of geology, hydrology, geophysics, geochemistry, civil engineering, and desert ecology.

For the next three days the committee's mandate was to listen, ask questions, take notes, and not take sides. They were to follow the science and let others make the decision about site suitability. Dave Prudic and two other hydrologists from the USGS Carson City, Nevada office were also attending in this capacity. From the earliest days under its first Directors, Clarence King and John Wesley Powell, the US Geological Survey was chartered to be an objective, fact-finding scientific agency.

The three USGS geologists, who were the reason for this investigation, unaccustomedly sat apart from their Survey colleagues. In his sixties, Howard Wilshire had a reputation for being against any development whatsoever on fragile desert ecosystems. As an expert on off-road vehicle impacts on desert environments, Wilshire was a champion of the endangered desert tortoise, and development destroys habitat.

Dr. Thompson opened the meeting with a summary of the concerns they were meeting to address. Two scientific questions dominated the discussion. First, can enough rain penetrate the burial trenches to carry the radionuclides 650 feet (200 m) down to the water table? Second, could groundwater then carry a dangerous dose of radiation to the Colorado River?

Groundwater hydrologists study this unseen world through modern high-tech instruments, mathematical models, and challenging detective work. Groundwater hydrologists who study questions related to low-level radioactive waste devote most of their time studying the *unsaturated zone* (UZ) – the zone between the land surface and the water table. The UZ can be likened to a sponge of varying thicknesses. In wet climates, the UZ might only be a few inches to a few feet thick. In arid climates, such as Ward Valley, the UZ is hundreds of feet thick. Nevertheless, unsaturated does not mean *dry*. There is virtually no place on our planet where the unsaturated zone is completely dry all the time. We're a wet planet.

The Wilshire Group's concern about Ward Valley low-level waste working its way to the water table and then to the Colorado River was based largely on the area's, albeit rare, soaking winter rains. The Colorado River is 18 miles (30 km) east, as the crow flies, from Ward Valley, but the underground path would be a much longer distance. Groundwater doesn't travel like crows fly. The most plausible route for contamination of the Colorado is a meandering 70-mile (110 km) trip with an ETA of 1,900 to 4,600 years. Additionally, the waste wouldn't all just make a beeline for the river, but would spread out as the water took different routes among minute paths between particle grains. Some contaminants would adhere to the particle grains along the way. Finally, Danby Dry Lake, not the Colorado River, is the main groundwater discharge area within Ward Valley, lying south of the proposed site. [9]

When presented with these facts, the Wilshire Group rebutted that there could be some *preferred pathways* down there that go right from the burial site to groundwater and then to the river. In dry desert soils some water moves through preferred pathways, such as roots and root tubules, but it is mostly absorbed by the dry soil around the pathways. The kind of preferred pathway the Wilshire Group was suggesting would have to be something like an HOV express lane. And this express lane would be impossible to prove or disprove without digging up the whole valley.

Fortunately, there was a simpler solution. If appreciable amounts of water don't get past the waste burial trenches constructed more than 25 feet (8 m) below the surface, the whole chain reaction of possibilities would become basically a moot point. This is where Dave Prudic came in. He had distinguished himself in many areas of hydrology, including as a key developer of the world's most popular groundwater computer model. While one's immediate impression upon meeting Prudic is that he's a nice guy, he's also a stickler for detail and scientific rigor. When it comes to arid zone hydrology, Dave Prudic is a person people listen to.

Eighteen years earlier, in 1976, the USGS began studying the geohydrology of LLRW disposal sites to better understand the potential for radioactive movement through the unsaturated zone. One of the long-term studies was at a field laboratory outside the fence of the LLRW facility near Beatty, Nevada. For years, Prudic and his associates had been studying UZ water movement at this field lab in the Amargosa Desert, 200 miles (300 km) northwest of Ward Valley. The Beatty data were one of the best records of water movement through deep unsaturated zones in the Southwest. It was logical that these data would enter the Ward Valley debate.

The Beatty and Ward Valley sites were considered analogous in several key ways. First, they were both located in extremely hot, arid climates. Beatty, Nevada is known as the Gateway to Death Valley. Ward Valley is 15 miles (24 km) from Needles, the Nation's record-breaker for heat. Secondly, both places receive almost all their annual precipitation during the winter months. Finally, the water table at both sites is hundreds of feet below land surface. Nonetheless, the Wilshire Group contended that the Beatty site did not provide an appropriate analog, because Ward Valley is at a higher elevation and may have more precipitation. Because of limited rainfall records for the area, no one knew the exact difference.

While the hydrology of the two sites may or may not be similar, the waste disposal practices most certainly were *not* going to be the same. The Beatty disposal facility had a checkered history. In the mid 1970s it was known to local residents as "the store," because anyone could (and did) purchase from site workers items such as contaminated tools and other materials slated for burial. At one point, employees borrowed a concrete mixing truck used to mix LLRW with cement and used it to pour concrete for local construction projects – including a patio behind the town saloon, and a floor for the town jail and courthouse building. When these practices were exposed in 1976, the saloon patio floor was found to emit excessive gamma radiation. It was torn up and returned to the site for burial. State health inspectors then went door to door with radiation detectors, collecting a motley assortment of radioactive items – military clocks, watches, shipping containers, tools, and so forth. Probably the biggest surprise was when inspectors found more than 20 metal tanks, once containing radioactive waste, that had been converted to septic or water tanks. It took 25 truckloads to return all of these hot items to the burial grounds. [13–15]

In the end, the US Environmental Protection Agency concluded that no one appeared to have received significant exposure from the radioactive grab-bag, but no one knew for sure. The Beatty site was temporarily shut down and the operator fined $10,000. After finally reopening in the early 1980s, the site was shut down two additional times after it was discovered that the dump was continuing to accept leaking containers *and* had illegally buried waste outside the fence. After many years of effort, in 1992, the State of Nevada finally prevailed in permanently closing the Beatty site from accepting any more LLRW.

Figure 10.2 Worker checking buried low-level waste for radioactive leakage.
Source: US Department of Energy.

The site, however, had been operated by the same company that would operate Ward Valley, running up a definite red flag for Ward Valley opponents.

There was another key difference between Beatty's and Ward Valley's disposal philosophy. Liquid wastes were dumped at Beatty, with the trenches being left open often for a year or more, collecting water from rainstorms. All wastes at Ward Valley would be solidified, then put in steel or concrete containers, and trench segments would be closed as they were filled. The Wilshire Group omitted these important differences from their report. [7]

On the first day of the meeting, Dave Prudic presented the results of his long-term studies of water percolation at Beatty, then compared these findings with the limited data available from Ward Valley. He stressed that movement of water through the UZ is very complex, particularly in arid climates. To begin with, when precipitation infiltrates the soil, some of this water evaporates while some is *transpired* by plants. Because it's hard to determine where evaporation stops and transpiration begins, these two processes that return water to the atmosphere are lumped together as *evapotranspiration* – also known as ET and not to be confused with the cute little outer-space guy!

The UZ in arid climates has an immense storage capacity. Before rain soaks down to the water table, it first has to fill up all those thirsty pores – and then ET kicks in. When the amount of rainfall exceeds what can be stored and returned to the atmosphere through ET, water works its way deeper into the UZ. If sufficient, it recharges the water table. If rainfall is marginal, the water table is not recharged. When investigating a proposed LLRW disposal site, the trick is finding out which is the more likely scenario.

Other processes add even more complexity. Liquid water moving through the UZ is affected by two primary forces – gravity and capillary action. Gravity pulls the water downward, while capillary action sucks the water from wet to dry places – like setting a dry sponge in a puddle of water. The balance between gravity and capillary action in the UZ is very sensitive to *water content* – how much water is down there. When water content is low, capillary forces hang on to it. When water content is high, gravity overcomes capillary action and pulls the water downward. And finally, because of the predominantly dry soils in arid climates, much of the water movement occurs in the vapor phase (as a gas) rather than as liquid water.

Measurements at Beatty indicated that water movement deeper in the UZ is primarily an upward flow of water vapor, driven by temperature gradients, the aridity at the surface, and the thirsty vegetation. After not getting a drop of water for many months at a time, these tenacious desert plants are masters at filling up during the winter rainy season. Studies at the Beatty field laboratory demonstrated that water has a much better chance of percolating downward when the vegetation is removed. This lesson had been learned; unlike at Beatty, the trenches would be revegetated at Ward Valley.

Groundwater hydrology in arid climates involves determining how all these elements combine into an integrated system. It's a huge challenge compounded by the fact that water movement in the UZ cannot be measured directly, but has to be inferred from other methods. Developed over time, there are a number of techniques for inferring this movement. One technique useful in arid areas is to measure the amount of chloride in the UZ. Chloride is very common – for example, chloride and sodium make salt. Chloride is everywhere on the planet, including the atmosphere. When rain falls to the ground and soaks into the UZ, it takes the chloride along for the ride. After the water evapotranspires, the chloride is left behind. Using measurements of the amount of chloride falling from the atmosphere, groundwater hydrologists can estimate how long it took for a chloride deposit to accumulate in the UZ.

Prudic found that the chloride concentrations at Beatty and Ward Valley were very high within about the top 30 feet of the UZ. After determining the amount of chloride, he calculated that water percolation was restricted to the upper 30 feet (9 m) during the past 16,000 to 33,000 years. It took that long to accumulate that much chloride at that depth. Chloride easily dissolves in water. If rainwater had soaked deeper into the UZ, it would have carried the chloride with it. This piece of detective work provided the State of California with very good evidence that very little, if any, rainwater *over the past 16,000 to 33,000 years* had soaked deeper than 30 feet (10 m) into the UZ. The time estimates had substantial uncertainty, but the existence of a long-term accumulation of chloride was undeniable. *And* this chloride band was well above the water table which lay hundreds of feet below. [16]

There are other ways to study liquid and gas water movement through the UZ. Today's groundwater hydrologists use sensitive instruments, like thermocouple psychrometers and neutron probes, that can measure very small amounts of water and infer its movement. At Ward Valley, six boreholes were drilled to a depth of 87 feet (almost three times deeper than the chloride deposits) to collect soil samples and run additional tests. The results showed that the UZ is very dry and deep percolation of water is extremely small. Unfortunately, there was evidence that this sensitive equipment was not properly used at Ward Valley, throwing the results into doubt.

Finally there's tritium, an isotope of hydrogen. Being the first element on the periodic table, hydrogen is the lightest and simplest element on Earth or, for that matter, in the universe. Hydrogen has three isotopes. One of these, tritium, is radioactive with a half-life of 12 years. During aboveground thermonuclear testing from 1952 to the mid 1960s, huge amounts of tritium were released into the atmosphere. These radioactive isotopes of hydrogen bonded with oxygen and joined the hydrologic cycle as part of the water molecule, H_2O. Tritium found above certain levels in the UZ or groundwater tells us that the water is younger than 1952. When atmospheric tritium peaked in around 1963, there were 1,000 or more tritium units (TUs) in rainfall *sampled anywhere on the continent*. Because of its short half-life, by 1994 the average amount of tritium found in precipitation had declined to around 5 to 15 TUs. [9]

But there's a catch. Tritium testing in the UZ of arid and semi-arid regions is extremely difficult because you have to collect enough water from these dry soils to run the test. Most UZ water in arid regions comes in the form of water vapor, or gas. As a result, collecting enough water vapor to run a tritium test involves pumping huge volumes of air *out* of the UZ without pumping *in* any tritium from the atmosphere. Given the difficulties, it's quite possible for tritium results to come back higher than estimated because some atmospheric tritium got into the sample.

A sample at Ward Valley, taken 100 feet (30 m) down, found about $1\frac{1}{2}$ units of tritium. One unit of tritium translates into one tritium atom for every 1×10^{18} hydrogen atoms; that is, one for every billion *billion* hydrogen atoms. By almost anyone's measure, this was nothing to get worked up about. A few tritium atoms found at such a depth could easily be explained as a sampling or laboratory error. And even if it wasn't an error, everyone considered it too small an amount to worry about. Everyone, that is, except the Wilshire Group, who maintained that the tritium sample was the smoking gun that Ward Valley water is moving through the UZ much faster than anyone had supposed. They also dismissed the chloride band 70 feet (21 m) above the tritium – regardless of the fact that if much water had percolated to a depth of 100 feet, the chloride would have been carried down with it. [9]

For three days, the NAS Committee listened and asked questions. The meeting ended with a plea for patience. Dr. Thompson understood that the State of California was on a holding pattern until these questions were resolved. Nonetheless, the Committee members had a huge task ahead of them. After a follow-up public meeting in August, they would review some of the 5,000 pages from the administrative record, dozens of documents of site characterization and monitoring data, various technical reports from US Ecology, new documents generated by the Wilshire Group, and several reports from pro and con organizations. It would take months to work their way through all this material without compromising the integrity of the review. Then they would write their report. Dr. Thompson reminded everyone that, on top of everything else, they all had day jobs. Their findings would be released as soon as possible, but realistically it would probably be sometime next winter or spring. Good science moves slowly. [9]

THE BEATTY CONTROVERSY

At the time of the NAS meeting, Dave Prudic was just beginning another research project. A few months earlier, he and fellow USGS scientist, Rob Striegl, had collected samples of soil gas at a test hole at about 350 feet (100 m) from the Beatty LLRW facility fence. They sent the samples to the lab for analysis of tritium and carbon-14. Prior to the meeting in Needles,

Prudic learned that the tritium results had come back and were larger than expected, but he hadn't seen the results.

When Prudic later met with Striegl to discuss the results, they considered several possible explanations. Sampling error was the most logical explanation. Aside from the usual delicacy of conducting the test, they had used equipment from the Nevada Test Site where atomic weapons testing had taken place. It was possible that their sampling equipment, or possibly the lab, had contaminated the samples. And then again, the Beatty site, itself, was a possible source of contamination due to all the tritium buried at the site. Yet, it was unprecedented that tritium gas would work its way through the deep UZ to 350 feet *outside* the fence. But then again, it was rumored that some waste had been dumped illegally off-site. Given Beatty's track record, anything was possible. The most likely scenario, however, and the first thing to check was whether their samples had been contaminated. They decided to set aside the tritium data until the carbon-14 results came back.

In May 1995, the NAS committee released their 200-plus page report on Ward Valley. In short, the committee recommended additional monitoring and some other precautionary actions, but concluded that the Wilshire Group's concerns didn't hold up under close scientific scrutiny. Agreement was not completely unanimous. Two of the 17 committee members had concerns about the $1\frac{1}{2}$ units of tritium found at depth. One member had concerns that the "raised cap" design could hinder revegetation. [9]

Perhaps most significant, none of the committee members viewed contamination of the Colorado River as a possible problem. To support their conclusion, they calculated what would happen if somehow *all* of the expected plutonium in the waste, over the entire 30-year operating period, leached instantaneously from the steel or concrete drums into the water table and traveled directly to the Colorado River, with no delay. This totally unrealistic worst-case scenario was equivalent to pouring the wastes directly into the Colorado for three decades. Even then, the NAS Committee calculated that the total amount of radiation would be small compared with the other alpha emitters already present in the river.

After release of the NAS report, Secretary Babbitt agreed to the land transfer, but with strings attached. The State of California had to accept Department of the Interior authority to enforce compliance with all of the NAS recommendations. The terms satisfied neither Governor Pete Wilson, a proponent of the site, nor the environmentalists who opposed it. "The Department of the Interior apparently believes that although it has no expertise, experience or legal role in radiation safety, it should second-guess the responsible State agencies," charged a frustrated and furious Wilson. Environmental groups opposed to the Ward Valley site accused Babbitt of reneging on a pledge to hold a public hearing first. [17]

The timing couldn't have been worse for the USGS. Two weeks after release of the NAS report, Dave Prudic and Rob Striegl received the carbon-14 results back from the lab. As with the tritium, the carbon-14 was surprisingly high. The Beatty waste-burial site moved higher up on the suspect list, yet contamination of the sampling equipment remained a prime suspect that had to be tested. However, if these results were accurate, the implications of finding two radioactive isotopes in such high concentrations hundreds of feet outside the Beatty fence was lost on no one. Rob Striegl sent an email to Bob Hirsch, Chief Hydrologist at USGS headquarters, warning him of the *"can of worms* [emphasis added] that this will open if we are actually seeing 14-C (carbon-14) movement from an arid radioactive waste site." Hirsch promptly ordered a second sampling to confirm the results, writing: "My sense is that this is very important for us to get the story both correct and fast."

The resampling was done in July, and the samples sent to the lab with a red flag to put them on the fast track for analysis. The results for tritium came back a month later, again with high concentrations. Samples collected from additional sites right along the fence were even higher. By now, the possible explanations had been winnowed down to *one*. For the first time, scientific lingo described what the test results had revealed: a plume of unexpectedly high levels of tritium had been detected in the unsaturated zone 350 feet outside the Beatty, Nevada LLRW disposal facility. In simple terms, a tritium plume was on the loose.

This was not the first time radionuclides had escaped from an LLRW facility, as witnessed by the now closed sites in Kentucky, Illinois, and New York. In addition, government LLRW facilities that took waste generated by the defense industry had experienced problems with leakage. Also, it wasn't the first time problems had developed as a result of dumping liquid waste. The NRC now strictly enforced solidifying the waste before dumping, and Ward Valley would be no exception to this rule. However, this was the first time radionuclides had escaped from an LLRW facility in the arid Southwest.

The carbon-14 results came back in September, again with high concentrations. As with the tritium, carbon-14 concentrations were highest along the fence. The amounts of tritium and carbon-14 in the soil gas did not represent a health hazard. Any gas that reached the water table would quickly be diluted, while most of the radionuclides would gradually and harmlessly vent into the atmosphere. Nonetheless, the findings became the trump card for Ward Valley opponents. Previously, Beatty had been decried as an inadmissible analog with Ward Valley. It wouldn't be long before they morphed into *near twins*.

On October 20, 1995, Gordon Eaton, Director of the USGS, sent a letter to the Department of the Interior informing them of the tritium and C-14 results and explaining that the concentrations were too small to pose a health risk. He added: "The purpose of this memorandum is to alert you to the fact that this discovery may be used by opponents of the proposed new low-level radioactive waste disposal facility at Ward Valley, California, to attempt to block construction of the facility and to challenge the transfer of land for the site from the Department of the Interior to the State." Soon accused of being pro-Ward Valley, the USGS Director was simply stating the obvious. [18–19]

A few days after Eaton notified Interior, Dave Prudic received a phone call from the Committee to Bridge the Gap. The caller wanted to know if there were any new findings at the Beatty research site. Prudic explained the tritium and carbon-14 data and provided the results. In spite of his cooperation, it was soon widely publicized that "an environmental group had forced USGS to make public the Beatty dump leak."

The first newspaper article appeared on the last day of October 1995. Frank Clifford of the *LA Times* wrote: "The discovery of tritium and carbon-14 near the Beatty dump site could have important implications for Ward Valley, where opponents of the proposed dump fear that long-lived waste particles could leach into the water table and ultimately to the nearby Colorado River, a source of drinking water for millions... [the results were] initially found in April 1994 but were not revealed to the National Academy panel that assembled in July." Daniel Hirsch, of the Committee to Bridge the Gap, was quoted as saying, "This appears to be an astonishing cover-up of a matter that could affect the safety of millions of people." [20]

The *LA Times* article kicked-off a several year media blitz fueled by Public Employees for Environmental Responsibility (PEER). Howard Wilshire was a member and later Chair of the Board for PEER, which bills itself as a watchdog organization of public resource professionals. Through Freedom of Information Act requests, PEER soon had in-depth knowledge of the sampling history at Beatty, the timing of confirmatory sampling, and plans to prepare reports documenting the findings. PEER and the Committee to Bridge the Gap ignored these facts.

In late 1996, PEER formally charged that scientists at the USGS secretly knew that the nuclear waste at Beatty was leaking, but had kept that critical information from the NAS Ward Valley committee. PEER went on to report how a USGS scientist had warned of the "can of worms that this will open if we are actually seeing [radioactive] movement from an arid radioactive waste site." According to PEER, information on the Beatty leaks may have been withheld for nearly a year and a half. [21]

As the media's accusations gained momentum, the "can of worms" email was no longer reported as coming from a scientist in the field, but rather from USGS headquarters to the scientists, presumably to shut them up. Caught in the middle, the Department of the Interior struggled with what to do. Deputy Secretary John Garamendi summoned Prudic and others from the USGS to his office in downtown Washington, DC. It was time for an explanation.

Garamendi had been an Eagle Scout, Peace Corps volunteer in Ethiopia with his wife, All-American offensive guard in football, and Pacific Coast wrestling champion. A picture of him riding a horse, on the cover of a western magazine, was prominently displayed on his office wall. The magazine story touted Garamendi as a family man who grew up on his family's California cattle ranch. Reinforcing the family-man image, his daughter was dutifully working on her homework in a corner of his large office when the USGS scientists arrived.

Prudic was eager to discuss the technical details and nuances of the Beatty investigations. He brought along a set of photographs showing how the field instrumentation had been emplaced, along with other features of the work. Garamendi was more interested in the political nuances of the whole mess. When the official meeting ended, Garamendi finally turned to Prudic and said, "Now, let's take a look at those pictures you have." With all the facts in hand, Garamendi took the safe route by ordering more sampling and a second supplementary environmental impact statement. These major undertakings created a serious impediment to moving forward with the site. [22]

Congressman George Miller (D-CA), senior Democratic member of the House Resources Committee, was drawn into the controversy. Miller told the press that reports of a USGS cover-up concerning radioactive material at the Beatty site "raises the most serious questions about the safety of the site and appears to give additional credence to criticisms of the proposed Ward Valley disposal plan." Congressman Miller also cautioned that, "We should not draw any conclusions until the investigation is completed, but the evidence disclosed by PEER is very shocking." Miller summoned the USGS scientists charged with the cover-up to his office. [23]

A Congressman from the San Francisco Bay area since 1975 and a long-time advocate of the environment, Miller began the meeting by explaining that he had always trusted both the USGS and PEER, so he was confused about where the truth lay. The USGS scientists explained that sound scientific practices were used at the Beatty research site to carefully collect and interpret the data, confirm results through resampling, and make the results available in a responsible manner. Although this may have looked like suppression of information to the general public, proceeding carefully and responsibly is a crucial part of scientific practice.

Miller brought up the "can of worms" email, and was presented with the full email. Following the oft-repeated sentence from Rob Striegl to Bob Hirsch, "Neither one of you need to be informed of the can of worms that this will open if we are actually seeing 14-C movement." Miller now read for himself the rest of the communication dropped by PEER, the Committee to Bridge the Gap, and the media. Striegl had gone on to say, "Because of this, I believe we need to proceed quickly and openly to resample borehole UZB-2 and to see if we can identify a gradient off of the site." Followed by Bob Hirsch's response: "Thanks for the info. I've spoken to Newell Trask [Chief, USGS Branch of Nuclear Waste Hydrology] about the situation and asked him to develop the strategy with all of you. Please keep him in the loop on all communications. My sense is that this is very important for us to get the story both correct and fast." Miller now realized he had been sorely misled by the opposition and the press.

But the damage had been done. By now, the scientific issues had been eclipsed by politics and lawsuits. The State of California and US Ecology were suing the Department of the Interior for exceeding its authority, arguing that Interior had neither the criteria nor the technical expertise to independently assess the suitability of the site. By law, the Low-Level Radioactive Waste Policy Act had turned site selection over to the States. Interior remained adamant that there would be no land transfer until California complied with all of the Department's conditions.

Also being challenged in court was Interior's alleged noncompliance with the Federal Land Policy and Management Act and the National Environmental Policy Act, and their alleged failure to protect native desert tortoises under the Endangered Species Act. The Department of the Interior also needed to address the effect of the Ward Valley facility on Native Americans in the region because of Executive Orders on Accommodation of Sacred Sites and Environmental Justice in Minority and Low-Income Populations. California and US Ecology viewed these Executive Orders as newly contrived blockades. They reminded everyone that, back in 1991 while preparing the original EIS, they had conducted an archaeological survey of the site, which included a walkabout with tribal representatives who did not identify any unique cultural resources. In addition, the proposed Ward Valley site had once been used for military tank maneuvers and was currently crossed by electric power lines. [24]

In 1997, the Senate Energy and Natural Resources Committee held a hearing on a bill introduced by Chairman Frank H. Murkowski (R-AK). The bill proposed bypassing Interior to transfer federal land to the State of California for developing the Ward Valley facility. Senator Murkowski opened the hearing by admonishing both the Department of the Interior and the State of California for allowing this process to drag on. He called the current waste-storage sites a threat to public health and gave examples of near misses – a storage site that caught fire during the 1994 Northridge earthquake and an attempted burglary at a UCLA waste facility. Murkowski emphasized that the government is obligated to protect citizens, and the best way to do this is to locate radioactive waste facilities away from people. He also pointed out that 149 biotech companies had recently left California – presumably some over waste-disposal issues. Calling the problems between California and Interior unconscionable, Murkowski closed by emphasizing the simple fact that nobody wants high-level or low-level waste, but we created it so we must do something with it. Murkowski's bill never passed. [22]

After several more contentious years – including a more than 100-day occupation of "ground zero" at Ward Valley by local tribes and the Save Ward Valley Coalition – the Department of the Interior terminated all land transfer actions at the end of 1999. The Ward Valley facility was never built.

On July 1, 2008, Barnwell, South Carolina closed its doors to all States that were not in its three-State Atlantic Compact.

For three decades after passage of the Low-Level Radioactive Waste Policy Act in 1980, only one new site opened in the United States. It accepts only Class A waste.

In December 2011, an LLRW facility opened for business in west Texas that accepts Class A, B, and C waste. While originally to be shared with Vermont only, many States are eager to sign on. No land transfer from Interior was necessary for approval.

The US Geological Survey continues to study radionuclide movement through the unsaturated zone at Beatty, Nevada. In spite of years of study, the exact cause of the tritium gas in the unsaturated zone remains unknown.

11

WIPP

You cannot say, or guess, for you know only

A heap of broken images, where the sun beats . . .

I will show you fear in a handful of dust.

T.S. Eliot, *The Waste Land* [1]

Ten thousand years *into the future*, on the baking windswept sands of what is today southeastern New Mexico, a huge earthen berm bears witness to a previous civilization. Twice as old as the Egyptian pyramids now are, the berm is an eroded fortress over a half-mile long and nearly as wide. The 100-foot (30 m) -wide base slopes upward, in places almost four storeys tall. Excavation into the berm uncovers small markers of granite, aluminum oxide, and fired clay, carved with hieroglyphs in seven dead languages. The berm also bristles with magnets and radar reflectors that transmit unique "signatures." [2–3]

Flanking the berm's inside walls there once stood huge stone pillars weighing over 100 tons each, now mostly toppled and broken. The same hieroglyphs cover the pillars' surfaces. Carved among the ancient words are human faces – unquestionably expressing terror or horror, or both.

In the center of the fortress there still stands part of a small granite enclosure, open to the blazing sky. The crumbling walls, once almost half the height of the berm, are covered with the same messages, the same terrorized human faces.

Patient excavation eventually reveals two buried rooms made of solid granite slabs – one directly under the berm, the other just outside. The granite surfaces are covered with the same hieroglyphs in the same combinations, the same order exactly.

CURIOUS. *VERY* CURIOUS

If human beings are still around 10,000 years from now, will they figure it out? Will the people of AD 12,000 be able to decipher the hieroglyphs to learn that once upon a time more than 800,000 barrels of radioactive waste were buried here? Will they *get it* that the messages and pictographs are trying to warn them away. Keep out! Don't dig here. Don't drink the water. Don't mine for oil or natural gas or potash or salt. Or will they be typically curious humans and dig deeper?

Thousands of years ago, Stone Age man constructed tremendous open-air enclosures and other structures formed by huge pillars and slabs of rock. Scientists have named them megaliths, meaning *great stones*. The most famous megalith is Stonehenge on England's Salisbury Plain, but in the Brittany province of France there still stands an ancient megalith, two and a half miles long and made of almost 3,000 great stones. Best guesses abound, yet archaeologists have no clear idea why these structures were built. [4]

And so it was that in the last years of the twentieth century AD, the Department of Energy in the United States of America developed a system of Passive Institutional Controls, designed to warn and inform future civilizations about the location, purpose, and danger of the WIPP site where huge volumes of radioactive waste were laid to rest. In full knowledge that it's impossible to send a message to humans 400 generations down the road, this warning system was developed out of the imperative to, at least, *try*.

THE STORY OF WIPP

In January 2002, the Waste Isolation Pilot Plant (WIPP) near Carlsbad, New Mexico received its 500th shipment of nuclear waste. When WIPP eventually is closed, 56 rooms, each about the size of a football field, will be packed with long-lived radioactive waste. Costing about $2 billion to build, the final price tag for waste disposal may be as high as $29 billion. Money, however, was the least of the problem. [5]

Prior to its opening, the site was studied for decades to determine its suitability as a permanent deep repository for transuranic waste. The challenges in constructing WIPP involved an unprecedented pioneering effort in the assessment of geological site suitability, while dealing with one political crisis after another. Although WIPP accepts only transuranic

Figure 11.1 Transuranic waste containers emplaced at WIPP.
Source: US Department of Energy.

waste from the US defense program, it is the only deep geologic repository in the world expressly built for, and currently accepting, long-lived radioactive waste. WIPP is the ground-breaker.

In simple terms, transuranic waste is not high-level waste but still has dangerous levels of long-lived transuranic elements, most notably plutonium. In the United States, transuranic waste is defined as having more than 100 nanocuries per gram of transuranic elements with half-lives longer than 20 years. Anything less is regarded as low-level waste. Most transuranic waste comes from reprocessing nuclear fuel, and virtually all reprocessed waste in the United States comes from the defense program. Although some transuranic waste is in liquid or sludge form, most of it consists of items such as protective clothing, rags, and equipment that have become contaminated with transuranics.

Transuranic waste is about a thousand times less radioactive than spent nuclear fuel and generates much less heat. Most transuranic waste can be handled when properly stored in containers. The main problem is its *longevity*. As a result of the long half-life of many transuranic elements, the waste requires secure containment for more than 10,000 years – far longer than any civilization has survived. [6]

Of all the transuranics currently being buried at WIPP, plutonium-239 is of greatest concern because of its high concentrations in the waste. Plutonium has 15 isotopes with half-lives all over the map, ranging from

20 minutes to 76 million years. By this measure, plutonium-239 has a moderate half-life of a mere 24,100 years.

Plutonium is relatively harmless outside the body. As an alpha emitter, it can't penetrate the outer layer of one's skin. To cause harm, plutonium must enter the body through a cut or by eating, drinking, or through breathing it. Drinking water containing small amounts of plutonium is a health hazard, but reduced because the gastrointestinal tract does not readily absorb plutonium into the body. The primary danger of plutonium comes from breathing it. The body's respiratory system is extremely vulnerable to even the smallest amounts of this long-lived transuranic. If inhaled, plutonium lodges in the respiratory system and eventually finds its way into the bone marrow. The large quantity of plutonium being buried at WIPP increases the possibilities for dangerous amounts to eventually make their way into the air, human drinking water, or livestock via local wells. All of these pathways must be considered. [7]

For almost three decades, transuranic wastes were disposed of along with low-level waste in shallow land-burial sites. From the beginning of the Manhattan Project, over one million curies were buried at Hanford, Idaho National Lab, Savannah River, Los Alamos, and Oak Ridge. Hanford once again had the lion's share, burying more than all the other sites combined. Most of these wastes remain in place to this day, although litigation to force their removal continues. [8–9]

In 1970, the AEC finally faced up to the dangers of transuranics and tightened disposal requirements. For the first time, transuranic waste had to be specially packaged and stored until a geologic repository was opened. Belgium, France, Germany, India, Japan, and the UK started to do the same. The world community was beginning to catch-on to the long-term dangers of transuranics. [10]

Much of the AEC's new-found motivation for addressing transuranic waste resulted from ongoing controversies involving the shipment of radioactive waste from the Rocky Flats Plant, near Denver, to Idaho's National Reactor Testing Station (NRTS). Beginning in 1954, trucks and trains had been hauling plutonium-contaminated waste from Rocky Flats to the NRTS. The modus operandi involved packing the waste in cardboard boxes or steel barrels and, upon arrival at the NRTS, dumping it in shallow, unlined pits. The operation attracted little attention until 1969, when a plutonium fire gutted the Rocky Flats Plant. Suddenly, hundreds of railroad cars bringing transuranic waste from the fire-damaged plant to Idaho unleashed a fury. Under mounting pressure, AEC Chairman Glenn Seaborg promised that the waste would be removed from Idaho by 1980. With the studies on a potential repository at Lyons, Kansas moving along, this seemed an easy promise to keep.

The scenario quickly changed in 1972, when hope for the Lyons repository went up in smoke. Suddenly, the AEC was left without a geologic repository even on the planning board. With Idahoans watching and the clock ticking, the AEC commissioned Oak Ridge National Laboratory and the US Geological Survey to search for other promising salt beds for burial of wastes principally from the defense program.

The Delaware Basin in southeastern New Mexico seemed to offer excellent prospects. After decades of extensive potash mining around Carlsbad, the area was geologically well understood. Another plus was that oil and gas development had come relatively late to this region, with better record keeping – a hard lesson learned from Lyons. Hopefully there would be fewer surprises. In other respects, the area seemed almost too good to be true. There was no agriculture because the soils were poor and water for irrigation was in short supply. And last but not least, that ubiquitous battle cry of "Not in my back yard!" would hardly apply to a windblown desert of red sand and withered scrub.

To cap off this promising scenario, the Carlsbad potash industry was in decline. In 1968, after a long-running monopoly in the Western hemisphere for this key ingredient in fertilizer, the US Potash Company shut down its Carlsbad operations. New discoveries in Canada were causing prices to plummet. In a year's time, over a thousand empty homes and commercial buildings blighted the town. Smaller potash companies in the area were scraping the bottom of the easily obtainable reserves. The backbone of the local economy was crumbling. [11]

When word got around that the AEC was looking for a new repository site, US Potash wrote to the Commission with the idea of using the old mine galleries. In what looked like a classic win–win, the AEC would have a place to build the long-promised repository and the increasing numbers of unemployed potash miners would have new jobs.

The AEC was interested but cautious. Still living down their humiliation from Kansas, the Commission was determined to do it right this time. Interested parties were informed that the AEC would not come to New Mexico unless invited by the Governor and, in that event, they would be coming "with geologists, not bulldozers." In 1972, with the encouragement of local businessmen and politicians, Governor Bruce King formally invited the AEC to come and have a look. [11]

In these early days, the basic plan was that the repository would be a permanent burial site for military transuranic waste and a "pilot" facility for commercial spent fuel. There was little opposition. On the whole, people were mostly worried about jobs. In addition, many locals felt it was pointless to try to stand up to the mighty AEC. Roxanne Kartchner, the wife of a potash company worker, led an anti-repository initiative but was only able to collect 2,000 signatures. [11]

There was also a prevalent attitude that New Mexico, Nevada, and Washington were just a lot more comfortable with all things nuclear. Each State in this nuclear triad had its own special claim to fame, but New Mexico took the prize. Los Alamos was where the first atom bomb had been developed, and the Trinity site (not far from Carlsbad) was where the first atomic bomb had been detonated in 1945. It just seemed logical that New Mexico should, and would, host the first geologic repository.

Scientists soon concluded that the abandoned potash galleries were not suitable, yet they remained confident that an acceptable site could be found in the Delaware Basin. This huge basin is roughly 150 miles (240 km) long by 100 miles (160 km) wide and is nearly surrounded by the Capitan Reef, a large limestone formation shaped like a giant horseshoe. Salt beds hundreds of meters thick had remained undisturbed since their formation more than 200 million years ago. Groundwater had carved out the spectacular Carlsbad Caverns near the western outcrop of Capitan Reef (an obvious poor location for a repository), yet the northern part of the basin, which included the Carlsbad potash district, apparently had not been significantly affected by dissolution. It was here that the search took place.

After such an idyllic start, the honeymoon quickly ended. While drilling a test hole in 1975, a reservoir of salty brine was breached just 200 feet (60 m) below the proposed repository site. Brine, hydrogen sulfide gas, and methane began shooting to the surface, almost asphyxiating one of the workers. The scientific implications were disturbing. A 200-foot buffer between the repository and brine was probably sufficient, yet who was to say the buffer was that thick everywhere? Years earlier, an oil company hit a brine reservoir in the Delaware Basin that gushed 36,000 barrels a day. Most troubling for the immediate future was worker safety. If the reservoir had been breached while miners were excavating deep underground, it could have been a catastrophe. [5, 11–12]

The site was moved six miles (10 km) to the southwest, considerably closer to oil and gas wells. In order to expand the number of potential sites, the requirement that WIPP had to be two miles from oil and gas boreholes was reduced to one mile. Not surprisingly, this fueled suspicions that the rules were being made to fit the circumstances. [11]

While studies continued to find a suitable location, determining what the Waste Isolation Pilot Plant would actually isolate became, at

Figure 11.2 Mining operations at WIPP.
Source: US Department of Energy.

times, a guessing game, and at other times all out war. In 1975, the new Energy Research and Development Administration (ERDA) cancelled the idea of including a pilot program for commercial high-level waste and redefined WIPP as an unlicensed facility for military transuranic waste. While many were relieved about the high-level pilot study being abandoned, the implications for an "unlicensed" nuclear-waste facility were a bit unsettling. For those involved in the project, however, there was a lot to be said for being unlicensed – they wouldn't have the newly formed NRC breathing down their necks. [11]

Within a year, another unforeseen complication arose when California banned construction of any new nuclear power plants until the problem of what to do with the spent fuel had been solved. Other States soon followed suit. Finding a home for commercial spent fuel suddenly took on much greater urgency, just when WIPP had been changed to military transuranic waste *only*.

In the fall of 1977, the newly formed Department of Energy (DOE) changed the game plan once again, then followed up with a series of political missteps. It began when Senator Pete Domenici (R-NM) received a letter from an official at DOE telling him that they were "considering" having WIPP licensed by the NRC so that they could use the repository for defense high-level waste. Domenici discovered that a few days earlier the same official had informed the NRC that WIPP was being expanded to accommodate high-level waste, and DOE definitely planned to have the facility licensed accordingly. [13]

Pete Domenici was a leading proponent of nuclear power. In addition, because of his support in Congress for the national labs, scientists at Los Alamos and Sandia National Laboratories affectionately called him "Saint Pete." But saint or not, Domenici had no tolerance for being deceived. As a former minor-league baseball pitcher, he knew how to play hardball.

Up to this point, the State had worked in good faith with the feds. When New Mexicans learned of DOE's behind-the-scenes maneuvering, all trust was gone. This new Energy Department was behaving like the old AEC in Kansas only a few years earlier. Senator Harrison "Jack" Schmitt (R-NM), scientist and the last Apollo astronaut to set foot on the moon, wryly commented that "the Land of Enchantment would not like to become known as the Nuclear Garbage Dump State." [13]

While Carlsbad's power structure consistently supported the project, the citizens of Santa Fe and Albuquerque had little, if anything, to gain. Within a few weeks, the State legislature was hotly debating whether to put WIPP on the ballot, in the form of a constitutional amendment that would ban storage of any radioactive waste brought into the State. Polls indicated that if New Mexicans were asked whether a nuclear waste repository should be built in the State, the answer most likely would be no. With NIMS (not in my State) threatening to strike again, the ballot proposal lost in the legislature by three votes. [13]

To calm the situation, James Schlesinger, the first Energy Secretary, met with Domenici and other members of the New Mexico congressional delegation. While having virtually no choice, Schlesinger acceded that WIPP would not be built over the State's objection. The immediate crisis simmered down, but Domenici knew that honoring this agreement was at the discretion of the Energy Secretary – and Schlesinger's successors might see it differently. On the Hill, Domenici continued to press for a strong State's role in any decisions about siting a nuclear-waste disposal facility. [5]

Only three weeks after Domenici's meeting with Schlesinger, the Carter Administration reinforced suspicions when a task force on radioactive management headed by John Deutch, an MIT chemistry professor, recommended a demonstration test in which 1,000 spent fuel assemblies would be placed at WIPP. Calling it a "modest demonstration," this recommendation was none-too-subtly linked to the California moratorium and the growing urgency to get some spent fuel into the ground. When Deutch was asked if the study phase of placing 1,000 spent fuel assemblies at WIPP would satisfy the California law, he responded, "In my view, it will meet those requirements." [14]

What had begun as a way of bringing new jobs to the Carlsbad area had suddenly escalated into something the State didn't need or want. A year earlier, citizen opposition to WIPP was minimal. Now it was organizing and growing rapidly. New Mexicans were afraid that if they allowed a few hundred tons of spent fuel through their door, much more would follow.

In an attempt to quell fears, or at least to even the playing field, DOE agreed to provide funding for an independent evaluation group. Soon titled the Environmental Evaluation Group (EEG), it was understood that neither the State of New Mexico nor the Department of Energy could try to bias or interfere in the group's technical conclusions. With a staff of seven and funds for outside consultants, the EEG soon proved competent to press everyone involved – both for and against WIPP – to substantiate and defend their positions. [9, 11]

After release of the Deutch Report, Carter appointed an Interagency Review Group (IRG) to lay the groundwork for a national waste-management policy. In March 1979, they came out with their report. All IRG members, except those from the Energy Department, recommended that WIPP be terminated in its current form. They gave two primary reasons. First, natural gas and potash in the area might invite future intrusion. Second, WIPP should be licensed by the NRC because the long-term dangers are comparable to high-level waste. This recommendation went against the strong *no-licensing, hands-off* position of the House Armed Services Committee and its chairman Melvin Price (D-IL). Price would almost single-handedly write the next pages of WIPP history. [15]

Hailing from the industrial world of East St. Louis, Melvin Price was elected to the House of Representatives when Franklin D. Roosevelt was President. He was one of the earliest congressional champions of nuclear power. An original member of the Joint Committee on Atomic Energy and its last chairman, Price was best known as co-sponsor of the Price–Anderson Act. Passed in 1957, and making legal history with its provisions, the Price–Anderson Act limited the liability of the nuclear industry in the event of an accident. A fund of $560 million was set up to be apportioned to the victims. According to the Act's other sponsor, Senator Clinton Anderson (R-NM), the fund was big enough to indicate that something meaningful would be done, but not so large as to "frighten the country and the Congress to death" by revealing the magnitude of risk. Opponents of the Act argued that it was a massive subsidy for the nuclear industry, and that it effectively repealed every citizen's right to sue for damages in the event of an extraordinary nuclear accident. Proponents pointed to the large pool of funds to provide prompt and orderly

compensation to those who incurred damages from a nuclear accident. Both groups recognized that the landmark legislation was essential to launch the nuclear-power industry, since no insurance company
would underwrite a nuclear power plant to the full extent of possible
damages. [16]

Melvin Price knew how to look out for the interests of his committee, and the Armed Services Committee did not want to see commercial
spent fuel being buried at WIPP. They wanted to keep this project simple, so that it might actually happen. Burying commercial spent fuel
at the site would give the NRC a role, and nothing was ever simple
for the NRC. In a surprising move, Price ramrodded Public Law 96–
164, the 1979 WIPP Authorization Act, through Congress before everyone went home for Christmas. Caught unprepared, New Mexico's two
House members were attending a dinner in Albuquerque and missed the
vote. [5]

The WIPP Authorization Act, once and for all, prevented licensing
by the NRC and firmly established WIPP for defense wastes only. The
Act expanded New Mexico's voice in the project, but also took away
their right of veto. The Armed Services Committee didn't pander to such
blatant non-cooperation. Instead, the Energy Department and the State
were directed to negotiate a "consultation and cooperation" agreement –
and were given nine months to do it.

Confusion was now added to anger. New Mexico's Secretary of
Health and the Environment, George S. Goldstein, spoke for pretty much
everyone when he said, "None of us knew what the agreement meant."
New Mexico insisted on an agreement that would be legally enforceable
and subject to judicial review. The Energy Department didn't want anything legally binding. They envisioned something more along the lines
of a gentleman's agreement. [17]

The spotlight was now on New Mexico's Attorney General, Jeff
Bingaman. While Governor King was lamenting that "There isn't much
the State can do about WIPP now," Bingaman was explaining to anyone who would listen that, without a legal pact, Congress could change
its mind and make WIPP a repository for high-level waste. Larry Har-
mon, the DOE project manager for WIPP in Washington, DC, tried
to quell these fears with classic political on-the-fence talk: "It is *absolutely unlikely* [italics added] that WIPP will become a high-level waste
repository." [5, 17]

First the environmentalists and then the Governor got behind
Bingaman. On May 14, 1981, Attorney General Bingaman filed a suit
against the Department of Energy, charging that it had violated States'

rights, had failed to adequately consult with State officials, and had refused to agree to a legally enforceable document to resolve these issues. For good measure, he threw in violation of the National Environmental Policy Act, the Federal Land Policy and Management Act, and the WIPP Authorization Act. Two months later, DOE and the State reached a compromise agreement and consented to stay the lawsuit. Before the suit, "it was obvious we were not getting the timely, accurate information about WIPP that we felt Congress intended," Bingaman said, but since filing the lawsuit he was happy to report that "communications have improved substantially." [17]

In exchange for having the lawsuit dropped, the Energy Department agreed to share information and address the State's concerns. A State-federal task force would begin addressing issues like emergency-response preparedness, road upgrades, and health studies of people who would be living near WIPP. The Department also acknowledged New Mexico's right to go to court and stop the project if dissatisfied. For the first time, the State had a substantial role in WIPP.

While the never-ending politics simmered and boiled in the background, somehow scientists had continued studying the site. Four months after the new agreement was signed, drillers again struck a brine reservoir with a whopping 350 gallons (1,300 liters) a minute gushing to the surface. This reservoir was less than a mile north of the proposed repository. Opponents soon dubbed it "Lake WIPP." The Energy Department planned to start construction in just over a year's time – *and now this*. One concern was that the brine reservoir might indicate that the salt beds were dissolving. Another was that future drilling for oil or gas might hit a pressurized brine reservoir, causing brine to enter the repository and transport radionuclides into the overlying aquifers and to the land surface. [17]

The Energy Department agreed to build the repository to the south of the central shaft, yet there was still a possibility that the reservoir extended beneath the relocated site. The problem was, the only way to be sure the reservoir was not directly beneath the site would be to drill boreholes. If they drilled boreholes and hit the reservoir, they might ruin the site. And if they drilled boreholes and didn't hit the reservoir, the site might be ruined anyway because of the boreholes. Everyone waited for the EEG to have their say. After studying "consequence analysis" scenarios, the group determined that even if brine were to flow up through the repository to the surface, the consequences for public health would be acceptable. The concentration of radionuclides in the transuranic waste was low enough to avoid large concentrations reaching the human environment. Lest no one forget, they added that the consequences would be much more serious if the repository contained high-level waste. [11]

In the spring of 1983, after a decade of tumultuous site-characterization studies, DOE was ready to begin construction. New Mexico's new Governor, Toney Anaya, wanted several conditions met before construction began. First (and once again), Anaya wanted strong assurances that the project would not be a foot in the door for high-level waste. Next, he wanted these assurances backed up, and trust rebuilt, by having the State's right of veto restored. Finally, Anaya wanted WIPP licensed by the NRC. This time even Senator Domenici and now-Senator Bingaman warned the Governor not to meddle with the WIPP Authorization Act, or heaven only knows where they could end up. But no matter. The Energy Department was now three years late in fulfilling their promise to get the Rocky Flats waste out of Idaho. On July 1, 1983, DOE turned down the Governor's requests and announced they were ready to begin construction. [18]

Considerable mining already had been done in order for scientists to study the site. A 12-foot (3.6 m) -wide exploratory shaft had been drilled to the repository horizon 2,150 feet (645 m) below, and from there almost two miles of tunnels had been mined. Now the Department was bringing in a monster mining-machine, complete with rotating head, that would eat its way through 120 acres (48 hectares) of salt, carving out a maze of tunnels and waste-storage rooms. [18]

Everything went fine during the initial construction phase, and then bad news struck again. After a decade of politics threatened to destroy the project, it began to look like the site itself was going to self-destruct. The salt was creeping, cracking, and leaking water faster than anyone had predicted.

The ability of salt beds to eventually self-seal any openings had long been viewed as one of the strong points in going with salt. Computer models indicated that the salt would cooperate nicely – it would "creep" slowly enough to allow for retrieval of transuranic waste for at least five years, but not so slowly that radionuclides could escape. A good understanding of the rate of closure was essential for the design of an effective repository. Now the scientists discovered that the salt was creeping several times faster than predicted – more at the rate of a lively crawl. [6, 11]

Tests within the repository also indicated that the salt contained more moisture than had been assumed. During construction, brine was observed seeping into new openings made in the salt. This seepage

Figure 11.3 Underground tunnel at WIPP.
Source: US Department of Energy.

suggested the possibility that the disposal rooms might eventually be flooded, resulting in all kinds of possible havoc. At a minimum, the metal canisters could be damaged.

Wendell Weart, a geophysicist at Sandia National Laboratories, felt these concerns were overblown. He was confident that "within 70 years, long after we are through down there, the mine will look like a perfectly solid rock." Monitoring eventually validated Weart's position, showing that after the initial disturbance, the brine inflow declined rapidly and finally tapered to almost immeasurable amounts. [5–6]

Nonetheless, the State's growing anti-WIPP movement pounced on the brine issue. After being assured that the salt bed was dry, suddenly a *little* water was supposed to be alright. Obviously the WIPP scientists were just the politicians' puppets. Years later, Weart noted in hindsight, "The biggest mistake we made was . . . telling people that one of the things we liked about salt was that it's dry. The minute the first beads of moisture appeared, we had to start telling [outsiders], 'It's still a dry rock in mining terminology; it's just not bone dry.'" [5]

Wendell Weart was the lead scientist on the project for more than two decades. Having grown up in the Midwest, Weart was in his element in rural New Mexico. He also possessed that rare talent for knowing how to explain scientific complexities to everyday people. On top of his demanding job, he traveled all over the State and spoke at community gatherings, inspiring confidence in the project through his honest, calm, and authoritative manner. Over time, Weart earned the nickname, "The Sultan of Salt." [5]

Regardless of these problems, when construction was completed in September 1988, DOE went before Congress to seek permission to begin limited operations. Scientists, members of Congress, and environmentalists argued that DOE was trying to proceed before serious scientific and engineering issues were satisfactorily resolved. Senator Domenici had warned for months that the October goal for opening WIPP was unrealistic. Energy Department officials stuck to the date, arguing that they needed to establish a fixed deadline just as a highway engineer does for a highway. Under withering cross-examination by Congressman Mike Synar (D-OK), the Energy Department's own engineers said they were not sure the repository was safe for operation. Minutes after the hearing, DOE announced they were closing the repository *indefinitely*. [5, 19]

This was not what Idaho needed to hear. Almost two decades had passed since AEC Chairman Glenn Seaborg promised that the Rocky Flats waste would be removed from Idaho's National Reactor Testing Station. Since that time, an average of 55 shipments a year of Rocky Flats waste had been added to the ever-growing mountain of waste. Idaho's Governor Cecil Andrus was out of patience. He was no longer accepting any of the *excuses* that were keeping the waste on a holding pattern in *his* State. Andrus summed up the problem this way: "A nation that can send a man to the moon and bring him back safely can find a solution to this problem. It's not a question of scientific ability. It is a question of political will." [5]

Having served as President Carter's Secretary of Interior, Andrus knew how things got done in Washington. As he wrote in his 1998 autobiography, "I had learned one basic lesson as a Cabinet secretary. The government in our nation's capital reacts only to crises. If we wanted action, we would have to create a crisis and force the Department of Energy to give us its attention." And that he did. On October 19, 1988, Governor Andrus announced that his State's borders were officially closed to waste shipments from Rocky Flats. He defiantly told the press, "I'm not in the garbage business anymore." Soon after this declaration, the *New York Times* ran a photograph of a stranded railroad car just inside Idaho's State line, with a State trooper car blocking the track. The Energy Department was forced to send the railroad car back to Rocky Flats. [5, 20]

Colorado's Governor Roy Romer was far from happy with this turn of events. If waste piled up at Rocky Flats, he warned, they might have to shut down the Nation's only facility that manufactured plutonium triggers for nuclear bombs. *USA Today* ran a front page photograph of Romer with his fist raised, telling the press, "I'm saying to the federal government, 'Get off your duffs.'" This Governors' feud set off a gleeful torrent of news coverage, dubbed by *The Washington Post* as "a high-stakes game of plutonium poker." [5]

Faced with the Governors' lists of demands, including federal money for highways and environmental cleanup, it was up to DOE's Deputy Secretary Joseph Salgado to solve the crisis. Meeting with the Governors of Idaho, Colorado, and New Mexico, Salgado worked some kind of magic. Following the meeting, Governor Carruthers of New Mexico told reporters, "We came together with DOE; I think we have a good-faith effort now." Romer emerged saying, "I am willing to go home and say, 'I think there's a solution.'" Even Governor Andrus was now willing to *reconsider* allowing Rocky Flat's waste into his State, telling the press, "I believe that DOE has put in motion a schedule that can work. Whether it works or not, I'm not yet prepared to say." A year later, the Energy Secretary announced that production at Rocky Flats was halted indefinitely after a raid by the EPA and FBI revealed serious environmental violations at the plant. For the time being, at least, the Idaho problem was diffused, but confidence in DOE sank to new lows. [5]

Back at WIPP, the technical challenges never seemed to end. During a standard inspection, the EEG discovered large cracks in the ceilings and floors of two huge waste storage rooms. The Energy Department had known about the cracks for several years but had neglected mentioning this latest problem in any of their reports, including a two-volume assessment of the site. Dr. Lokesh Chaturvedi, deputy director of the EEG, pointedly explained: "They should have told us about it. We are here to know everything about the site." James E. Bickel, an assistant manager at DOE's field office, argued that engineers did not consider the cracks important enough to make public because common mining techniques could make them harmless. The technique being used was a million dollar engineering system, anchoring the ceiling with 13-foot (4 m) metal rock bolts that were electrically monitored and grouted with a special epoxy to increase their holding power. The system worked. [21]

In 1989, while still on their holding pattern, DOE held public hearings in Albuquerque and Santa Fe to listen to reactions to their latest supplemental environmental impact statement. The hearings quickly broke down into angry denunciations. Protestors compared DOE to the People's Republic of China and WIPP officials to Darth Vader. The hearings in Santa Fe resembled a carnival, where a stream of dancers, singers, poets, and storytellers took center stage. Eighty people had testified in Albuquerque – 545 showed up in Santa Fe. Virtually the only support came from about 30 Carlsbad residents, who said they were spat upon as they entered the hearing. [22]

With the public turned against the project and any number of scientific questions still not resolved, the late 1980s was perhaps the darkest hour for WIPP. Few people would have predicted that this site would ever open. Indeed, another decade would pass before the Department of Energy met all of the safety requirements and was finally authorized to open WIPP. In addition to the many scientific questions, there were numerous other challenges. As with Ward Valley, a complicated land transfer with the Department of the Interior had to be worked out. In addition, a permit from New Mexico was needed because of toxic chemicals contained in the waste. The Environmental Protection Agency also had to finalize its controversial standards for WIPP. The Department of Energy battled in the courts with another New Mexico Attorney General, Tom Udall (nephew of Morris Udall).

Although the 1979 WIPP Authorization Act limited WIPP to defense waste only, it remained unclear whether this might include military spent fuel and high-level waste. The 1992 WIPP Land Withdrawal Act finally resolved this issue by banning spent fuel or high-level waste, confirming the repository's restriction to disposal of defense transuranic waste only.

Transportation and emergency response issues were also of paramount importance. To New Mexicans, hauling transuranic waste along the largely run-down "blue highways" that crisscrossed the population centers of this mainly rural State was just asking for trouble. In the end, by-passes were built around Santa Fe and several other cities, along with upgrading the roads that would be used.

Additional transportation requirements were enacted in response to the pressure from New Mexico. Before a WIPP-bound truck left a transuranic waste site, it would be inspected to Level VI, the industry's highest safety level. The containers approved for carrying the waste, costing almost half a million dollars each, had undergone rigorous testing, including a 30-foot (10 m) drop onto a steel surface and a 30-minute burn test in jet fuel to ensure that the containers would remain leak-tight. Drivers would travel in pairs so that they were rested, as well as ensuring that truck and payload would be attended at all times. Waste shipments would be tracked by satellite 24/7 and continuously monitored from a control center at WIPP. Drivers would be required to stop and check their trucks and payload every 150 miles (240 km) or three hours, as well as to notify State officials two hours before entering each State. WIPP would also train thousands of emergency response professionals along the routes to assure an effective response in the event of an accident. [23]

By the late 1990s, almost everyone believed that natural processes, by themselves, would not disrupt the WIPP site. The only probable threat was disturbance by human activity. What would humans be doing in the vicinity of WIPP during the next 10,000 years and beyond? Would they heed the warning signs? What about future exploration for oil and gas? There were many such possibilities. In the end, Sandia scientists concluded that there is a reasonable expectation that WIPP will contain all but a small fraction of its transuranic wastes for the next 10,000 years. On May 13, 1998, the Environmental Protection Agency announced that DOE's 100,000-page application had met all their safety standards and WIPP would be certified. [24]

Finally it happened. On the evening of March 25, 1999, a tractor-trailer carrying three dumpster-sized steel containers left Los Alamos to begin the 260-mile (400 km) journey to WIPP. "It was one of the most anticipated garbage pickups in U.S. history," wrote Chuck McCutcheon in his book, *Nuclear Reactions*:

> As the truck rumbled away from the place where the atomic age had been conceived 54 years earlier, jubilant Los Alamos workers lined the road to greet it. "There's a lot of pride in this, a lot of pride," laboratory manager Dennis Rupp told a reporter. But when the truck reached the outskirts of Santa Fe, about 15 miles south, the primary onlookers became environmental activists who had fought for years against the prospect of ever witnessing such a sight. One tried to block the vehicle's path with his car before New Mexico state police interceded; others waved signs and beat Tibetan shaman drums in protest. "You're evil!" one woman yelled at the driver. The truck continued south, a state police escort and television news crews in tow. Finally, at around 4 a.m., nearly 500 weary but excited bystanders cheered as the caravan pulled up to a cluster of stark white buildings rising out of the barren expanse of scrub oak, coyote trails, and mesquite upholstering the southern New Mexico Desert. The truck had arrived to deposit its cargo at the world's first permanent deep underground burial site for nuclear materials: the Waste Isolation Pilot Plant (WIPP). [5]

Almost three decades after the AEC first sent "geologists, not bull-dozers" to New Mexico, and after weathering six different presidential administrations, the world's only deep repository for long-lived nuclear waste was up and running. With this achievement the Department of

Figure 11.4 First shipment of transuranic waste arrives at WIPP, March 26, 1999.
Source: US Department of Energy.

Energy hoped that WIPP would bolster public confidence in their ability to solve the much more intractable problem of high-level waste. One of the main conclusions from the Waste Isolation Pilot Plant is that when it comes to burying nuclear waste, no site is perfect. Perhaps the accomplished fact of WIPP would help pave the way for solving the much thornier problem – 1,000 miles to the northwest – at Yucca Mountain.

Part II The mountain

The search for a geologic repository

I wonder how one finds the words to talk about a man who has achieved so much,

who has served with such distinction and who has touched the lives of so many.

Only two words keep coming back to me, over and over again – thank you.
Senator John McCall, in a tribute to Congressman Morris Udall [1]

Following the Atomic Energy Commission's public embarrassment at Lyons, Kansas in 1972, the search for a geologic repository took many twists and turns until the field was narrowed, in 1987, to a single candidate – Yucca Mountain. In the intervening decade and a half, the federal government searched in vain for what Daniel A. Dreyfus quipped was a "technically appropriate subsurface with a politically compliant Governor on top." Dreyfus, then staff director for the Senate energy committee, was later to take charge of the Yucca Mountain project – where a politically compliant Governor was certainly not to be found. [2]

In 1970, nuclear power plants provided the Nation with less energy than it derived from firewood. Only a dozen or so reactors of modest size were operating. This would soon change. During the middle and late 1960s, the AEC had authorized the building of nearly a hundred large reactors. Practically overnight, a major construction boom had begun. [3]

AN ENERGY CRISIS AND A NEW AGENCY

The vulnerability of the United States to energy supply disruptions would come into full focus in the early 1970s. On October 6, 1973, Syria and Egypt launched a surprise attack on Israel during the Yom Kippur religious holiday. Less than two weeks later, the Arab members of the Organization of Petroleum Exporting Countries (OPEC) placed an oil embargo on the United States. The embargo was quickly extended to Western Europe and Japan.

The US decision to resupply Israel with arms during the Yom Kippur War united the OPEC countries in their embargo. However, the root causes, which had been festering for some time, were economic as well as political. Consumption of oil by the West and Japan had steadily increased, while prices paid to petroleum producing countries remained low. Adding insult to injury, devaluation of the American dollar was causing goods imported by the OPEC countries to become ever more expensive. By the early 1970s, the situation had become intolerable. Before the outset of hostilities, Saudi King Faisal had secretly made a deal with Egypt. Emboldened by the West's increasing reliance on Middle Eastern oil, Faisal had agreed to use the "oil weapon" to support Egyptian President Anwar Sadat. Saudi Arabia was now the swing producer for the entire world (a position once held by Texas) and was in the driver's seat. M. King Hubbert's prediction that oil production in the Lower 48 would peak around 1970 had come to pass. [4]

As oil suddenly came to a trickle from the Middle East, the price quadrupled from $3 to $12 per barrel – bargain-basement prices by today's standards. The oil embargo affected almost every American. Gasoline was rationed. Huge lines formed at the pumps. Many stations ran out of gas. Drivers with license plates ending with an odd number could purchase gasoline only on odd-numbered days, and vice versa for even-numbered plates. A national maximum speed limit of 55 miles per hour was imposed to increase gas mileage, and the Nation was encouraged to observe "gasless Sundays." The embargo continued until March 1974.

The Arab oil embargo increased the urgency for a broad-based energy policy. Several months before the embargo, President Nixon had directed Dixy Lee Ray, chair of the AEC, to lead a review of energy research and development activities and to recommend an integrated energy program for the Nation. Many people believed that nuclear energy would be the primary beneficiary of this review. Three-quarters of every federal dollar for energy R&D was already going to nuclear energy, and most of it for the breeder reactor.

"Nature did not make me willowy," Dixy Lee Ray once observed. This was true in physique (she was heavy set) as well as spirit (she was blunt and confrontational). With a Ph.D. in zoology, Dixy Lee Ray taught at the University of Washington and became head of the Seattle Pacific Science Center, when it was formed after the Seattle World's Fair in 1962.

Figure 12.1 Dixy Lee Ray, ca. 1980, Portraits of State Governors,
1889–2004, Washington State Archives.

In spite of her training in the natural sciences, Dixy Lee Ray spent much
of her political career battling environmentalists. [5]

Nixon appointed Dixy Lee Ray to the Atomic Energy Commission.
Ray's academic background made her an unlikely candidate for the AEC
chair, but her deep concern about the Nation's energy supply drew President Nixon's attention. The women's movement was also forcing new
inroads for hiring women into high federal posts. She drove her mobile
home all the way to Washington, DC, stopping to inspect nuclear power
plants on her way. Not your run-of-the-mill bureaucrat, Ray bucked the
strict Washington, DC dress code by wearing tweed skirts, knee socks,
and sensible shoes. She also caused a stir by bringing her miniature

poodle and huge Scottish deerhound daily with her to AEC headquarters. Six months after joining the commission, Dixy Lee Ray became the
seventh, and final, AEC chairperson.

Though an outspoken advocate of nuclear energy and a true
believer that it could be made completely safe, Dixy Lee Ray proved
to be an independent-minded chairperson. She opened up access of AEC
reports to antinuclear activist Ralph Nader. She ordered Consolidated
Edison Company to build a costly water-cooling system at a nuclear plant
along the Hudson River to protect fish. She gave reactor safety research
new prominence.

Dixy Lee Ray also rattled the cages in the AEC hierarchy by removing responsibility for reactor safety research from Milton Shaw, the controversial director of the civilian reactor development program. A protégé
of Admiral Rickover, Shaw had been the project leader for the *Nautilus*, the
first nuclear submarine, and the *Enterprise*, the first nuclear aircraft car-
rier. He had long been criticized for postponing critical work on reactor
safety in favor of directing as much funding as possible toward breeder
reactor development. Upon removal of his responsibilities for reactor
safety research, Shaw quit. Ray also refused to back reappointment of
James Ramey to the AEC Commission. For more than a decade, Ramey
had been the liaison to the Joint Committee on Atomic Energy, a relationship that had become all too cozy. Dixy Lee Ray's reining in of Shaw
and dismissal of Ramey came in the face of substantial opposition by the
AEC establishment and the Joint Committee. [6–7]

On December 1, 1973, Dixy Lee Ray submitted her report, *The
Nation's Energy Future*. The report called for increasing domestic oil and
gas supplies through oil shale and other development; a massive shift
from oil and gas to coal during a transition to heavy reliance on nuclear
power; continuing research on the breeder reactor; and conservation
through more efficient building designs and other means. Solar energy
research would increase but would remain a relatively minuscule part of
the overall energy program. Other alternative energy sources like wind
and biofuels were not on the radar screen. In the long run, it was thought
that nuclear fusion would save the day. [8]

The Nixon Administration also unveiled Project Independence, an
ambitious plan to make the US energy independent by 1980. On November 7, 1973, in a major Presidential address on energy, Nixon announced,
"Let us set as our national goal in the spirit of Apollo, with the determination of the Manhattan Project, that by the end of this decade we
will have developed the potential to meet our own energy needs without
depending on any foreign energy source." The cover on *Time* magazine

that week was a picture of Nixon, but the heading was, "The Push to Impeach." [4, 6]

Like similar efforts in later years, the lofty goal of energy independence far exceeded the political will to achieve it. The goal of independence would not only require substantial investment in research and development, but also many tough decisions. As Robert Gillette commented in *Science* magazine at the time:

> Nor, it seems, does anyone but President Nixon seriously regard this R&D effort as an analog to the Manhattan and Apollo projects, except perhaps, in terms of cost. In this case, creation of new technology is only half the battle; commercial application of the new technology depends on myriad policy decisions – bearing on things from oil shale leasing to power plant siting – that fall outside the realm of R&D. The success of Project Independence thus depends as much on politicians as on technicians. [9]

In short, the Nation could not simply buy its way to energy independence. Within a couple of months, Nixon changed his policy from "self-sufficiency" to "reducing" the Nation's reliance on "potentially insecure sources of foreign energy." [10]

To make headway toward even basic energy goals, a single entity was needed to bring order to the process. To accomplish this, Congress passed the Energy Reorganization Act of 1974. The Act was signed by President Ford on October 11, 1974. Ford inherited the Act from Nixon, who had resigned a couple of months earlier after the long Watergate siege. In one fell swoop, the Act accomplished what many had long sought: it abolished the AEC and replaced it with two separate agencies. The regulatory functions for atomic energy were incorporated in the newly formed Nuclear Regulatory Commission (NRC), which would serve as an independent "watch dog" regulatory agency. The remainder of the AEC, along with parts of other government agencies responsible for energy development, became the Energy Research and Development Administration (ERDA). In effect, the fox was no longer guarding the henhouse with the same agency responsible for both promoting and regulating the nuclear industry.

In her report to Nixon, Dixy Lee Ray had supported dividing the AEC, thereby helping to write herself out of a job. For a short while, she moved to the State Department as an Assistant Secretary for oceans and the environment. Frustrated with Washington bureaucracy and lack of attention to her office by Secretary of State Henry Kissinger, Dixy Lee Ray got in her mobile home and drove back to Washington State. She later became Governor and soon became enmeshed in nuclear controversies

once again. After a single term in office she lost decisively in the Democratic gubernatorial primary, largely because of her outspoken support
of nuclear energy and waste disposal at Hanford.

MOVING PAST LYONS

In late 1975, the newly formed ERDA announced a reinvigorated plan
to address disposal of high-level radioactive waste. The Nuclear Waste
Terminal Storage Program was soon up and running. The program was
ambitious. Six repositories were to be identified to accommodate the
waste generated by the several hundred reactors planned to be operating
by year 2000. The first two repositories would be built in salt and would
start operating at a pilot scale by 1985. The other four would be built in
other kinds of rock, such as granite and shale. All six would be operating
by the mid 1990s. [11]

The number of repositories was based on projections of future
growth by the nuclear power industry. These projections drastically
ignored slowing growth in electricity demand since the oil crisis in
1973. Soaring oil prices had caused electricity rates to increase com-
mensurately. Subsequent economic "stagflation" produced interest rates
exceeding 20 percent. With less demand for electricity and the cost of
financing increasing, utilities began scaling back their planned increases
in energy generation capacity.

Meanwhile, to meet its ambitious goals, ERDA decided to greatly
expand the search for potential disposal sites. Thirty-six States were targeted for evaluation. At least 13 of these States would be selected for
drilling and other exploratory fieldwork. ERDA decided to try the opposite approach from the AEC's non-communicative practices of the past.
In November 1976, ERDA Administrator Robert C. Seamans, Jr. wrote
to the Governors and legislators of the 36 States, telling them that
they would be searching for repository sites within their borders. The
letter offered to work closely with the States and to keep the Governors informed of how the efforts were progressing. Seamans committed to terminating a project if the State raised technical issues that
could not be "resolved through mutually accepted procedures." Scores
of messengers hand delivered the letters to the State officials almost
simultaneously. [12]

The response was swift and mostly negative. Some States banned
ERDA from even exploring potential repository locations. The ERDA letter
also needlessly antagonized States that were not under serious consideration for a repository. Over the next few years, more than a dozen States

enacted laws that either prohibited or made the establishment of repositories extremely difficult. A task force later noted, "what began as a new initiative, a fresh start in the area of waste management, soon got mired down in the reluctance of State officials even to contemplate a facility on their soil." [12–13]

A representative example of the States' reluctance occurred in the Salina Basin. As the name 'Salina' implies, the basin contains extensive salt beds. Underlying a broad swath from the Finger Lakes of New York across western Pennsylvania and eastern Ohio to the Lower Peninsula of Michigan, it was here that the newly formed ERDA turned after the failure in Kansas.

In 1975, ERDA and Oak Ridge National Laboratory proposed drilling test holes near the town of Alpena, Michigan, on Lake Huron's Thunder Bay. The test drilling required a permit from the State. Governor William G. Milliken was reluctant to agree to any such studies. Strong negative responses at a statewide series of public hearings reinforced his doubts. In May 1977, Milliken informed ERDA that they were to cease all activities in Michigan. [2]

Ohio and New York followed in Michigan's footsteps. Ohio took particular exception to the suggestion that the most promising geologic site was in densely populated northeastern Ohio. New York was already wary of nuclear waste as a result of their experiences at West Valley. When ERDA tacitly suggested that federal negotiations with New York about the West Valley cleanup might also cover the possibility of a repository in the Salina Basin, the State was outraged and cried "nuclear blackmail" to the press. In the end, nothing beyond the preparation of literature surveys was ever accomplished in the Salina Basin. [2]

Colin Heath, newly appointed ERDA director of geologic disposal, was told by his superiors that the siting effort was "too widespread" and should be focused only on the most promising States. In 1977, Heath proposed to press ahead in six *salt States* for the first two disposal sites. He also accelerated work at Hanford and the Nevada Test Site, having concluded that these sites offered the best possibilities for retrievable storage of spent fuel. There was a long-standing interest in using Hanford due to its convenience and supportive local community. Vincent McKelvey, the Director of the US Geological Survey, had highlighted the Nevada Test Site as a potential site for a repository the year before. Heath prepared a letter for his superiors to send to the Governors, letting them know that some States were off the hook. The letter was never sent, as ERDA was in the process of being absorbed into the newly created Department of Energy. Nonetheless, Heath's general

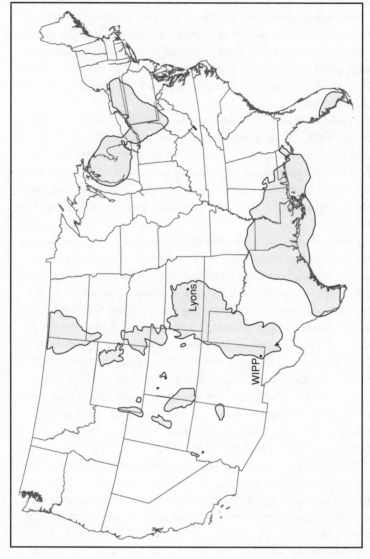

Figure 12.2 Map of salt deposits in the United States. The shaded areas show regions of major salt deposits. Compiled by J.B. Epstein, US Geological Survey, based on maps from Reference [28].

strategy of focusing on a combination of salt sites, Hanford, and the Nevada Test Site, would be followed over the next several years. [2]

TIME TO REASSESS

When the Department of Energy was formed in October 1977, geologic repositories had been the leading choice for high-level waste for more than two decades. The almost single-minded focus had been on salt formations. After more than 20 years of work, it was hard to make a case that anything of note had been accomplished. Clearly, it was time to reassess the situation.

California had already begun the reassessment with a moratorium on building any new nuclear power plants until the State energy commission was able to certify a federally approved method for permanent disposal of spent nuclear fuel. The moratorium was overturned by the courts. Upon appeal, the US Supreme Court upheld the moratorium on the grounds that California had a legitimate right to question the economic viability of future nuclear plants, in the absence of a concrete plan for waste disposal. This moratorium still stands today. Connecticut, Wisconsin, and other States soon followed California's example, making progress toward a permanent waste disposal facility a pre-condition for constructing any more nuclear power plants. [2]

Scientists were also asking questions. In late 1976, a top ERDA official declared that the six repositories would require only "straightforward technology and engineering development." Less than two years later, Luther J. Carter, writing an article in *Science* magazine entitled, "Nuclear Wastes: The Science of Geologic Disposal Seen as Weak," described "an emerging consensus among earth scientists familiar with waste disposal problems that the old sense of certitude was misplaced." [14]

Scientific concerns were coming from several directions. Five senior USGS scientists published a report on the uncertainties connected with geologic disposal of high-level radioactive waste and identified some key geologic questions that remained unanswered. According to these scientists, the challenge was more complicated than just selecting the right host rock. The behavior of the disposal system as a whole – the waste package, repository design, and hydrogeologic environment – needed to be evaluated. More attention should be devoted to interactions between the heat-generating waste and the disturbed host rock and its water. The authors were all widely respected earth scientists. Their report, published as a USGS Circular, received considerable attention.

The White House Office of Science and Technology Policy, along with a panel of scientists convened by EPA, expressed similar concerns. The science community basically remained confident that a technical solution could be found but felt that the science case was, after all these years, surprisingly weak. [14–15]

Scientists were also raising questions about the wisdom of using salt formations for disposal of high-level waste. Like the Atomic Energy Commission before it, ERDA continued to keep the faith that salt was the best emplacement medium. In 1976, ERDA's first comprehensive report on alternatives for managing radioactive wastes contained virtually no information on options other than salt. Salt continued to be viewed as a simple, almost foolproof, approach to waste disposal. [16]

The test program at Lyons, Kansas had demonstrated that the small brine-filled cavities in salt tended to migrate toward the heat source resulting from the wastes. This process was thought to be manageable by using a steel sleeve around the waste packages to protect their integrity from the corrosive brine. The USGS Circular expressed concerns that small amounts of brine might cause a substantial decrease in the salt's mechanical strength and possible movement of waste. Dave Stewart, one of the authors of the USGS report, told Congress in 1979 that a decade of laboratory evaluations and tests inside a repository might be necessary before the scientific questions about disposal of high-level wastes in salt could be adequately answered. [2]

About this same time, the American Physical Society, a leading professional organization for physicists, established a study group to look into nuclear fuel cycles and waste management. In a departure from the prevailing DOE view, the study group recommended examining not just salt but other geologic media like granite and shale as well. [17]

President Carter's decision in April 1977 to ban reprocessing of high-level waste was not good news for salt. Suddenly retrievability became a big issue, as this energy resource might be needed down the road. One of salt's most positive features for waste isolation is that it creeps over time and closes up openings from the pressure of the overlying rock mass. While great for isolating the waste, this self-sealing mechanism all but negates future retrievability.

When announcing the reprocessing ban, Carter also called for a review of radioactive waste management policy by an internal task force led by the DOE director of research, John Deutch. In March 1978, the task force released their report. The Deutch report reiterated DOE's position that salt should be used for the first repository site. The report also acknowledged that the planned opening of the first repositories by 1985 was unrealistic. [18]

Nuclear waste policy was in disarray, with widely differing views within the Carter Administration. In an effort to come to some consensus, Carter announced formation of a high-level Interagency Review Group (IRG) on Nuclear Waste Management. The Group comprised representatives from 14 governmental agencies, with strong representation by environmentalists. It was chaired by the Secretary of Energy. Amid the ongoing controversies surrounding nuclear power, the Group decided they would stick to waste management and remain neutral on the future of nuclear power. The IRG would neither attempt to "shore up the nuclear option [nor] to undermine it." [19]

When the IRG released their report in March 1979, certain key disagreements remained. DOE continued to favor salt as the geologic medium, and pushed for selection of two or three candidate sites. Sticking with salt would facilitate construction *finally* getting underway. Most of the other agencies favored a slower approach. They wanted selection of more candidate sites, including some in geologic media other than salt. These internal controversies about waste policy were exacerbated by two external events – the Three Mile Island accident in March 1979 and the Iranian hostage crisis that began in November 1979. These crises resulted in almost a year's delay of Carter's final policy statement.

The vacuum created by this delay was readily filled by members of Congress. Senator Gary Hart (D-CO), chair of the Environment and Public Works Subcommittee on Nuclear Regulation, proposed phasing out nuclear power over a 10-year period, beginning in 1985, unless the NRC could attest that an adequate plan was available for the permanent isolation of spent nuclear fuel. At the other extreme, Senator Bennett Johnston (D-LA), who sat on the powerful Committee on Energy and Natural Resources, sponsored a bill calling for spent fuel to be stored indefinitely in retrievable surface storage facilities. *Indefinitely* meant up to 100 years or longer. By passing the problem to future generations, Johnston was conveniently able to maintain his pro-nuclear stance while keeping a geologic repository away from the salt domes of Louisiana. The Nevada Test Site, located 1,300 miles (2,000 km) west of Louisiana, happened to be one of Senator Johnston's "particular favorites" for a long-term storage facility. Johnston, a masterful politician known as *Mr. Energy* in the Senate, forcefully pushed the long-term storage concept. [2, 13]

The Carter Administration favored some form of centralized interim storage, but with a much shorter time horizon. In October 1977,

President Carter proposed construction of a large *away-from-reactor* (AFR) storage facility. The AFR would serve as a temporary measure to keep utilities from shutting down because of storage pools filled to capacity with spent fuel. The government would take title to the fuel and have responsibility for it until it was permanently laid to rest. There would be a one-time charge for these government services. Unlike the long-term interim storage envisioned by Senator Johnston, an AFR was viewed as a very temporary measure to be used for as short a timeframe as possible. The away-from-reactor storage concept continued to be batted around for a while. It largely dissipated, however, as dry casks became available for on-site storage at reactor sites at a cost comparable to, or less than, storage at a federal AFR facility. [11]

On February 12, 1980, President Carter finally issued his nuclear waste policy statement. Largely ratifying the conclusions of the IRG, he viewed his policy as an interim strategy pending a full environmental review under the National Environmental Policy Act. Carter reiterated the long held position that the first disposal facilities should be in a geologic repository. Conservative engineering practices and multiple independent barriers would be used to reduce risks and compensate for uncertainties. Interim storage of spent fuel was supported, but the waste problem should not be deferred to future generations. Carter also addressed the issue on which the IRG had been unable to come to consensus – the scope and diversification of geologic investigations. The President adopted the more conservative view: "When four or five sites have been evaluated and found potentially suitable, one or more will be selected for development as a licensed, full-scale repository." [20]

Carter's policy statement, the IRG report, and continuing debates in the years to follow focused on a principle that came to be known as "consultation and concurrence." This loosely defined principle had been promoted by the National Governors Association for the States to have a bigger role in the decision-making process. Accordingly, State and local agencies, tribes, and the general public should be closely involved in all stages of the development of facilities for high-level waste disposal.

While "consultation" is a relatively straightforward term, "concurrence" was open to wide interpretation. To most Governors, the nuclear industry, and the Carter Administration, concurrence meant that a potential host State would have a major role in decision-making. The startup of each new phase of operation would require that the State be satisfied with what had gone before. The protocol for what would happen in the case of an impasse was never nailed down.

Environmental and antinuclear groups equated concurrence with the right of each State to veto any proposed waste repository. The obvious problem with this interpretation, noted Oregon Senator Mark Hatfield, was that "Simply allowing the State to say 'no' [would be] inviting nothing but no's." Further complicating matters, DOE had already given assurances to Louisiana, New Mexico, and New York that they had the right to veto potential waste-disposal sites. Louisiana's veto was promised in return for the State going along with the development of the Strategic Petroleum Reserve in their salt domes. New Mexico's veto power was part of the WIPP negotiations. New York got its veto because the federal government still had not cleaned up the abandoned high-level wastes from reprocessing at West Valley. [21]

As the Carter Administration came to an end, Congressman Udall (D-AZ) and Congressman Dingell (D-MI) oversaw passage of legislation in the House of Representatives that largely reflected the Carter philosophy, insisting on geologic disposal as the first priority. The Senate passed a similar waste bill but, instead, emphasized development of Senator Johnston's envisioned long-term storage facilities. The differences between these two bills proved insurmountable. In addition, the House and Senate were unable to agree on the rights of a State to veto a repository for military wastes. Like the Carter Administration, nuclear waste policy came to an inglorious end as Congress adjourned in 1980. [11]

CONGRESS GETS ITS ACT TOGETHER

The new Reagan Administration promptly declared support for nuclear power and promised swift action to develop an acceptable solution for commercial high-level waste disposal. The change in administration brought a new set of policies. The ban on commercial reprocessing was lifted, though DOE efforts to encourage private investment in reprocessing proved unsuccessful. The Carter Administration's offer to provide federal storage facilities for spent fuel (the AFR concept) was withdrawn, leaving utilities with the primary responsibility for storing spent fuel until the long-promised reprocessing facilities were developed. Perhaps most significant in the policy shift, three geologic repository sites would now be studied prior to selecting the first site, rather than the four or five proposed by Carter. The Nevada Test Site, Hanford, and a salt site not yet selected were the leading contenders.

Unlike Nixon, Ford, and Carter before him, Reagan saw little need for government funding of energy research. He took a budget ax and dramatically slashed government research in general, and research on alternative energy sources in particular. Reagan had better uses for the money, such as his futuristic defense program that came to be known as Star Wars.

By now, "the nuclear industry badly needed some good news on the waste front because certainly there was none coming from elsewhere," notes Luther J. Carter, author of *Nuclear Imperatives and Public Trust: Dealing with Radioactive Waste*. By the end of 1981, utilities had cancelled 77 reactor orders. They cancelled another 18 in the following year. The decline in electricity demand and economic issues were partly responsible for the cancellations. Lack of public confidence was a further cause. The Three Mile Island accident, in particular, had greatly undermined public confidence in nuclear technology. And the pools at nuclear power plants were filling up with spent fuel. Great frustration in the industry was matched by the public's growing distrust that the nuclear waste problem would ever be solved. [2]

The utilities continued to view geologic disposal as the only feasible long-term solution, but saw an immediate need for the federal government to step in and provide interim storage for their growing accumulation of spent fuel. The utilities and nuclear industry were also concerned that the lack of a designated disposal site might cause the NRC to refuse to license any more reactors – a moot point, as no new reactors were ordered anyway. Along with growing concerns in the public sector, the industry was losing faith that the federal government would ever meet a schedule or stick to a policy. The cry was on for a nuclear waste law to bring stability to the situation.

Environmentalists were of mixed opinion. They were concerned about the waste problem and wanted to make sure that spent fuel would not be reprocessed. At the same time, they did not want to aid legislation that might help expand nuclear power. Environmentalists adamantly opposed interim storage and seriously doubted that the federal government would deal adequately with safety.

A dozen States that were potential, and largely unwilling, hosts for a geologic repository or interim storage facility took an active part in congressional debates. These included the six States currently under investigation for a repository: Washington, Nevada, and the four remaining salt States of Louisiana, Mississippi, Texas, and Utah. Three Midwest States – Michigan, Minnesota, and Wisconsin – were among potential host States for a second repository in their Precambrian granite formations. Finally, Illinois, New York, and South Carolina feared that the defunct commercial reprocessing facilities already in place in their States might be retrofitted for interim storage. Senators Charles Percy

(R-IL), Alfonse D'Amato (R-NY), and Strom Thurmond (R-SC), all self-described supporters of nuclear power, adamantly opposed interim storage. [2]

University of Washington professor, Kai N. Lee, summed up two vastly different stereotypical viewpoints at the time. The first viewpoint was held by *technical rationalists*, an optimistic bunch. In their view, enough was known to proceed with an orderly program of waste disposal. Estimates of waste confinement based on models and analyses were sound and reassuring. Public fears, while politically troublesome, were really nothing more than misinformation compounded by antinuclear demagoguery. Strong presidential leadership, clear decisions, and effective implementation by the federal government were needed. Furthermore, delay threatened the economic well-being of the Nation. [22]

The second viewpoint favored *cautionary consultation* and emphasized uncertainty. In this view, the unease of the general public had a sound basis in technical uncertainty. Although safe disposal of wastes is important, there was no need to rush, as scientists and other credible experts couldn't even agree on how to get started. Highly simplified models of the behavior of radioactive materials under geologic conditions remained untrustworthy. Perhaps most importantly, bureaucratic momentum must not be allowed to force a premature choice in an inappropriate medium or location. In the meantime, vigorous conservation and development of alternative energy resources could adequately meet the Nation's needs for electric power. Their bottom line – nuclear power would have to wait until these issues were satisfactorily resolved.

Although both viewpoints had merit, they were totally incompatible. Technical rationalism had traditionally guided federal waste management policy, while the politics of nuclear waste had increasingly become those of cautionary consultation. With such a tug-of-war in progress, compromise and consensus seemed impossible. In spite of these difficulties, Congress continued to work toward a solution. One person, in particular, had the right stuff to see it through – a one-eyed, liberal Democrat from conservative Arizona whose good friends included Barry Goldwater and John McCain.

Morris Udall (D-AZ) came from a prominent Mormon family. Mo, as he was known, succeeded his older brother, Stewart, in a special election for the congressional seat after Stewart was named Secretary of the Interior by President Kennedy. He served in the House of Representatives for three decades. Mo was easy going, civil, and touted a bipartisan spirit, which soon earned him respect among his congressional peers. He was 6 feet 5 inches tall and athletic. In spite of a glass eye (he lost

Figure 12.3 Morris K. Udall and his love for nature. Credit: MS 325 Morris K. Udall Papers, Box 738 Folder 1, Courtesy University of Arizona Libraries, Special Collections.

an eye accidentally to a pocket knife when he was six), Mo played for a year with the Denver Nuggets professional basketball team. Udall was famous for his wit. A political commentator called him "too funny to be president," when he ran for the Democratic nomination against Jimmy Carter in 1976. Udall later entitled his autobiography, "Too Funny to be President." [23]

Udall was a highly productive legislator. He championed the rights of Native Americans and Alaskan Natives. His love of nature resulted in numerous pieces of environmental legislation. Chief among his accomplishments was the Alaska Lands Act of 1980, which doubled the size of the national park system and tripled the size of the national wilderness system. He also authored legislation on campaign reform and congressional ethics.

Mo Udall was diagnosed with Parkinson's disease in 1979, yet this did not stop him from taking a leadership role in the 1980s in developing radioactive waste policy. The disease forced him to resign from the House in 1991, and he was eventually confined to a veteran's hospital a few miles from the Capitol. His main visitor from the Hill was fellow Arizonan, Senator John McCain, who paid Udall a visit every few weeks until his death in 1998. Udall had befriended McCain when he first ran for Congress and helped show him the ropes around Washington. Though of opposite political parties, they both became leading voices for reform and, through their actions, demonstrated a willingness to work across party lines. McCain never forgot his friend. [24]

As Chairman of the Interior Subcommittee on Energy and the Environment in the mid 1970s, Mo Udall acquired oversight authority in nuclear energy – previously the sole domain of the Joint Committee on Atomic Energy. Acting more as an advocate of the nuclear industry than as a regulator, the Joint Committee's role began to diminish when the NRC was created in 1974, and was finally abolished in 1977. Noting that the Joint Committee had long been a "closed club," Udall welcomed the chance to get a fresh start and provided a refreshingly open and much more balanced approach. His credibility with environmental interests served as a moderating influence with the contentious issue of State's rights in the siting process. [25]

On April 29, 1982, the Senate passed a nuclear waste bill introduced by Senator James A. McClure (R-ID), chairman of the Senate Committee on Energy and Natural Resources. The McClure bill passed with little fuss by a vote of 69 to 9. The nuclear industry was pleased with the result, but environmental and antinuclear lobbyists were adamantly opposed to several provisions in the bill, including an emphasis on interim storage, restrictions on environmental assessment, and the accelerated schedule for siting and licensing. [2]

In a remarkable balancing act, Mo Udall tenaciously shepherded a parallel nuclear waste bill through the House. He managed to keep the environmental lobbyists engaged in the process, despite the ambivalence of their constituency. He addressed the interim storage issue, while maintaining geologic disposal as the preferred approach. Finally, he successfully addressed the concerns of potential host States, while avoiding special amendments to exclude a particular site. After many gyrations, the House passed the bill on December 2, 1982. Its fate now depended on negotiations with the Senate. [2]

The two houses of Congress carried out intense negotiations, progressively ironing out their differences in an effort to complete the bill before the end of the 97th Congress. In the final hours, the legislation almost failed when Senator William Proxmire (D-WI) threatened to filibuster. Proxmire insisted on a proviso that a Governor's veto of a proposed site would stand unless overridden by both the Senate and the House. This provision would strengthen the State's hand beyond what was currently in the bill. Proxmire's concerns stemmed from the fact that granite formations in his home State of Wisconsin were of interest to DOE. Senator McClure yielded to Proxmire's threat, removing the final barrier to passage of the bill. *Consultation and concurrence* had now been codified into law. [2]

The legislative process had not been pretty. The old adage applied quite well – there are two things best not directly observed; one is the making of sausage, the other the making of laws. Udall described the nearly 100-page bill as a "delicate fabric of agreements." A spokesman for the Atomic Industrial Forum, an industry group, called it a "masterpiece of compromise." Sierra Club lobbyist Brooks Yeager aptly noted, "There's an awful lot of politics in this bill." David Berick of the Environmental Policy Center quipped, "One of the reasons the bill went through was because people were just tired of working on it." [26]

On December 20, 1982 – 31 years to the day after the lighting of four light bulbs in Idaho with a nuclear reactor – the House and Senate passed the Nuclear Waste Policy Act (NWPA). The Act became law when it was signed by President Reagan on January 7, 1983. The NWPA established the current geologic disposal program, including a comprehensive national policy for management and disposal of spent nuclear fuel and high-level radioactive waste. Along with amendments, it remains the statutory framework for the US high-level waste disposal program.

The Act embraced geologic disposal as the highest priority and set a schedule for siting two waste repositories. As always, the schedule was optimistic. Selection of the first repository site was mandated for 1987. Repository construction would start as early as 1989 – just six years away. The President would decide on a second site by March 31, 1990. DOE would begin accepting wastes by January 31, 1998. The repositories could be used for both commercial and military waste, with a provision to allow the President to decide by January 1985 if a separate repository for military waste was required. Responsibility for waste disposal was partitioned among DOE, NRC, and EPA. Responsibility to implement the Act went to DOE. The NRC has responsibility to develop the regulations, while the EPA sets the standards that repositories must meet to assure public and worker safety and health.

The Act deferred the decision about whether long-term interim storage facilities were needed. The Department of Energy was directed to present a site-specific proposal for long-term interim storage within two-and-a-half years. Senator Bennett Johnston wanted the proposal within one year, while Mo Udall wanted to wait five years. They basically split the difference. [27]

One of the most important features of the Act was the provision for a Nuclear Waste Fund, that would be financed by a fee on the utilities of 0.1 cent per kilowatt-hour of nuclear-generated electricity. This fund would ensure that the full costs of the federal waste disposal program are paid for by the nuclear utilities – although the consumers are actually footing the bill. The option to increase the fee also held out the possibility of a handsome "reward" for a host State. In return for the fee, DOE signed binding contracts with nuclear utilities to take legal charge of the spent fuel by January 31, 1998.

President Reagan declared mission accomplished. "The Act," he proclaimed, "provides the long overdue assurance that we now have a safe and effective solution to the nuclear waste problem." [2]

13

Nevada wins the lottery

Democracy is the worst form of Government except for all those other forms

Winston Churchill [1]

After nearly 30 years of scientific, congressional, and public debate about what to do with the Nation's high-level radioactive waste, the Department of Energy was suddenly under the gun. A daunting schedule was now written into law under the Nuclear Waste Policy Act (NWPA) to get a geologic repository up and running in a few short years. If DOE failed in this mandate, they could find themselves besieged with lawsuits.

The aggressive schedule established by the NWPA meant that the nine sites already under consideration when the Act was passed, automatically formed the basis of site screening for the first repository. By 1984, these were narrowed to five sites – three in salt formations in Mississippi, Texas, and Utah; one in basalt at Hanford; and one in the volcanic tuff of Yucca Mountain at the Nevada Test Site. The process of selecting the first repository from among these candidates became known as the *First Round*.

Regardless of the patchwork of compromises that shaped the final result, the NWPA was worded very carefully. It did not talk in terms of a perfect place to dispose of the waste because, as most everyone was catching on by now, there was no perfect place. After studying the many possibilities, the Act stated that the *preferable* method of long-term disposal was in a geologic repository. There was not, because there could not be, a guarantee of 100 percent safety. Not surprisingly, none of the five candidate sites was perfect. Several were far from perfect. All three salt sites had serious conflicts with overlying or nearby land uses.

The Hanford site had major geologic issues. Yucca Mountain was in a tectonically active region.

The Richton salt dome near the town of Richton, Mississippi, not far from Hattiesburg, was among the five finalists. It was one of several salt domes in the Gulf Coast region that had been considered for a geologic repository. Salt domes form when deep layers of bedded salt, being relatively light and buoyant, push up through weak places in the overlying rock. The domes sometimes rise thousands of feet toward land surface in enormous columns several miles across. Hundreds of salt domes occur in the Gulf Coast region. Some lie beneath the land surface; others are offshore.

Salt domes have their drawbacks for waste disposal. They are far more geologically complex than salt beds. Many salt domes are also associated with petroleum deposits, making them more likely targets of future breaching by humans. The Richton salt dome was relatively large and its potential for future petroleum exploration seemed low.

The flashpoint for opposition to the site was its proximity to the town of Richton, a community of just over 1,000 people. During the NWPA legislative debates, the "Richton amendment" was a regular feature championed by Senator Trent Lott (R-MS). The amendment would eliminate any sites (namely, Richton) exceeding a certain population density from hosting a high-level repository. The first version was so poorly crafted it could have eliminated all sites under consideration. Of course, given the time scale of geologic disposal, the idea of a population-based test is easily challenged. Today's remote area might be a thriving urban center several hundred years from now. [2]

Some of the Richton locals remembered their experience with Project Dribble a couple of decades earlier. In this oddly named AEC project, a nuclear explosion was set off in a nearby salt dome in 1964 to see how seismic wave patterns could be used to detect underground nuclear testing. In spite of assurances by the AEC to the contrary, the explosion rocked the area, breaking windows and cracking masonry in Hattiesburg, over 20 miles (30 km) away. More than a thousand residents successfully filed damage claims totaling over $650,000 – a sizable sum at the time. [2–3]

The second of the three candidate salt sites was located in Deaf (pronounced "Deef") Smith County in the Texas Panhandle. The county was named after a partially deaf soldier and scout for General Sam Houston. Deaf Smith was the first to reach the Alamo after its fall. Many people know the county more by its former brand of natural peanut butter.

The salt deposits in Deaf Smith County were part of vast Permian-age salt beds extending from Kansas (including Lyons) through western Texas and eastern New Mexico. The proposed repository horizon was more than 2,000 feet (600 m) below the wheat fields of this flat farming country. Thick shales and other low permeability rocks lie above the salt beds, creating a natural barrier for radioactive waste. The flashpoint for opposition, in this case, was that the salt beds of Deaf Smith County are below the High Plains (Ogallala) aquifer, the lifeblood of the region. Irrigation water from the High Plains aquifer has transformed a large part of the Great Plains into one of the major agricultural regions in the world. Even if the disposal plan were technically sound, public perceptions of feed crops being irrigated with water located near nuclear waste did not sit well with the agricultural industry.

The third salt candidate was in the Paradox Basin near the southern border of Utah in the Four Corners area. Paradox Basin got its name from a valley *paradoxically* crossed perpendicular to its axis by the Dolores River. The residents were already quite familiar with the nuclear industry, as the basin contains uranium ore deposits and the leftover mill tailings from earlier mining years. The decline of the uranium mining industry had put many people out of work, so local residents were generally supportive of the idea of a new industry coming to the area. However, the vast majority of residents in the State of Utah were not so pleased, particularly when test boreholes drilled at four locations in the basin identified the best site, geologically speaking, to be within a mile of Canyonlands National Park and not far from the Colorado River. Selection of a site so close to one of the Nation's most treasured scenic areas easily mobilized opposition by environmentalists and the National Park Service.

The fourth candidate, the Hanford Reservation, sits atop one of the world's largest accumulations of lava. From 17 million to 6 million years ago, hundreds of overlapping lava flows engulfed much of Washington, Oregon, and Idaho. The flows did not come from your classic volcano vent; instead *basalt flows* poured out of long fissures onto the landscape. The extensive sequence of sheet-like basalt lavas accumulated to a thickness of more than 10,000 feet in places and formed a lava plain known as the Columbia Plateau.

Almost everyone assumed that DOE chose Hanford as a proposed repository site solely because of its politics. The local economy is almost completely dependent upon the DOE operations at Hanford. Richland, the biggest metropolis in the area, is a town long accustomed to nuclear activities. The local high school, "Home of the Bombers," has used a mushroom cloud as a school mascot since 1945 to celebrate the region's role in ending World War II. Remediation and management of the defense wastes had provided many millions of dollars annually toward employment opportunities at Hanford. A high-level waste repository would only add to these well-paid job opportunities.

Repository interest at Hanford centered on the Cohassett flow, a 200 to 250-foot (80 m) -thick lava flow more than 3,000 feet (900 m) below land surface. The Cohassett, like other lava flows in the Columbia Plateau, is contained within the Columbia Plateau aquifer system, an extensive regional system of groundwater flow within the layers of basalt. The general direction of groundwater movement is from recharge areas near the edges of the Columbia Plateau, including the Cascades, toward regional drains such as the Columbia River. Although the proposed repository was less than five miles (8 km) from the Columbia River, the DOE contractor claimed it would take tens of thousands of years for radionuclides to reach the river.

Groundwater movement through the Columbia Plateau aquifer system is very complex. The geologic system consists of dense basalt lava flows interspersed with narrow layers of sedimentary rocks that were deposited on the flows between eruption periods. The tops of the lava flows are rubbly and full of cavities formed by gas bubbles that emanated from the hot lava as it cooled. As a result, the tops of these lava flows and the sedimentary interbeds act like sluiceways for groundwater. Further complicating the picture, faults and vertical cooling joints potentially connect the dense, impermeable centers of the basalt layers (where the waste would be buried) to these permeable zones. Minerals deposited over time have closed some of the openings along the interfaces, faults, and joints, but not all of them.

These geologic features and their hydrologic significance were impossible to characterize adequately to assure the conservative travel times postulated by the DOE contractors. The USGS described the system as "very leaky" and thought that much more rapid transport was possible. Travel times to the accessible environment were quite possibly less than 1,000 years. If this were true, the site would not meet licensing criteria. According to the USGS, "Available data are insufficient to conclude much of anything with regard to groundwater travel time or direction." Moreover, since the area was already contaminated by wastes from the plutonium factories, it might be difficult to tell if future contamination was from the repository or from the former military waste burial grounds. [2]

Prolific water-bearing beds above the Cohassett also created a major engineering problem for construction of a repository several thousand feet underground. The large diameter shafts that would be needed to access and ventilate the proposed repository might also provide conduits for catastrophic flooding. The expense of constructing a repository at Hanford would be much greater than at the other sites, and it might not even be possible with "reasonably available technology." Notwithstanding these technical difficulties, Hanford would prove remarkably resilient in the selection process. [2]

The fifth, and last, candidate was Yucca Mountain, which will be discussed in depth in the following chapters.

NARROWING THE FIRST-ROUND CHOICES

The Nuclear Waste Policy Act mandated that no later than January 1, 1985, the Secretary of Energy would recommend three of the nominated sites as candidates for a geologic repository. In late 1984, DOE announced the findings from their environmental assessments of the sites. The dubious winners were Deaf Smith County, Hanford, and Yucca Mountain.

The response from the Governors of the *winning* States was predictable. "Arbitrary, capricious, uncaring, and unreasonable," fumed Texas Governor Mark White. "Before the people of Deaf Smith County glow in the dark, sparks will fly." "Nevada has already done its share in the nuclear arena," declared Governor Richard Bryan. The possibilities of "earthquakes and groundwater contamination," worried newly elected Governor Booth Gardner of Washington. [4]

Representative Edward Markey (D-MA), Chairman of the House Subcommittee on Energy Conservation and Power and a self-described nuclear watchdog, challenged the selection process. DOE responded to Markey that it could not give the subcommittee its working files on the decision, because they had been thrown away. The idea that DOE made its selection without external involvement did not sit well with anyone. And the possibility that they had intentionally ditched the files destroyed any remaining credibility DOE might have had in the selection process. Under pressure, DOE once more turned to the National Academy of Sciences, asking them to set up a new and more methodical selection procedure. When the Academy completed this task, Hanford came in last among the five sites – instead of third as before. It appeared that Hanford would be dropped from the three runner-ups when DOE came back with their revised list. [2, 5]

On May 28, 1986, Energy Secretary John Herrington announced in a press conference that three sites had been chosen for exploration in the West. To most everyone's surprise, Hanford, Deaf Smith County, and

Yucca Mountain were once again the winners. DOE argued that Hanford may have been ranked last of the five sites according to the Academy's site-ranking methodology, *but* if the cost of repository construction and of spent fuel transportation *were not* taken into account, and the DOE estimate of tens of thousands of years of travel time to the Columbia River *was* taken into account, then Hanford came out on top. The DOE argument not only ignored their previously solicited NAS input, but also ignored concerns by the USGS about the possibility of much shorter travel times to the Columbia River. To pretty much everyone who was paying attention, DOE seemed intent on selecting Hanford, regardless of what studies might indicate. [6]

At this point, the structural weaknesses of the Nuclear Waste Policy Act itself were becoming obvious. During the congressional debates leading up to passage of the Act, and among the many sticky problems to be worked out, one central problem persisted – no State wants this stuff in their backyard. Congress was acutely aware of this problem. As a result, many provisions were written into the Act in an effort to deal equitably with a State's concerns. After all, passing a bill dealing with nuclear waste without also dealing with the public's resistance would only serve to provoke greater resistance down the road. Yet, built into the NWPA is the implicit recognition that any and all the States would likely refuse, under any and all circumstances, to host a repository. Any legal loophole or potential safety problem would be taken advantage of to shut down the project.

This reality could be overcome only by giving the federal government the ultimate and final say about where a high-level waste repository would be built. Clearly, everyone would benefit from finally facing up to the national problem of radioactive waste disposal. In the Constitution, our Founding Fathers laid out the rights of States and under what general circumstances States' power is superseded by the power of the federal government.

This chain of power is reflected in the Nuclear Waste Policy Act. The federal government, through the Department of Energy, has all the real clout in this Act. DOE selects the five sites for study and narrows it down to three sites by 1985; the President approves or disapproves of these sites within two to six months; DOE then studies the sites and comes up with one site for a repository; DOE then gets (or doesn't get) the President to recommend the site to Congress. If Congress signs off and the Nuclear Regulatory Commission authorizes construction, DOE gets the repository up and running. The States have no authority to halt site exploration and characterization during this process.

There is, however, one significant *check and balance* built into the Act. In the event of huge resistance on the part of the host State and/or a very poor choice, the State and affected tribes were given the opportunity to object. This objection becomes effective unless overridden by both Houses of Congress.

The possibility of a State or tribe objection is only intended for hardball. The NWPA tries to avoid a rough-and-tumble endgame by giving the States enough involvement so that everyone feels a sense of fairness in the process. For example, after each of the major DOE decisions, the Act specifies public involvement through comments and/or public hearings. After DOE finalizes site characterization for each of the three sites, and before it announces one site for the actual repository, the State (and affected tribes) may provide an impact report that goes to the President. Nevertheless, the bottom line is that talk is talk and input is input. In the end, there are no guarantees that a State or tribe will have any influence over the final decision. Everyone was well aware of this fact.

Compounding this problem, by the time of the passage of the NWPA, federal credibility had been seriously eroded concerning all things nuclear. As the stories came out over the years – the contamination at Hanford, Savannah River, Oak Ridge, Idaho National Lab, and West Valley; fires and plutonium contamination at Rocky Flats; the mismanagement of uranium mill tailings in the West, the fate of uranium miners working in poorly ventilated mines; the effects of radioactive atmospheric fallout downwind of the Nevada Test Site; the Three Mile Island reactor meltdown and the befuddled response by officials. The list goes on. The federal government's nuclear track record was a mess. If this was arguably true in reality, it was most assuredly true in perception.

By September 1982, around the time the NWPA was becoming the law of the land, this eroded trust was blatantly obvious. Throughout the United States, approximately 160 State laws, initiatives, and resolutions and 250 local laws pertaining to high-level radioactive waste had been passed. The next spring, these laws were being pre-empted by higher court decisions. One such overruling took place in Virginia, when a US District Court voided Louisa County's ban on storing spent fuel that came from facilities outside the county boundaries. In issuing the decision, the judge cited NWPA as "clearly giving the Federal Government the authority over storage of radioactive material." Then, in May 1983, the US Supreme Court decided not to review lower court decisions that declared as unconstitutional, restrictions on the transportation and storage of radioactive materials in Illinois and Washington. This did nothing to reassure a growing number of people that the federal government was acting in their best interests. [7]

The States had a number of legitimate issues. Naturally, at the top of the list was fear of high-level nuclear waste. No matter how remote the possibility, the dangers to public health and safety are understandable concerns. In addition, there were serious issues of equity, with some States potentially bearing all the costs and risks while everyone else reaped the benefits. States and localities were also concerned about other impacts of this new industrial complex in their backyard. Historically, States have been accorded primary responsibility for protecting the property, health, and general welfare of their citizens. Police and fire departments would have increased responsibilities to prepare for, and possibly deal with, nuclear accidents. There could be losses in property values, directly translating into losses in tax revenue. Especially in western States, there was concern about demands on the already scarce water resources. The State would also be unable to exploit mineral and other resources on the large tracts of land surrounding a repository. Finally, there were possible *boomtown effects*. Construction and operation of radioactive waste management facilities results in an influx of new residents and transients to that area. Such a population surge often leads to social disruption with rising rates of crime, alcoholism, and divorce. In some rural areas of western States, boomtown effects have been severe. Although many of these effects would fall on local governments, the NWPA did not provide for participation in siting decisions by units of local government – leaving it up to the State to decide their role. [7]

With the inherent State–federal conflicts as a backdrop, the decisions that led to the three runner-ups for the first geologic repository were already controversial enough, but it was the *Second Round* that would truly stir up a hornet's nest.

THE SECOND ROUND

Most of the first-round sites were in the West. Although not explicitly stated in the NWPA, it was expected that the second-round choice would focus on States in the East to provide geographic equity. This idea of siting the second repository in the East was part of the delicate balancing act DOE needed to perform to maintain any semblance of fairness and credibility. Without it, most westerners were not inclined to take any waste from the East, where most nuclear waste is generated.

To comply with the implicit agreement for a regional distribution of repositories, DOE focused second-round studies on igneous and metamorphic rocks (granite and gneiss) in the eastern USA. Known collectively as *crystalline rocks*, they form beneath the Earth's surface at very high temperature from a molten liquid (igneous rocks) or conversion of deeply buried rocks (metamorphic rocks). It follows that crystalline rocks would be very stable in the presence of the heat from high-level waste. A second advantage is that crystalline rocks, when mined, form self-supporting caverns. A common feature of crystalline rocks is very low total porosity (open space in the rocks). Water flows principally through fractures that range from small cracks to large shear zones. The challenge for a repository in crystalline rocks is to find a relatively unfractured section where groundwater movement would be retarded.

Crystalline rocks are prevalent throughout much of the world, though in many places are covered by a blanket of sedimentary rocks. Many countries have directed their research efforts on disposal of high-level waste toward use of crystalline rocks, including Canada, the Czech Republic, Finland, India, Sweden, and Switzerland.

In January 1986, 12 sites in the eastern USA were selected as tentative candidates for a second-round repository. The sites were distributed among five States along the East Coast (Georgia, North Carolina, Virginia, New Hampshire, and Maine) and two States in the upper Midwest (Minnesota and Wisconsin). A few States had two sites; Minnesota had three. The process of choosing an eastern site had received little attention until then. As one activist said, "When they finally put the pins on the map, the intensity of the response took everyone by surprise." DOE collected 60,000 comments, most of them negative. [5]

Nowhere was the protest louder than the choice of the Sebago Lake area in Maine. The proposed site was within a few miles of the Portland and Lewiston-Auburn metropolitan areas, Maine's two largest communities. Sebago Lake is the second largest lake in Maine and the source of Portland's drinking water. From DOE's perspective, Sebago Lake overlaid a large granitic body, known as a batholith, in a tectonically stable region. The local populace saw it differently. When DOE officials went to Maine to brief the public on how the screening had been done, some 3,000 anxious and angry people showed up, including the Governor and both US Senators. The meeting lasted until 3:30 a.m. [8]

A second candidate site in New England was in Hillsboro, New Hampshire – a rural town of about 3,000 residents and birthplace of Franklin Pierce, the 14th President of the United States. In this case, the town was directly on top of the proposed repository site. Many residents could face the loss of their homes, farms, and businesses. One can only imagine the feeling of waking up one day to discover that your home may lie directly over a future high-level radioactive waste dump. Governor John Sununu came out strongly against siting a repository in the Granite State. At public meetings, Sununu had difficulty explaining how he was for the controversial Seabrook nuclear plant, but against a waste site in New Hampshire. "The two are unrelated," Sununu replied. [8]

Hearings on the Hillsboro selection were held around the State. One story that illustrates the gap between citizenry and the government involves a young boy, about 10 years old, who stepped up to the microphone and asked "What if it leaks? There'd be no more me." The DOE specialist replied, "That wouldn't happen for thousands and thousands of years." "Oh," said the boy. "Don't you care about the future?" [9]

Vice President George H.W. Bush got an earful when he was in Maine for a fund raiser. It was no secret that Bush planned to run for the Presidency in 1988. The brewing nuclear waste controversy could affect the outcome of the presidential primary, which kicks-off in New Hampshire. Republicans were also concerned about the upcoming congressional election in which Republican control of the Senate was at stake.

When Energy Secretary John Herrington held a press conference on May 28, 1986, in which he announced the three sites chosen in the West, he dropped a political bombshell. Herrington announced that for the indefinite future DOE was unilaterally dropping all plans to explore sites in the East. Herrington insisted that he did not change the waste program to suit the Vice President's plans for 1988, or the needs of the Republican candidates running in the East that fall. Rather, the decision was based on revised figures showing that a second repository was not needed until 2020. "It is not prudent to spend hundreds of millions of dollars on site investigation and identification now," he said. Herrington claimed he was just trying to save money. Virtually no one believed him. [5]

Herrington's announcement clearly violated the spirit, and probably the letter, of the Nuclear Waste Policy Act. The NWPA repository selection process, with its provisions for technical integrity and its measures to assure fairness, was seemingly abandoned. Several key sponsors of the Act and all the Senators from the three selected first-round States immediately denounced Herrington's suspension of the search. More than 50 pieces of legislation to amend the Act were introduced during 1986 and 1987. One bill would place a moratorium on field research at potential repository sites pending recommendation from a blue-ribbon commission and the enactment of new legislation by Congress. When the idea of a moratorium won Mo Udall's support in July 1987, it ended

Figure 13.1 Aerial view showing crest of Yucca Mountain.
Source: US Department of Energy.

"a dark, dark week" for the nuclear utilities, according to one lobbyist. An alternative bill, sponsored by Senators Bennett Johnston and James McClure, would simplify and speed up site selection by giving a large reward to any State willing to serve as host. The prize for a repository would be $100 million a year. This became jokingly referred to as the "bribe Nevada plan." [10]

In late 1987, facing an upcoming presidential election year and an immediate Christmas deadline, Senator Johnston ramrodded amendments to the NWPA through Congress. Johnston used his in-depth knowledge of the Senate's labyrinth of procedures to adeptly steer the amendments through. He confined substantive discussions to committees who favored Yucca Mountain, and attached his plan to a money bill that included pork-barrel projects – making it difficult to drum up opposition.

On December 21, 1987 (now 36 years and 1 day after the lighting of the four light bulbs in Idaho), President Reagan signed the Omnibus Budget Reconciliation Act of 1987, which included the NWPA amendments as an add-on. Yucca Mountain at the Nevada Test Site suddenly became the *only* candidate for a repository. The prize for the host State had been reduced from $100 million to $20 million per year, provided the State agreed not to exercise their right to disapprove the facilities. Nevada expressed no interest in accepting the $20 million per year "bribe" and chose to fight. The "bribe Nevada plan" had now become what is known in Nevada as the *Screw Nevada Bill*.

Representative Mo Udall summed up the result: "We created a principled process for finding the safest, most sensible places to bury these dangerous wastes. We were confident that while no State wanted a nuclear waste repository, the States ultimately chosen would accept the outcome because the selection process would have been fair and technically credible. Today, just 5 years later, this great program is in ruins. To help a few office seekers in the last election, the administration killed the eastern repository, shattering the delicate regional balance at the heart of the 1982 act. Since then, the Western States have felt they are being treated unfairly, and they no longer trust the technical integrity of the Department of Energy's siting decisions." [11]

Senator Johnston proclaimed mission accomplished. "I think we've solved the nuclear waste problem with this legislation," he said. A congressional aide was more circumspect. "We have reason to believe it will work out, but if it doesn't, *man* we're in trouble." [12]

14

The Nevada Test Site

How come dumb stuff seems so smart while you're doing it?

Dennis the Menace [1]

The Trinity test of the first atomic bomb in New Mexico and the bombings of Hiroshima and Nagasaki were horrific displays of shock and awe. Yet, at war's end military officials still knew very little about the overall effects of nuclear weapons. The Joint Chiefs of Staff requested and received presidential approval to conduct a series of tests during the summer of 1946. The radiological hazards of atomic bombs *were* known, and strongly influenced the decision to locate the tests in the middle of the Pacific Ocean at Bikini atoll in the Marshall Islands.

Under an agreement with the United Nations, the Marshall Islands were a new trust territory of the United States. The agreement allowed for military use of the islands, along with responsibilities for native welfare. It was hard to make a case that relocating the natives and turning their island into a nuclear weapons test site was to their benefit. Nonetheless, the Bikini islanders were moved to Rongerik atoll, which was too small and barren to support them. The USA did little to help. When the dismal record of American stewardship became known in the fall of 1947, it stirred up worldwide protest. Regardless of the international ramifications, the Joint Chiefs had no good alternatives. Even these islands had serious drawbacks. The distance from the United States made for extraordinary logistical and security challenges, while the humid climate wreaked havoc with the sophisticated electronic and photographic equipment. [2]

As the Cold War intensified, so did demand for nuclear weapons and a full-scale testing program. After a series of tests on Enewetak atoll near Bikini, the myriad problems with logistics, weather, security, and safety concerns initiated thinking about a continental test site. As an Air Force official later put it, one of the pluses of a continental site was the "advantage of educating the public that the bomb was not such a horrible thing that it required proof-testing 5,000 miles from the United States." [2]

When the Joint Chiefs brought this proposal to the Atomic Energy Commission in September 1948, Chairman David E. Lilienthal responded that the Commission was willing to cooperate in a "preliminary survey," but he believed that psychological considerations were strongly against testing atomic weapons inside the United States. Nevertheless, in his formal written response Lilienthal stated that a continental site could have certain advantages over Enewetak for some tests and gave the go-ahead for a clandestine study of possible sites. [2]

The Armed Forces Special Weapons Project (AFSWP), established in 1947 from the military remnants of the Manhattan Project, was mandated with overseeing nuclear weapons doctrine, training, and logistics for the entire military establishment. The secret study fell within their jurisdiction. Code-named *Project Nutmeg*, AFSWP selected Navy Captain Howard B. Hutchinson to conduct the study. As a highly qualified meteorologist who had been at Enewetak, Hutchinson collected data and other information from prior tests and extrapolated how radioactive debris would behave in the meteorological environment over the USA. Captain Hutchinson concluded that "at properly engineered sites, under proper meteorological conditions" continental testing would "result in no harm to population, economy or industry." [2]

The Project Nutmeg report proposed no specific location as a test site, nor did it consider problems involving real estate, public relations, and security. The report simply targeted the best regions in the continental USA for detonating nuclear weapons. These turned out to be the arid southwest and the southeastern coast. Both came with drawbacks. The prevailing winds in the US are eastward, so a testing site in the southwest would carry radioactive fallout over population centers. In targeting the southeastern coast, somewhere between Cape Hatteras and Cape Fear, the radioactivity would be harmlessly blown out to sea. The problem was, nearly all the land that could be used as a coastal test site was inhabited and the region's considerable ocean-going shipping would have to be curtailed during test periods. After considering the problems with both regions, the AEC concluded that, excepting a national emergency, a continental site was not desirable. [3]

A national emergency soon arrived. In August 1949, the Soviets tested their first atomic bomb. Responding to this threat, President

Truman began deliberating the move to the next generation of nuclear weapons – what AEC Commissioner Lewis Strauss called a *quantum jump* – which would increase the explosive yield of the bomb 100- or even 1,000-fold. Such a thermonuclear weapon, known as the hydrogen bomb or the "Super," would restore the absolute advantage over the Soviets. Following intense internal debate on the possibility, morality, and wisdom of taking this quantum jump, President Truman approved development of the thermonuclear weapon in January 1950. Nuclear testing would be critical for every stage of this new Super bomb. [4]

Los Alamos scientists envisioned a series of preliminary tests to be conducted at Enewetak in the spring of 1951. With these plans almost complete, North Korean troops suddenly stormed into South Korea on June 25, 1950. Truman's decision to commit American ground troops to the conflict caused severe strains on military shipping and air transport. Los Alamos Laboratory Director, Norris E. Bradbury, was dumbfounded to discover that Enewetak might not be available for testing. "Just as one wants and needs it the most," he observed, "the chances of using it decrease alarmingly." [2–3]

The delays revived Project Nutmeg. AFSWP was charged to recommend at least one site in the continental US for emergency nuclear weapons testing. The AFSWP initially favored Alamogordo, New Mexico, where the Trinity device had been tested. Second choice was the Las Vegas Bombing and Gunnery Range. The search parameters included avoiding a site within a 125-mile (200 km) downwind radius of a major population center. El Paso, with a population of 130,000, was just outside the radius for the New Mexico site. Las Vegas, with a population of 25,000, was well within the Nevada site's 125-mile radius. [2]

The Nevada site, however, had significant advantages. Immediately to the south of the bombing and gunnery range was a government-owned airfield, complete with adequate runways and housing for several hundred people. Convinced that they had found the right place, the AEC hired a contractor to find a specific testing site. The area chosen encompassed two large valleys, Frenchman Flat and Yucca Flat. A range of hills provided natural barriers from public roads. There was easy access to the airfield's facilities and transportation. And the area would be easy to secure. The one serious drawback was that Frenchman Flat was only 65 miles (100 km) from downtown Las Vegas, with Yucca Flat not much further away.

A group of experts, that included Edward Teller and Enrico Fermi, met to discuss the distance problem. Assuming that meteorologists would pick no-wind and no-rain days for the detonations, there still would be measurable off-site fallout. Fermi suggested that anyone subject to exposure should be warned to stay indoors and take showers. Everyone agreed there was no risk "that anyone will be killed, or even hurt," but there was a "probability that people will receive perhaps a little more radiation than medical authorities say is absolutely safe." [2]

On August 7, barely six weeks after North Korea invaded its neighbor to the south, Secretary of Defense Louis A. Johnson took the Nevada proposal to the White House. President Truman requested time to think about it. The explosive yield of the proposed tests hardly suited a continental site. As autumn passed, the solution became to talk it up, as though there were no downsides. On November 22, Los Alamos officials recommended the Nevada site in glowing terms. The military concurred. General James McCormack, director of military applications, concluded that the Nevada site "most nearly satisfies all of the established criteria." He noted that high safety factors must be established and "the acceptance of these factors by the general public must be insured by judicious handling of the public information program." The AEC and National Security Council accepted the recommendation. On December 18, 1950, President Truman signed off on the selection. [2–3]

The following day, representatives from the Departments of State and Defense and the AEC met to discuss the ticklish issue of how they were going to break the news to the public. All agreed that the press release should emphasize that nuclear weapons testing was a routine activity, and radiological safety was under control and nothing to worry about. They also would remind the public that continental testing had been done successfully with the Trinity test in New Mexico. The group agreed that these simple, straightforward assurances would "make the public feel at home with atomic blasts and radiation hazards." [3]

The first series of tests, dubbed *Ranger*, were scheduled to begin in just a few weeks' time. Carroll L. Tyler, as lead commissioner for these first tests, wanted to see Ranger and the new Nevada Test Site go forward without public panic. He advised that any national reaction could be "conditioned" by composing all public announcements around certain "primary themes." The most important theme was that the testing would speed up the Nation's weapons development program, which was of major importance to national defense and security. Noting that Las Vegas was "highly aware of national publicity angles," Tyler recommended that the AEC should play on "local pride in being in the limelight." He also advised approaching the tests "rather matter-of-factly and not stimulate sensational attention by making too big a thing of them . . . too much reiteration may come under the category of the lady doth protest too much." [2]

The draft press release emphasized heavily that radiological safety requirements had been given *full consideration*. It stressed that extensive monitoring would be done, then listed various committees and panels that had given their seal of approval to the Nevada Test Site. Finally, it listed individuals, such as Enrico Fermi, whose names would lend considerable authority towards reassuring a skitterish public. When Fermi found out about it several days later, he told them to remove his name from the press release. [2]

When the Joint Chiefs read the draft press release, they wanted to do some serious editing, eliminating all reference to radioactive danger and making no mention of the fifth Ranger test. The Chiefs had promised Truman there would be no big tests at the continental site, and it turned out that the fifth test was three or four times too big. The AEC Chairman acquiesced to rewrite a "somewhat misleading" press release. [2]

The press release announcement of the Nevada Test Site and the Ranger series was scheduled for January 11, 1951. The day before, the AEC commissioners worked the phones. They discussed the matter with Nevada's two Senators and one Congressman, then with Nevada's Governor, State legislature, and any local officials who "if not taken into our confidence, might misinterpret the whole program." No one raised an objection. All 18 members of the Joint Committee on Atomic Energy were informed by hand-carried memorandums. William L. Borden, executive director of the Joint Committee, reported to the commissioners that some were "glad that it isn't where I live." There was some concern about the risks, but no one raised a strong objection. In lieu of its proximity to the new test site, the commissioners reasoned that California might be worried about contamination of water supplies. Unable to reach the Governor, they hastily contacted Los Angeles Mayor Fletcher Bowron and told him that they were going to "perform a few explosions" at the new Nevada site, but "there couldn't possibly be any damage" to Colorado River water from fallout. To totally put his mind at ease, the mayor was assured that tests would only be conducted when the wind blew away from California. Bowron replied that "he would see from Los Angeles that there is no one who gets the wrong idea." [2]

The press release went off without a hitch. The AEC's public relations staff in Nevada reported an overwhelmingly favorable reaction at the city, county, and local levels. Their two-hour press conference in Las Vegas was "largely a get-acquainted session." [2]

With the public now informed, the Ranger series moved full steam ahead. Los Alamos was in charge of most of the experiments. The military got on board with several "weapons effects" experiments to determine how much radiation troops would be exposed to at incremental distances from ground zero. The AEC added Operation Hot Rod, to determine whether cars would provide shelter during a nuclear attack. After the blast, Hot Rod clearly demonstrated that anyone trying to shelter in a car half a mile away would be killed twice – once by the blast and fire, the second time from radiation. At two miles or more, "chances of survival without injury were very good." [2]

The first test of the Ranger series was detonated in the early dawn of January 27, 1951. Named "Able," the one-kiloton bomb was dropped from a B-50 bomber out of Kirtland Air Force Base in Albuquerque. Twenty-four hours later, the eight-kiloton "Baker" was dropped. Baker's flash and shock wave were much stronger, managing to wake up the entire city of Las Vegas, except for people who were already awake in the casinos. The city's homes were jarred by several strong shocks that rattled dishes and shook windows and nerves. When a few people started talking about moving away from the area, Las Vegas *Morning Sun* publisher Hank Greenspun tried to quell the "irresponsible and hysterical utterances" by admonishing residents to "feel proud to be a part of these history-making experiments." A few days later, Baker-Two broke several store windows downtown and uncooperative winds carried the radiation cloud to the immediate west of the city. The wind also didn't cooperate with the one-kiloton Easy, when part of the radiation cloud veered off over southern California. A few days after the tests, the AEC received reports of "radioactive snow" falling in the Midwest and northeastern United States. The first reports came from Eastman Kodak whose Rochester, New York, plant detected radioactive particles in air filters, raising concerns about its sensitive photographic film. [2–3]

Fox was the last shot in the Ranger series. Of the first four tests, Baker and Baker-Two, at 8 kilotons each, had the biggest yields. Fox was the one the Joint Chiefs wanted to keep hush-hush, with its anticipated yield of 33 to 35 kilotons. Test officials took a brief time-out to ponder what a blast four times as powerful as the Baker bombs might do, then pushed ahead. Just to be on the safe side, however, they issued a public announcement urging people to stay away from windows at the time of the blast.

In the early dawn of February 6, Fox went off. Observers commented that the visual show was much more spectacular than the previous four

Figure 14.1 Small Boy Event, Frenchman Flat. View looking north from official observer point. Photograph courtesy National Nuclear Security Administration/Nevada Site Office.

detonations. Twenty to 50 miles away, the mountains were "illuminated by blinding whiteness." Gamblers in Las Vegas dived under the tables.

The Ranger series was just the opening ceremony for the Nevada Test Site. From 1951 to 1962, 100 nuclear bombs were dropped from planes, detonated at or near ground level, placed on towers, suspended from hot-air balloons, and shot from a 280-mm cannon. During those 11 years of aboveground testing, Las Vegas experienced an unprecedented influx of tourists who didn't come to gamble but to witness a detonation. The city became caught up in an ongoing Fourth of July mood, as though the nuclear blasts were just a more spectacular form of fireworks. The desert was "blooming with atoms," according to Governor Charles Russell. And you didn't even have to live in Las Vegas to experience the thrills. More than 200 miles (320 km) from the site, Los Angeles residents occasionally could watch the eerie "false dawn" from a blast. [5]

The Nevada Test Site also became a Mecca for the press. In the spring of 1952, a craggy knoll across from the Control Point was set up for reporters and photographers to witness, snap pictures and scribble off their stories. Dubbed "News Nob," Walter Cronkite broadcast to the world from this spot. During the ensuing years, hundreds of international journalists and photographers came to The Nob, turning the Nevada Test Site into one of the most photographed and heavily reported places in the world. [6]

The AEC's hastily contrived and misleading public relations program turned out better than anyone could have wished. When Americans sat in their living rooms to watch the first televised atomic blast, it wasn't long before "atomic fever" swept the country. Designers of everything from lamps to clocks to corporate logos adopted what became known as "atomic style" in their creations. The Franciscan China Company launched the wildly popular Starburst dinnerware pattern. KIX Cereal advertised an Atomic Bomb Ring, offering it to kids for 15 cents and a cereal box top. There was Atomic Fireball candy and even an atomic brand of sewing needles. The Washington Press Club offered the "Atomic Cocktail," as "a mixed drink for modern times." The word "atomic" became synonymous with anything powerful and modern. [7]

The movie industry quickly jumped on the atomic bandwagon. *Them!* told the story of ants mutated to huge proportions by an atomic blast. *The Beast From 20 Thousand Fathoms* was the tale of a prehistoric monster released from its icy prison by a nuclear bomb detonated at the pole. The most famous atomic movie was *Godzilla*, in which a 400-foot prehistoric monster made a serious statement about the dangers of nuclear bombs and radiation.

Musicians also embraced the atomic hullabaloo, turning out hundreds of popular songs that featured the bomb. Jackie Doll and his Pickled Peppers sang *When They Drop the Atomic Bomb*, enthusiastically calling upon General MacArthur to use the bomb against North Korea. Sultry Fay Simmons crooned *You Hit Me Baby Like an Atomic Bomb*. Dr. Strangelove and the Fallouts came up with *Love That Bomb*. Teenagers danced the atomic-bomb boogie.

There were even beauty pageants. In 1952, Candyce King, a Vegas showgirl, became the first Miss Atomic Blast. Candyce was chosen and crowned by US Marines, who had participated in atomic maneuvers at Yucca Flat – only to discover in Candyce another kind of "Big Bang" who radiated "loveliness instead of deadly atomic particles." After a series of stunning beauty queens, Lee Merlin, a Copa showgirl, became the last and probably most famous Miss Atomic Bomb in 1957. Lee became a pin-up sensation, wearing a cotton mushroom cloud pinned to the front of her swim suit. Her publicity photo appeared in hundreds of publications worldwide. [8]

Suddenly, in 1958, the Soviet Union unilaterally declared a moratorium on nuclear testing. A few months later, President Eisenhower responded to this propaganda coup by declaring a moratorium on nuclear testing in the United States. When the Soviets unexpectedly resumed testing in 1961, growing concerns about radioactive fallout, particularly strontium-90 in mothers' milk and baby teeth, spurred international debate and negotiations. Edward Teller, that manic Dr. Strangelove of all things nuclear, argued that radiation from fallout "might be slightly beneficial" and that any mutations it might induce should be welcomed for accelerating the evolutionary process. Regardless of his Promethean viewpoint, the 1963 Limited Test Ban Treaty prohibited all atmospheric nuclear testing. [9]

While the treaty was cause for international celebration, Teller was seriously dismayed by this turn of events. The treaty not only shut down the nuclear version of the Wild West but, even more importantly, was getting in the way of his latest nuclear obsession – Project Plowshare. Harking back to Isaiah's vision of an earthly paradise – "they shall beat their swords into plowshares and their spears into pruninghooks: nation shall not lift up sword against nation, neither shall they learn war anymore" – Teller's brave new world of Plowshare was going to elevate the Atoms for Peace program to undreamed heights.

The basic idea had all the glibness of a slick sales pitch. For a fraction of the time, headache, and cost of conventional methods, Plowshare would use thermonuclear weapons for large-scale earthmoving and planetary engineering. With Teller at the helm, his precious nuclear devices would carve harbors, dig canals and mines, even move mountains. "If your mountain is not in the right place," he boasted at a press conference, "drop us a card." Plowshare applications seemed endless: blow up rapids to make rivers navigable, straighten the route of the Santa Fe Railroad, mine coal and minerals, free oil and gas reserves, melt ice to yield fresh water, close off the Strait of Gibraltar to make the Mediterranean a freshwater lake for irrigating crops in North Africa, cut a waterway across Thailand, excavate a harbor in Alaska. Teller even claimed that the AEC could "dig a harbor in the shape of a polar bear, if required." [10–11]

Plowshare would set the stage for "an altered public relations program," whereby nuclear projects were publicized rather than concealed. "We will change the earth's surface to suit us," Teller proclaimed. But why stop with earth? "One will probably not resist for long the temptation to shoot at the moon ... to observe what kind of disturbance it might cause," he wrote. [9–10]

Plowshare first gained support during the 1956 Suez Canal crisis. Scientists at AEC's Livermore laboratory proposed using nuclear bombs to blast a new canal across Israel. The crisis passed before the proposal got beyond the talking stage, but the idea survived. The Suez crisis later prompted interest in a second, sea-level canal in Panama, a "Panatomic Canal" that would increase shipping and minimize the security risk of relying on a single inter-oceanic canal. Marine biologists raised the alarm that a sea-level canal could create disastrous ecological imbalances by inducing contact between the separate Atlantic and Pacific marine species. Even AEC Chairman Glenn Seaborg, who was a Plowshare aficionado, conceded that there was another difficulty with the Panatomic – the nuclear excavation would require temporary removal of tens of thousands of people, at the very least. There would also be fears of radioactivity and of induced earthquakes that "could lead to considerable apprehension, regardless of the great saving to world commerce and substantial benefits a canal could bring to the area." After $17 million in feasibility studies, the Panatomic Canal was finally shelved in 1970. [9]

Private industries proposed some Plowshare projects of their own. One of the most ambitious ideas was the North American Water and Power Alliance, which would create a coast-to-coast waterway across the US by nuclear blasting, particularly through the Rockies. Nuclear-powered water pumps would sustain the flow, and the hot water from the pumps would keep the waterway ice-free. In 1963, the Santa Fe Railroad proposed using Plowshare derring-do to create a "Nuclear Right-of-Way" for rail lines across the mountains between Barstow and Needles, California. The plans called for simultaneously detonating 23 nuclear bombs. [9]

Suddenly, the Limited Test Ban Treaty threatened to nip "planetary engineering" in the bud. Throughout the negotiations, the AEC repeatedly tried to include an exemption that would allow for peaceful nuclear explosions, but the Soviets would have none of it. The treaty was signed without the exemption. Within months, however, both sides were violating the brand-new agreement. In the end, it didn't really matter. During its 15-year lifespan, the only Plowshare experiment actually tried in practice was using nuclear bombs to "stimulate" natural gas production in Colorado and New Mexico. The resulting gas turned out to be radioactive and no utility would buy it. [10]

The Soviets got into the plowshare business in an even bigger way, detonating more than 120 "peaceful nuclear explosions" to dig canals, create underground storage cavities, extinguish stubborn fires in gas wells, and promote oil and gas flow. In 1991, a Soviet trading company

Figure 14.2 Sedan Crater. Note roads in foreground for scale. Photograph courtesy National Nuclear Security Administration/Nevada Site Office.

proposed using nuclear explosions to incinerate toxic and radioactive waste. The cash-strapped and newly entrepreneurial Soviets were "willing to entertain all ideas." Fortunately, environmental concerns within the country had effectively brought the program to a halt. [10, 12]

An enduring testimony of Plowshare came from the first large-scale cratering experiment at the Nevada Test Site. In the summer of 1962, a 104-kiloton nuclear bomb was detonated 640 feet (200 m) below ground surface, to determine the feasibility of using nuclear explosions for large excavation projects. The test, named *Sedan*, lifted a dome of earth 290 feet (90 m) above the desert floor, displaced 12 million tons of soil, and resulted in a crater four football fields across by one deep. The blast caused seismic waves equivalent to a magnitude 4.75 earthquake. Sedan's "vast amount of fallout" helped put a coffin-nail into Plowshare by demonstrating there was no way to carry out large-scale earthmoving projects without contaminating the atmosphere. [13–14]

The Nevada Test Site had several other ambitious projects. From 1955 to 1973, scientists worked on developing a nuclear-powered rocket. In 1961, President Kennedy proclaimed this would someday allow space exploration "to the very end of the solar system itself" – a now largely forgotten goal in his famous speech to land a man on the moon (and safely return him to Earth) by the end of the decade. A second more nefarious effort, aptly named Project Pluto after the Roman God of the underworld, set out to develop a devastating, and virtually unstoppable, nuclear-powered cruise missile. The missile, dubbed "The Flying Crowbar," would be used to retaliate against an enemy nation in the event of a nuclear attack. It was designed to travel at three times the speed of sound and to discharge up to a dozen hydrogen bombs on widely separated enemy targets thousands of miles away. Flying near the ground, the supersonic shock wave and intense radiation emanating from the unshielded reactor would wreak havoc in its wake. The project was terminated in 1964, because the faster and less costly Polaris missile made it unnecessary. [5, 15]

Between 1951 and 1992, there were 928 nuclear detonations at the Nevada Test Site, with 828 of them underground. Large areas became covered with subsidence craters, giving the sparsely vegetated terrain a distinct moonscape appearance. Two decades have passed since the nuclear detonations finally stopped, yet for the State of Nevada the memory lives on.

The Nevada Test Site is, and will continue to be, an environmental disaster zone for hundreds of thousands of years. During the four decades of testing, the Department of Energy has estimated that more than 300 million curies of radiation were released into the soil and groundwater, thereby qualifying the site as one of the most radioactively contaminated places in the United States. Many of the underground tests were done near or below the water table. The groundwater mainly flows southward, to the east and south of Yucca Mountain. Local farming communities in these downstream areas, such as Amargosa Valley, rely on groundwater for their main water supply. The wells also provide water for agricultural and livestock needs. While there is no immediate risk to public health, the area's water-lifeline could eventually be seriously impaired or destroyed. [16–18]

The Nevada Test Site has yet another legacy. In response to growing public concern for radiation safety, the AEC mounted a campaign to assure "downwinders" that their health and safety were not in peril even while evidence to the contrary began to emerge. In 1979, a study reported in the *New England Journal of Medicine* found a 2.4-fold increase in leukemia rates among children who grew up in southern Utah during the above-ground weapons tests. In 1997, the National Cancer Institute reported that fallout from atmospheric tests at the Nevada Test Site had deposited high levels of radioactive iodine-131 across a large portion of

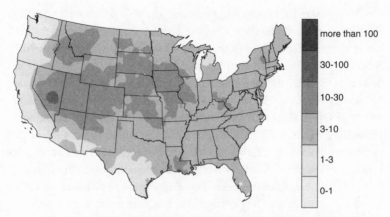

Figure 14.3 Total external and internal dose (in milligrays) to the thyroid of adults since 1951 from all tests at the Nevada Test Site. (1 milligray = 100 millirem)
Source: Reference [20]. Courtesy National Cancer Institute, National Institutes of Health, Bethesda, Maryland.

the contiguous United States, with doses high enough to produce tens of thousands of additional cases of thyroid cancer. While the increase in mortality and cancer rates caused by fallout from the atomic bomb testing remains debatable, the worry and anxiety to those who lived downwind is not. [3, 19–20]

For the citizens of Nevada and other highly affected areas, these consequences eroded a long-held trust in the federal government to protect the health and safety of its citizens. The government, of which the AEC was an all-too-active part, was guilty as charged on a number of counts. The government had seriously underestimated the potential environmental and health effects of radioactive fallout. The government had not arranged for adequate monitoring and follow-up testing of radiation exposure. The government approached radiation danger almost exclusively as a matter of short-term, external exposure – despite growing evidence of the long-term hazards of internal exposure through the food chain. And finally, the government attempted to suppress evidence of fallout-induced health problems among exposed persons and animals, and tried to discredit scientific studies suggesting strong evidence to the contrary. This legacy of irresponsibility and consequent distrust in the federal government set the stage for the controversies that would soon unfold at Yucca Mountain. [9]

15

Yucca Mountain

> And it is not our part here to take thought only for a season, or for a few
> lives of Men, or for a passing age of the world. We should seek a final end
> of this menace.
>
> Gandalf in *The Lord of the Rings* [1]

The year was 1972. The proposed salt repository at Lyons, Kansas
was in its death throes. Environmentalists were demanding a solution to
the waste problem or shutting down the entire nuclear industry. Caught
off guard, the AEC contracted with the US Geological Survey to study
other geologic media, look further afield, think outside the salt-box. Dr.
Isaac "Ike" Winograd, a research scientist for the USGS, had been doing
just that. At the annual meeting of the Geological Society of America, Ike
proposed using the thick unsaturated zone somewhere in the American
Southwest for long-term storage of solidified high-level waste. In the
world of high-level radioactive waste, this was quite the revolutionary
idea. [2–3]

For over a decade, burial in salt had been essentially the only game
in town. The USGS had a long-standing concern with this medium –
salt can behave in a highly unpredictable manner in the presence of
the extremely high heat generated by high-level waste. The other option
being considered for geologic disposal was to bury the waste below the
water table, deep in the saturated zone. Burying the waste *in* water would
protect the canisters from oxidation. Uranium in the spent fuel is also
more stable in the absence of oxygen. Yet, even then, eventually the
canisters would be breached and radionuclides released. A major challenge with a saturated-zone repository was how to build a solid, secure
repository surrounded by all that water, and then seal it back up.

Ike saw two key problems with salt and the saturated zone. First,
there was no way to monitor the waste, keep an eye on it, see how it's

Table 15.1 *Key events discussed in this and previous chapters.*

1951	First electricity production by nuclear energy illuminates four light bulbs in Idaho
1954	Atomic Energy Act makes possible the use of nuclear energy for civilian purposes
1957	First NAS report on geologic disposal of radioactive waste
1969	Fire guts Rocky Flats plutonium weapons plant
1972	AEC abandons Lyons, Kansas project
1976	USGS recommends Nevada Test Site for high-level waste repository
1980	Low-Level Radioactive Waste Policy Act
1982	DOE changes focus at Yucca Mountain to unsaturated zone
1982	Nuclear Waste Policy Act (NWPA)
1987	Amendments to NWPA identify Yucca Mountain as sole candidate
1999	First shipment of transuranic waste to WIPP
2002	President George W. Bush officially recommends the Yucca Mountain site to Congress
2004	US Court of Appeals rejects 10,000-year compliance period in EPA regulation

doing. Second, in the event of a problem there would be virtually no way to retrieve it. Ike cautioned that burying the waste in either salt or the saturated zone falls in the disposal category – out of sight and out of mind. On the other hand, a geologic repository in the unsaturated zone, between land surface and the water table, would allow for monitoring and retrieval. This would be more like long-term storage instead of disposal. [4]

The southwestern United States has some of the thickest unsaturated zones in the world. The region's arid and semi-arid climate also means a lot less rain would percolate downward into a repository. Another advantage is that the Southwest offers remote federally owned lands for consideration. As a possible repository location, Ike suggested somewhere in the 1,350 square miles (3,500 square kilometers) of desert and mountainous terrain of the Nevada Test Site.

Up to this time, the unsaturated zone had been studied mostly to depths of interest for agriculture – in other words, the root zone. In these early days, deeper and more complex studies of how water behaves in the unsaturated zone had seldom been attempted. There was also the considerable challenge of how to evaluate the effects of possible stresses when a major heat source was buried in all that rock. In short, if the AEC decided to explore Ike's revolutionary idea, it would have to be matched with some pretty revolutionary science.

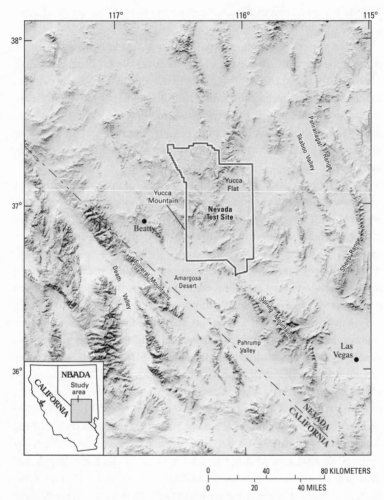

Figure 15.1 Location map of Yucca Mountain and vicinity.
Source: US Geological Survey, modified from reference [27].

In a 1976 letter, USGS Director Vincent McKelvey officially recommended the Nevada Test Site as a high priority, having "major geological
advantages, as well as obvious logistical, political, and economical advantages." The Nevada Test Site had been extensively studied for decades,
McKelvey added, producing 900 man-years of data collection and interpretation in the fields of hydrology, geology, and geophysics. In many
respects, the breadth and depth of this information was unequaled
elsewhere in the United States. In McKelvey's view, public acceptance
was "highly probable," due to the extensive radioactive contamination

already present, thereby eliminating the use of this area for most other purposes. [5]

The major disadvantage with the Nevada Test Site, McKelvey noted, is its tectonic setting. The test site is an area of major faulting, is susceptible to earthquakes, and is located within the Southwest Nevada volcanic field. The volcanism began around 15 million years ago, when nearby volcanoes began blowing their top on a scale that made the 1980 Mount St. Helens eruption look like a minor event. The material blown out of these ancient calderas formed thick deposits of a rock known as volcanic tuff.

Tectonic activity appeared to be the only significant drawback to the Nevada Test Site, but was not considered a show-stopper. The volcanism had long ago quieted down, and earthquakes primarily affect the land surface. Far underground, at repository depth, the effects would be greatly attenuated.

McKelvey's letter had its effect. A search began at the Nevada Test Site for high-level waste disposal sites deep below the water table in the, still favored, saturated zone. Within a few years, the repository search led them to the thick tuff formations at Yucca Mountain. After discovering problems with using the saturated zone, USGS scientists suggested studying the thick unsaturated zone beneath the mountain. In 1982, one decade after Ike Winograd first proposed the idea, the Department of Energy changed the focus of the Yucca Mountain investigation from the saturated to the unsaturated zone. [6]

Yucca Mountain is actually a low ridge that spans six miles (10 km) from north to south. To the west and south is the Amargosa Desert. The Funeral Mountains and Amargosa Range rise in the hazy, heat-baked distance. On the crest of these formidable mountains is Dante's View, where half-dead pioneers were treated to a bird's eye panorama of Death Valley. Many place names in the vicinity of Yucca Mountain testify to the perils of this region – Skull Mountain, Broken Limb Ridge, Fatigue Wash, and Busted Butte. To date, the area has seen little settlement or human activity. Beatty, with a population of around 1,000, is the closest town – lying about 15 miles (24 km) to the west of Yucca Mountain.

At places, the depth to the water table from the top of the ridge at Yucca Mountain measures more than 2,000 feet (600 m). Such a thick unsaturated zone (UZ) would allow a repository to be built about 1,000 feet (300 m) below land surface while also being about 1,000 feet above the water table. With the area's average rainfall only a few inches per year, and after factoring in evapotranspiration, a very negligible average of 0.2 inches (5 mm) per year soaks into the UZ. In turn, much of this water

Figure 15.2 Ike Winograd at Dante's View, Death Valley National Park.
Photograph courtesy US Geological Survey.

would be held by capillary forces around openings in the unsaturated rock, further reducing the possibility of water entering a repository. Scientists at the USGS tentatively concluded that a repository could be designed at Yucca Mountain in the thick UZ to allow very little, if any, water to come into contact with the waste canisters. This conclusion was of considerable significance, in that water is the most likely means by which the waste would be transported to the biosphere.

There were other natural barriers. Beneath the proposed repository "horizon" are rocks containing zeolites that grab and hold onto certain radionuclides. In addition, Yucca Mountain is in a hydrologically closed basin. No surface water leaves the area. The area's major river, the Colorado, is more than 100 miles (160 km) away. On top of these advantages, Yucca Mountain has a relatively uncomplicated geology and lacks mineral or energy deposits that might encourage future generations to mine here. To cap it off, building a repository over a thousand feet under the mountain was logistically feasible – welded tuff is suitable rock for mining stable tunnels that would hold up well over time. In many ways, the mountain seemed made to order.

POLITICS ENTER THE PICTURE

The 1982 Nuclear Waste Policy Act (NWPA) was intended to finally solve the problem of what to do with the growing amounts of spent fuel and high-level waste being stored all over the country. Repository possibilities in the eastern States soon succumbed to politics. Of the three western finalists, the Hanford site was in the home State of the House Majority Leader, and the Deaf Smith, Texas site was in the home State of Vice President George H.W. Bush *and* Speaker of the House Jim Wright. When the NWPA Amendments were passed in 1987 and Yucca Mountain was chosen as the only site to be studied, it escaped no one's attention that Nevada was a State with virtually no political clout at the time. Regardless of Yucca Mountain's scientific advantages, the public's perception became irrevocably clouded by politics.

Prior to 1987, Nevada's opposition to the use of Yucca Mountain was, for the most part, respectful and restrained. Almost instantaneously, the *Screw Nevada Bill* stirred up a hornet's nest of defiance, protests, and litigation. Gone were the days when the State was expected to have a patriotic duty to host all-things nuclear. The blatant political maneuvering was considered nothing less than a slap in the face for a State that, to this day, has no nuclear power plants. [7]

If Congress hadn't meddled with the Nuclear Waste Policy Act, almost assuredly Yucca Mountain would have been the top choice for study. But Congress did meddle. In the blink of an eye, the site was viewed as nothing more than the federal government running roughshod over a State's rights. All the arguments favoring geologic disposal in the unsaturated zone beneath Yucca Mountain were suddenly drowned out by shrieks of *foul play*. In the spirit of its State motto, *Battle Born*, Nevada was going to fight.

Foul play was far from the only issue. By the 1980s, Nevada's trust in the federal government was a rare commodity. Many people still remembered the above-ground atomic bomb tests. Brian Greenspun, president and editor of the *Las Vegas Sun*, had watched with his father: "He would take us up to the top of Mount Charleston when we were little kids, so that we could watch the blasts. You could see the mushroom cloud go off. And we thought that was the neatest thing in the whole world. And then, minutes later, this pink cloud would come over and we would get sprinkled with dust. No one ever thought anything of it. Thirty-forty years later, we are the thyroid cancer capital of the world." [8]

Las Vegas Mayor Oscar B. Goodman still kept a copy of a 1957 AEC handbook, in which it explained that fallout can be *inconvenient*. "That's the way they expressed it. So I'm not going to help the federal government lie to us again. Nope, not during my administration." Having moved to Las Vegas in 1964 with his wife and $87 in their pockets, Goodman had risen to become the self-proclaimed "happiest mayor in the universe." He was also one of the "15 Best Trial Lawyers in America," according to the *National Law Journal*. This popular three-term mayor soon became a formidable opponent of Yucca Mountain. [8]

No law can wipe out a legacy of distrust. Nevadans had finally had enough. No more nuclear guinea pigs and no more patriotic duty to be the dumping ground for all things nuclear. During their first meeting after the *Screw Nevada Bill*, the Nevada Legislature passed Assembly Bill 222, making it illegal to dispose of high-level nuclear waste in their State. Just to make sure the feds got it, the legislature also unanimously passed resolutions opposing the repository and denying the State's approval for ceding jurisdiction over the land required to build it. These resolutions became the foundation for Nevada's claim that it had exercised its veto of the repository under the NWPA. The problem was, the NWPA veto pertains to when all site-characterization studies are complete and DOE formally recommends the site to the President. And even then, Nevada's veto can be overturned by a simple majority vote in both Houses of Congress. [9]

While protesting loudly through the courts and the media, the State was not averse to making a grab for federal money. As compensation by the federal government, the county that was selected as the repository site would receive "funds in lieu of taxes." Depending on whether you asked the Department of Energy or the State of Nevada, the appropriate amount ranged from two to 40 million dollars. Suddenly Nye County, one of the largest, least-populated counties in the United States, was coming into a considerable windfall. With the strong backing of Clark County, home to Las Vegas, the State legislature had a brainstorm. They created a brand-new county, christened Bullfrog County, which completely surrounded Yucca Mountain. Bullfrog County had no residents, so the "funds" would go directly into the State coffers. Clark County, where most Nevadans live, would become the main beneficiary. Once word got out, Nye County sued the State and the whole scheme was overturned by the Nevada Supreme Court. [10]

By the early 1990s, lawsuits were flying. In January 1990, the United States sued the State of Nevada in the US District court, stating that Nevada's notice of disapproval (veto) was premature. The US also sued the State for not allowing the pertinent agencies to process the necessary permits for the study to commence. The United States Circuit Court of Appeals soon ruled against Nevada, finding that the federal government had the right to proceed with its study of the Yucca Mountain site. The Supreme Court elected not to hear the case and the decision stood. [9]

Nevada Governor Bob Miller warned, "Nevada will continue to oppose the imposition of this unwanted project on the State through any available lawful means." Louisiana Senator J. Bennett Johnston, the purported father of the *Screw Nevada Bill*, went on record as saying that he would do the same if he were Governor of Nevada. [7]

In this dry, hot, desert State, one of the most crucial battles was fought over water. State law governs water use, and the Energy Department needed a lot of it for drilling boreholes to study the site. Nevada refused to allow DOE a water permit, was overruled in the courts, and for the next two decades played hardball over when, and how much, water could be used at Yucca Mountain.

Transportation issues were also a continual rallying point for drawing national attention to what was otherwise Nevada's problem alone. Mayor Goodman reminded everyone that it wasn't just Nevada screaming "not in my backyard." The nuclear waste would have to travel through a lot of backyards before it arrived at Yucca Mountain. Nevada Senator Harry Reid raised the possibility of a *mobile Chernobyl*: "How are you going to haul the most poisonous substance known to man across the highways and railways of this country?" [8]

For a State where "all roads lead to Las Vegas," Nevadans were fearful that most of the waste would travel through the metropolitan

Figure 15.3 Burn test of transportation container at approximately 2,000 °F (1,100 °C). Source: US Department of Energy.

area on its way to the mountain. In a pre-emptive strike, Las Vegas passed a law making it illegal to haul nuclear waste through the city. In a 2003 interview on *60 Minutes*, Brian Greenspun voiced this concern: "Who wants to be the unlucky person who's here outside a hotel on the Las Vegas Strip when one of those trucks turns over and the nuclear waste spills? And you know it's going to happen. Accidents happen." [8]

Secretary of Energy Spencer Abraham assured Nevadans that the transport casks had undergone a whole range of strenuous tests. They had been smashed into concrete barriers, broadsided by roaring trains, and burned in jet fuel for 90 minutes. And they stayed intact. Nevertheless, the casks weren't designed to withstand *all* disasters. After two decades, studies demonstrated that the shipping casks were still vulnerable to anti-tank weapons or explosive charges. Senator Reid warned, "Every one of these trucks, every one of these trains, is a target of opportunity for a terrorist to do bad things . . . I mean, you talk about a dirty bomb." From Secretary Abraham's perspective, the more important issue was whether the current locations of nuclear waste lying around all over the country provided more vulnerable targets. "We think that it just stands to reason that consolidating the waste in one facility in a very remote part of America will make it much easier to protect on a long-term basis," he argued. [8]

In 2006, the National Academy of Sciences released a study on the safety of transporting spent nuclear fuel and high-level waste. The Academy concluded that, from a technical viewpoint, these shipments present "a low radiological risk activity with manageable safety, health, and environmental consequences when conducted in strict adherence with existing regulations." The history of waste shipment supports this conclusion. Since the early 1960s, more than 3,000 shipments of spent nuclear fuel have traveled over more than 1.7 million miles in the United States, and there has never been an accident that has resulted in release of radioactive material harmful to the public or the environment. The Academy also concluded that, compared to the significantly greater volume of other hazardous materials commonly transported in this country, the relative risk of transporting spent nuclear fuel is thousands of times less. In 2006, for example, American railroads transported hazardous materials in over one million rail cars. Much of this waste was in gaseous or liquid form. [11–12]

Although technically manageable, the Academy admitted that there remained a number of *social and institutional* challenges before such a large-scale radioactive waste shipping program could be implemented. Probably at the top of the list was fear – and in Las Vegas this fear was understandable. As the transportation issue dragged on, DOE finally decided to build a railway spur across the northern part of the State, well away from major population centers. Among Nevadans, however, these arguments largely fell on deaf ears. Any concessions were viewed as just the fed's latest trick in trying to hoodwink the State.

THE NEVADA INITIATIVE

A year after Governor Miller's warning that Nevada would oppose Yucca Mountain through any available lawful means, the anti-Yucca Mountain coalition had evolved from grass-roots protests into a well-organized political machine. The State continued to fight through the courts and hold up permits, while also launching a public-relations campaign to convince Nevadans that Yucca Mountain was another disaster waiting to happen – compliments of the federal government.

In 1991, the American Nuclear Energy Council (ANEC), an industry trade association, hired Oram, Ingram and Zurawski Advertising (OIZ) to conduct a three-year, $9 million advertising campaign to correct rampant misperceptions about Yucca Mountain and to build bridges with Nevada's political leaders. The ANEC reasoned that, through OIZ, the nuclear industry would provide the unadorned facts, based on sound and responsible science.

Kent Oram, the principal partner at OIZ, was a major player in Nevada's political arena. He had been instrumental in the campaign of Governor Miller and many of the State's politicians, whose careers now depended on defeat of Yucca Mountain. Originally, Kent Oram had opposed Yucca Mountain and had played a key role in Las Vegas passing a city ordinance opposing the site. After having researched the issue, interviewed key researchers, and visited nuclear facilities, Oram and his staff became convinced that nuclear energy was a necessary technology, and a solution to the waste problem was long overdue. While Oram understood that the Yucca Mountain account was a huge challenge, he believed it was a battle that could be won. [13]

If Oram thought his job was to run a simple information campaign, those thoughts were soon laid to rest. A few months after launching the campaign, an antinuclear group released confidential ANEC documents providing the unvarnished details of the ad campaign's strategies. Among these documents was a 22-page report dubbed *The Nevada Initiative*. Using military-like jargon, the report claimed that a "political beachhead" had been established and now "air cover" would be provided so that elected officials could negotiate benefits in exchange for the repository. To counter misinformation distributed by antinuclear forces, DOE scientists would be trained as "truth squads," to be backed by deployment of a "professional media attack/response team," using well-known news reporters. The report boasted that within 24 months the majority of Nevadans would support the repository. The result would be "checkmate for the antinuclear forces." [14]

The response to these leaked documents was outrage and derision. Newspaper and television coverage featured scathing attacks by State officials that continued for weeks, while a pair of Las Vegas radio disc jockeys began to spoof the OIZ ads.

Oram's first round of television ads featured popular former sportscaster, Ron Vitto, as the narrator. In an attempt to correct a widespread misperception that the waste would be in liquid or gaseous form, which is much harder to transport and store safely, Vitto held a "dummy" spent-fuel pellet in his hands. While merely trying to assure Nevadans that the spent fuel would be in solid form, critics screamed that OIZ was trying to trick Nevadans into believing that spent-fuel pellets are so safe you can hold them in your hand. The idea that anyone was suggesting such a thing was ludicrous, yet the opposition had a heyday. The disc jockeys sang scathing ditties about Vitto becoming a sellout. Citizen Alert published cartoons of Vitto dissolving into a skeleton. "Dump it in Ron Vitto's yard" became a best-selling bumper sticker. Vitto didn't know what had hit him. [15]

By 1993, the ad campaign ended as a spectacular failure. Ironically, this *simple information* campaign had actually managed to harden
the opposition. The public distrusted the ANEC and now viewed DOE
scientists as "hired guns" of the nuclear industry, pointing to the leaked
documents as proof of attempted public opinion manipulation. In addition, Oram had been continually hamstrung by a dithering and indeci-
sive industry group. The American Nuclear Energy Council had hired OIZ
because they were the big political guns in Nevada, with a proven ability
to get the job done, yet the ANEC officials nit-picked and micromanaged
almost every step of the process. In the words of Kent Oram, they "just
drove us nuts." [16–17]

SCIENTIFIC BATTLES

While the war continued through the media and the courts, the USGS was
having its own battle with DOE. Here was a very different style of combat.
On the whole, the scientific community is a civilized group that hold
themselves to a high behavioral standard. Disagreements, supported by
scientific evidence and careful thought, are respectfully aired at meetings
or in peer-reviewed papers. This is a world where people work slowly and
carefully and speak even slower – a world that would drive most people
out of their minds.

Nevertheless, by the early 1990s, the USGS was in sharp disagreement with DOE over two key issues. Repeatedly, in meetings, papers and
letters, the Survey had lobbied DOE for keeping the repository's temperature below the boiling point of water. Sustained temperatures above
boiling point would change the fractures and mechanical strength of the
surrounding rocks, increase chemical reaction rates, make new reactions
possible, and greatly complicate predictions of repository performance.
[18–19]

The second battle involved the ubiquitous fractures in the mountain. The State of Nevada and other opponents considered these to
be a major site liability, while the USGS argued that these numerous
fractures were, in fact, a major asset. In the event of a much cooler, wetter
climate during the geologic timeframe of high-level waste, the fractures
would drain the water from the repository. DOE engineers repeatedly
failed to recognize this natural feature and, by the late 1990s, were even
proposing to line the waste tunnels with concrete. [6, 18]

There was also broader controversy about using the unsaturated
zone for a repository. Critics noted that, except for the United States,
every other country was studying only the saturated zone. Proponents
countered that these other countries don't have the benefit of a deep

unsaturated zone. This general argument aside, when studies began at Yucca Mountain, the capability of predicting water and chemical movement through the UZ was in its infancy. Even less was known about how water behaves in the unsaturated zone of fractured rock. The effects of future climatic conditions also would be more unpredictable in the UZ than in the saturated zone, as the latter is already saturated.

An additional limitation of burying high-level waste in the UZ is the presence of oxygen in the rock pores and fractures, making the waste canisters more susceptible to corrosion. The solubility of many important radionuclides, including the uranium that comprises the bulk of the spent fuel, is also much higher in the presence of oxygen. [20]

These limitations of the UZ would be of less concern if it could be determined that only small amounts of water would pass through the repository. Demonstrating this hypothesis proved to be a much more lengthy and difficult process than originally believed.

ANOTHER INCONVENIENT TRUTH

For over three decades, Yucca Mountain became the most studied piece of real estate on the planet, involving *hundreds* of studies by federal, State, university, and industry scientists. Yet, for every study completed, new questions arose. It was like a dog chasing its tail – there would never be an end to it. This inconvenient truth meant that if the Yucca Mountain repository is ever constructed, it will be done in the face of uncertainty. In this respect, the Yucca Mountain project became a symbol for a much more encompassing truth – there will never be a perfect place, with 100% guarantee, where high-level radioactive waste can be safely contained over the hundreds of millennia required for it to be rendered harmless. The problem is just too big, too complex.

The Nuclear Regulatory Commission had ruled that a geologic repository must provide "reasonable assurance" that it would perform as hoped. However, increasingly during the 1990s, the battle lines were drawn around a guarantee of *total certainty*. With this impossible imperative gaining momentum, the strengths of Yucca Mountain – burial in the unsaturated zone where the waste could be monitored and, if necessary, retrieved – were lost in the hubbub of what questions hadn't been answered and what's wrong with the site. As the debates dragged on, Ike Winograd and Gene Roseboom summarized the problem:

> The more we learn about a given subject – especially one involving the interface of multiple disciplines over geologic time frames – the more complex it becomes. Another decade of study of Yucca Mountain will

likely provide the data needed to address some of the current questions about this site, but probably will also introduce new questions, as well as unearth surprises. Thus, there is unlikely to be complete closure. Nor will honest disagreements among scientists and engineers regarding some Yucca Mountain issues likely ever cease. This reality enables critics of this use of Yucca Mountain to ignore major attributes of the site while highlighting the unknowns and technical disputes for the press. Not surprisingly, the press, the public, and our elected officials are left with the impression of a flawed site. [21]

On February 15, 2002, President George W. Bush officially recommended the Yucca Mountain site to Congress. As permitted by the NWPA, Nevada's Republican Governor Kenny Guinn subsequently filed a notice of disapproval, thereby vetoing the site. He declared that the repository was "not inevitable" and that he would press the fight to "expose the Department of Energy's dirty little secrets about Yucca Mountain." Guinn made history by being the first Governor to veto a President, but his veto was short-lived. In July 2002, both the Senate and House voted to overrule the Governor. These actions bolstered the battle over how much uncertainty was acceptable. [22]

One of the most vocal critics trumpeting the uncertainties of Yucca Mountain was Dr. Allison Macfarlane. In May 2003, while working on MIT's Security Studies Program, she appeared before the Senate Energy and Water Development Subcommittee. Macfarlane began her testimony with a sweeping indictment of the scientists working on the Yucca Mountain project: "First, the science done at Yucca Mountain is produced by scientists mindful of the political goals of the agencies they work for, and the work they produce is evaluated by managers trying to meet these goals." After twice repeating this allegation, and without citing any evidence, she continued: "Now that we know that politics does indeed play a role in the science produced to uphold the Yucca Mountain site, we can ask the question, has politics limited some of the science done and the questions asked about the site? I would argue that the answer is yes." Macfarlane then proceeded to tear to shreds the science being done, and *not* being done, at Yucca Mountain. [23]

It is difficult for a political subcommittee to evaluate the fairness and accuracy of such testimony. However, this was not the case a year earlier when Macfarlane and Dr. Rodney Ewing, a well-known expert on nuclear waste at the University of Michigan, wrote a Policy Forum editorial for *Science* magazine. Citing Thomas Jefferson's advice to George Washington, "Delay is preferable to error," the authors charged that "The scientific basis for the selection of the Yucca Mountain site

Figure 15.4 Secretary of Energy Spencer Abraham visits Yucca Mountain on January 31, 2002, prior to recommending the site to President Bush. *Source:* US Department of Energy.

continues to be only a marginal consideration." They argued that, in spite of the scientific uncertainties about the site, there was "a surprising sense of urgency to move forward with a positive decision on Yucca Mountain." Ewing and Macfarlane advocated that the project "should not go forward until the relevant scientific issues have been thoughtfully addressed." [24]

Their article elicited strong responses. Scientists from the Colorado School of Mines, Harvard University, and the University of California at Berkeley wrote: "Insisting on comprehensive knowledge is neither possible nor necessary to assess the suitability of Yucca Mountain . . . Furthermore, insisting on scientific understanding of all possible processes only diverts limited resources from a few key processes that control the long-term performance and safety of a geological repository." A professor at Texas A&M University wrote, "The 40 years of study already invested in the disposal of nuclear waste, with some two decades focused on Yucca Mountain in particular, should not be cast aside with the implication that our government is acting in haste . . . There is indeed some risk associated with the Yucca Mountain site, as Ewing and Macfarlane point out. But there is far greater risk in not proceeding with the licensing process." Another argued, "If public policy required the level of certainty called for by the authors, we would have no vaccines or water treatment facilities." Yet another responded, "We are at a complete loss to account for the authors' failure to look at the relative risks or their conclusion that the project should not go forward. After all, there is already far more radioactivity under the ground – with no special containment – at the adjacent Nevada test site." [25]

In general, the commentators viewed Ewing and Macfarlane's editorial as a personal viewpoint rather than a persuasive, scientific rationale for delay in proceeding to the next step. While how much uncertainty was acceptable remained highly controversial, Macfarlane, Ewing, and many other scientists generally agreed on one point – too much of the science produced at Yucca Mountain was hidden away in obscure government publications or "gray" literature, hampering open scientific debate.

Key to many of the arguments to move forward was employing a *stepwise approach*. In short, the official site recommendation did not end DOE's obligation to continue the necessary scientific studies to confirm, or not confirm, repository safety all the way to final closure of the site a half-century or further down the road. This phased decision-making approach was endorsed by the National Academy of Sciences, the USGS, and the scientific community worldwide, thereby becoming the rationale for proceeding with the site recommendation. Yet, the inconvenient truth remained – resolving all uncertainty is not possible. [26]

Regardless, the uncertainty factor became the opposition's trump card. On July 9, 2004, the uncertainties took on an entirely new dimension when a federal court ruled that the safety of the proposed Yucca Mountain repository must be demonstrated, not for 10,000 years as recommended by the Environmental Protection Agency, but for *one million years* into the future. The quest for certainty had now fully entered the realm of science fiction.

16

How *long* is long?

I have never, I think, in my life,

been so deeply interested by any geological discussion.

I now first begin to see what a million means.

Charles Darwin in a letter to James Croll [1]

Modern *Homo sapiens*, with our great big brains, are really something special. We have walked on the moon. We have peered into the smallest particles of matter. We have developed vaccines and unbelievable technology. Our weapons have moved way beyond wooden clubs. We reason. We create. Yet, when it comes to dealing with *time*, we haven't changed much over the eons. We don't think or plan too far ahead. We have the ability to imagine the near future, yet when that time arrives it's so different from what we thought it would be. We do somewhat better with looking back and remembering past events, yet in the span of a lifetime our memories of childhood recede into an impressionistic haze. Without effort, most of us have a hard time remembering what we did two days ago. Like it or not, *Homo sapiens* mainly inhabits the present tense.

For many thousands of years our ancestors created an historical record in written and pictorial forms, yet these glimpses into the past have little meaning and no appreciable effect on our daily lives. Any child studying for a history exam will tell you that. Neil Chapman and Ian McKinley, eminent European geologists in the field of nuclear waste disposal, aptly stated the problem: "A pragmatic approach to time-scales would be that if you can really imagine what it would be like at that time, it is meaningful – if not, forget it. Computer models will happily predict releases after the expected lifetime of the Earth." [2]

What was life like even 100 years ago when the average life-expectancy was 47, six percent of Americans graduated from high school, and only a few thousand cars somehow managed to get around on 150 miles of surfaced roads and various mud-bogged tracks? Even with effort, it's hard to imagine living in a world where the primary mode of transportation was the horse or your own two feet. Yet that was our life just 100 years ago.

Three hundred years ago no one was making the world safe for democracy. Even the most basic tenets of equality and civil liberties had almost nothing to do with a world where kings still ruled by want and whim.

One thousand years ago people lived their entire lives within running distance from the walls of the feudal keep. Plagues mysteriously came and went. One thousand years is not even a blink of the eye in the life of high-level waste.

A mere 5,000 years ago the Sumerians were developing a more settled and complicated way of life in the Fertile Crescent. Eventually, their dazzling inventions included the wheel, writing, numbers, money, and smelting tin and copper to make bronze. [3]

Ten thousand years ago the last ice age was loosening its grip, but enough water was still locked in ice that Great Britain was not an island. Volcanoes were erupting in central France. And humans were beginning to make the astonishing discovery that if you put a seed in the ground, it will grow and produce more seed. Life was tribal, ruthless, and mercifully short. Before the court's 2004 ruling regarding the time required for isolation of the waste, 10,000 years was EPA's prediction standard for Yucca Mountain.

Twenty thousand years ago, an ice age was near its peak and sea level was almost 400 feet (120 m) lower. The coastline of California lay 25 miles (40 km) to the west of the Golden Gate Bridge.

Forty thousand years ago (we're now less than *one-twentieth* of the court-mandated one-million-year prediction timeframe for Yucca Mountain) the first *Homo sapiens* reached today's Europe. The place was bustling, and probably bristling, with Neanderthals who had ruled here for over 200,000 years (one-fifth of the million-year timeframe). For many thousands of years our ancestors lived among these SUVs of the ancient world, who required one-third more calories to lug their short, stocky bodies around. A mere 25,000 years ago, while eking out a meager survival on Gibraltar, the last Neanderthals succumbed to extinction. [4]

For that dim *one million year* mark on our timeline, we have only a few hard-won clues to guide us. *Homo erectus* – our earlier, clumsier, and mentally challenged predecessor – was wandering around Africa and dipping into Asia while the ages of ice came and went. Ice so thick it sank the continental plates by hundreds of meters. Recent DNA evidence suggests that, also around this time, our human predecessors were extremely close to extinction. [5]

And that's about it. One million years in the past is just too far back to have more than a vague idea of what was happening on the planet. Most people, of course, could not care less. Yet, for scientists working on the Yucca Mountain Project, *one million years* had suddenly taken on a whole new meaning. If you need to predict that far into the future, understanding that far into the past would offer some basic planetary clues of what that future might look like. And the most basic clue is ice. *Big time* climate change.

THE STORY OF ICE

While the twentieth century was a period of unlocking the smallest particles of matter, the nineteenth century witnessed huge breakthroughs into understanding the planet itself. Geologists were taking a fresh look at all those gigantic boulders stranded bizarrely out of place, the scraped-down rock beds and immense gravel deposits found all over the northern parts of Europe and North America. In such a religious age, all this geological rearranging had been easily explained as remnants of the biblical deluge. Read your Bible, here's your proof. It wasn't until the early decades of the nineteenth century that a few geologists, led by Louis Agassiz, began to question the explanation. This rock debris looked remarkably like deposits formed by the slow grind and slide of Alpine glaciers, but on a scale that staggered the imagination. Could *ice* somehow have caused this colossal geologic mess?

Goethe was perhaps the first to conceive of the idea of an ice age, popularized in his 1823 novel, *Wilhelm Meister*. Yet, this was just science fiction on the grand scale. Literary high jinks. It wasn't until the 1860s that most scientists accepted the incredible fact that, long ago, these northern regions had been buried in ice over a mile deep. Soon, another amazing whopper turned up. It appeared there had not been just one ice age but *four*, with each lasting many tens of thousands of years. [6]

Theories abounded to explain these glacial cycles. The Earth was simply cooling, or had passed through hotter and colder regions of space. The uplift of mountain ranges and other massive reconfigurations of the Earth's surface altered ocean currents and wind patterns, thereby changing climate. Others postulated that massive volcanic eruptions or extraterrestrial events, like the huge meteor that ended the Cretaceous period, could have brought on an ice age. The problem was that none of these theories could explain the *cycles*. Why had the ice sheets grown and retreated, again and again? What switch turned them on and off? It soon became clear that whoever solved the riddle of the ice ages would achieve lasting fame. [7]

While most geologists were busy looking at the Earth, two European scientists looked to the sky. Joseph Adhémar, a mathematician who made his living as a tutor in Paris, suggested that the precession of the equinoxes was the trigger-switch for an ice age. Precession affects the amount of solar radiation reaching the Earth's surface, and is caused by the Earth's axis of rotation wobbling like a slowly spinning top. As a result, the North Pole sweeps a roughly 24-degree circle in space. Today the North Star is the point around which the stars rotate, whereas 4,000 years ago the North Pole pointed to a spot midway between the Little Dipper and the Big Dipper. The precession cycle is very slow, averaging around 22,000 years before the Earth's axis returns to the same point in the circle. Adhémar believed that this precession cycle drives alternating 11,000-year hemispheric ice ages. His theory was soon shot down by Baron Alexander von Humbolt, who cited recent proofs that both hemispheres receive the same amount of heat each year. [8]

Twenty-five years later, Adhémar's theory was discovered and revised by James Croll – a most unlikely scientific visionary. Born in 1821 in a village in Scotland, Croll received little formal education. When he was 11, he came upon a copy of *Penny Magazine*, a British weekly aimed at disseminating useful knowledge to the working class. The magazine soon failed, but Croll's penny was well spent as he embarked on a rigorous course of self-education. [9]

At age 16, this brilliant farm boy became an apprentice wheelwright, followed over the next two decades by working as a carpenter, tea merchant, innkeeper, and insurance salesman. When he was 38 years old, Croll became the janitor at the Andersonian Museum in Glasgow. To his mind, the subsistence-level pay was amply compensated by having access to a fine scientific library and "a good deal of spare time for study." [1]

In a groundbreaking paper in 1864, Croll presented his astronomical theory of climate change. Since Kepler, it had been known that the Earth does not revolve in a perfect circle but in an elliptical orbit. Through his tenacious library reading, Croll was familiar with the recent calculations by the great French astronomer Urbain Leverrier, who demonstrated that the shape of the ellipse is constantly changing. Over the past

100,000 years, Leverrier discovered, the degree of eccentricity has varied from near zero (a circle) to about six percent.

While Earth's total heat budget each year is unaffected by changes in eccentricity, Croll discovered that *seasonal* solar radiation is strongly affected by eccentricity. Croll reasoned the critical season must be winter because that's when snow accumulates. Yet, this few degrees difference in eccentricity still didn't explain it to his satisfaction. Adhémar must have been correct – the precession of the equinoxes also plays a decisive role. Croll theorized that during the long periods of high eccentricity, when the effects of precession would be intensified, the precessional cycle results in ice ages alternating between the southern and northern hemispheres. [9]

Yet, these astronomical vagaries couldn't be the complete explanation. How could such subtle orbital changes bring on something as extreme as an ice age? With this concession, Croll broke new ground as an original thinker. James Croll was the first person to come up with the idea of *amplifying effects*, where a minor change in the orbital parameters triggers a major response in Earth's climate. He reasoned that the first amplifying effect would be sunlight reflecting off the snow, immediately causing colder temperatures and allowing more snow to accumulate. But wouldn't there be other, much slower, amplifying effects? He concluded that more snow at one of the poles means colder temperatures, eventually causing stronger trade winds that force the warm equatorial currents to shift towards the opposite hemisphere. This shift in ocean currents would result in a much greater heat loss at the pole, allowing a lot more snow to accumulate.

Croll's astronomical theory became a scientific *pièce de résistance*. Finally, here was a brilliantly reasoned explanation that could be tested against the geologic record. But there were some serious discrepancies. Croll had predicted glaciation in only one hemisphere at a time. There was little evidence to support this idea. It was finally Niagara Falls that brought Croll's theory crashing down. The Niagara River, which had formed concurrently with a glacial deposit, was eroding the lip of the falls at the astounding rate of about 3 feet (1 m) per year. By measuring the length of the spectacular gorge, geologists calculated that the glacial deposit had formed about 10,000 years ago. This was a far cry from Croll's prediction that the last ice age had ended 80,000 years ago. By the end of the century, it had become obvious that Croll's theory did not explain many of the geological facts. [10]

In the early 1900s, the Serbian mathematician Milutin Milankovitch took another look at Croll's astronomical theory. After working as an engineer in Vienna and obtaining several patents related to building with reinforced concrete, Milankovitch was offered a Chair of Applied Mathematics at the University of Belgrade. After settling into his new profession, he began to look around for a challenging scientific research topic. But there was a considerable obstacle. Being far from the scientific centers of Europe with their state-of-the-art libraries and budgets, competing on cutting-edge topics would put him at a severe disadvantage. Milankovitch decided to search for a problem in which he could use his mathematical abilities, and discovered the topic of ice ages. He would develop a mathematical theory capable of describing the Earth's climate, today *and* in the past. Here was an opportunity to make his mark. [11]

Milankovitch began with the fact that three orbital properties determine how the Sun's radiation is distributed over the Earth's surface – the 22,000-year precessional cycle, the 100,000-year cycle of eccentricity, and the tilt of the axis of rotation. While James Croll had been lucky in having Leverrier's eccentricity calculations for the past 100,000 years, Milankovitch inherited a similar goldmine. Detailed calculations had now been made for all three orbital parameters over the past one million years. Milankovitch decided to start with the present. If he could calculate the distribution of solar radiation as it is today, it would be possible to calculate past climates when the Earth's wobbling, tilting, and eccentric orbits were different. [10]

In the midst of these laborious calculations, World War I broke out and Milankovitch became a prisoner-of-war of the Austrian–Hungarian army. When a Hungarian professor learned that the talented Serbian mathematician had been imprisoned, he petitioned for his release in the "interest of science." Milankovitch was moved to Budapest where he worked completely undisturbed in the top-notch library of the Hungarian Academy of Science. By war's end he had achieved his goal of calculating, for each season and latitude, the amount of radiation received over the entire Earth. Meteorologists soon embraced the work as a major contribution to the study of modern climate. [11]

Completing this first step toward unraveling the ice-age riddle brought him another boon. The great German climatologist Wladimir Köppen and his son-in-law, Alfred Wegener, who had made his mark with his theory of continental drift, were working on a book about *paleoclimate* (climates of the geologic past). Upon discovering Milankovitch's climate studies, they invited him to collaborate. Milankovitch could hardly have found two people better equipped to help steer him into the complex labyrinth of paleoclimate.

The next problem was to figure out which latitude and season were critical to the growth of an ice sheet. Adhémar and Croll had thought the critical season was winter. Colder temperatures obviously meant more snow accumulating at high latitudes. Köppen suggested the opposite – the critical season was summer. Modern glaciers melt during the summer, and therefore, colder summers would inhibit melting and lead to glacial expansion. In addition, the *sensitive zone* most affected by a decrease in summer solar radiation would be where all that glacial debris had piled up – between the latitudes of 55 and 65 degrees north. Over the next three months, Milankovitch worked from morning until night calculating summer radiation at these latitudes for the past 650,000 years. He mailed a graph of his results to Köppen. After comparing Milankovitch's calculations with the geological evidence for the timing of past ice ages, Köppen wrote back that there was a good match. Convinced he was on the right track, Milankovitch continued work on his theory. In 1938, after a quarter-century of mathematical labor, Milutin Milankovitch put the final touches on his astronomical theory of the ice ages.

The Milankovitch theory was a profound breakthrough into understanding paleoclimate. The theory demonstrated that ice ages are triggered by astronomical variations in Earth's orbital properties, with the 41,000-year tilt cycle dominating. The theory also identified the critical season and latitudes when a decrease in solar radiation triggers glacial expansion. Amongst the theory's most important features were Milankovitch's *radiation curves*, demonstrating that not just four, but many ice ages had occurred over the past 650,000 years.

Milankovitch had solved his "cosmic problem." Now all scientists had to do was prove it with the geologic record. As it turned out, validating the Milankovitch theory would involve some of the most intriguing and difficult challenges in all of scientific history.

THE BIG DIG

Geologists began at the same old place, digging through glacial deposits and trying to figure out how many layers there were. The problem was that each glacial advance wreaked havoc with the previous deposits. When it became obvious that the land surface was not the place to be looking for a record of paleoclimate, scientists turned their sights to the ocean floor.

In Croll's lifetime, scientists knew more about the surface of the moon (amounting to precious little) than they did about the ocean depths. This would soon change. In 1872, the British government financed a round-the-world voyage of ocean study that lasted over three years. The expedition's six scientists took soundings, collected water samples and dredged the bottom at all depths. [8]

Their findings were fascinating. On the continental margins, the ocean floor was covered with sand, mud, and fragments of plants and other materials transported there by rivers, redistributed by currents, and resulting in a great big mix of well-churned muck. But away from the continental margins, and covering such a vast area of seafloor that it equaled all the continents combined, they dredged up fine-textured ooze that was largely composed of plankton fossils, collectively known as *foraminifera*. Over the eons, a steady rain of forams had built up into a deep and undisturbed sequence of sediments.

It wasn't until after World War II that this discovery was put to brilliant use by the Nobel laureate Harold C. Urey at the University of Chicago. Working on finding geochemical answers to some of the fundamental mysteries of Earth history, Urey came up with the idea of using oxygen isotopes as a proxy thermometer for reading ocean temperatures of the past.

Ocean water contains two distinct oxygen isotopes – the heavier oxygen-18 and the lighter oxygen-16. While both these isotopes are present in the fossilized skeletons of foraminifera, the colder the water, the more oxygen-18 in the skeleton. Urey proposed that it would be possible to calculate water temperature during the organism's lifetime by measuring the ratio of these two isotopes. However, in order to unlock these secrets of the dead, they needed long seafloor cores.

Ever since the British expedition's findings a half-century earlier, scientists had been slipping and slogging about on the wet decks of heaving and freezing ships trying to obtain sections of that tantalizing sediment pile way down below. This was no easy feat. The first attempts involved lowering a long steel pipe until it hung above the ocean floor, then releasing it and letting the force of gravity carry it into the muck below. These "gravity corers" retrieved cores about a meter long, giving them only a peek at paleoclimate. They next tried attaching lead weights to the pipe, but the extra friction with the water basically canceled out the extra weight. At one point, they experimented with using dynamite to blast the pipe deeper into the sediment, but (to no one's surprise) the resulting core was badly damaged. In the 1940s, the problem was finally solved by a Swedish oceanographer, Börge Kullenberg, who developed a core pipe with a piston that could suck up undisturbed cores 10 to 15 meters long. [10]

With a few longer cores becoming available, Urey formed a team whose formidable mission was to develop the instrumentation and procedures for accurately measuring the extremely delicate isotopic ratios. After several years of work, Cesare Emiliani, a young Italian post-grad, arrived on the scene. Urey realized that Emiliani's fossil knowledge would be extremely useful for applying their new technique on foraminifera. In 1955, after five daunting years spent analyzing eight cores, Emiliani published his results. His pioneering work proved to be a landmark in the study of the ice ages, with some highly unanticipated findings. Over the past 300,000 years, there had been *seven* ice ages. His results also showed a "reasonably good" correspondence with the Milankovitch radiation curves. [12]

Appetites were whetted for studying more cores, longer cores, cores from all over the world – and by this time, Doc was bringing them in. Maurice "Doc" Ewing, a Texas farm boy, became the indisputable leader of marine geology for more than three decades. As a senior in college, his first research paper was published in *Science*. After finishing his Ph.D. in geophysics in 1930, he taught at Lehigh University, a small engineering school in Pennsylvania coal country. He eventually moved to Woods Hole Oceanographic Institution on Cape Cod, where he began his lifelong passion of studying the structure, composition, and origin of ocean basins. [13]

With marine geology still in its infancy and many of the tools yet to be developed, Doc's can-do common sense often came to the rescue. At one point he had to figure out how to safely put geophones underwater, and solved the problem by sealing them in rubber condoms. Most solutions, however, were far from that simple. Doc invented (or greatly improved) many instruments and techniques for gathering geological and geophysical data at sea – the bathythermograph, heat-flow probes, sonar, hydrophones, gravimeters, deep-sea cameras, and a much improved piston corer that allowed him to drill cores faster and much deeper. [13]

When Doc moved to Columbia University, he took his precious cores with him and started the institution for which he is legendary – the Lamont–Doherty Earth Observatory (formerly the Lamont Doherty Geologic Observatory). From its beginning in the early 1950s, Lamont quickly developed into a world-renowned center for oceanographic and geophysical research.

Doc became known for his ability to coax, wheedle and, if all else failed, *bully* to get the money he needed for equipment – including an ocean-worthy research ship. In 1953 he bought the Vema, a retired

Figure 16.1 Maurice 'Doc' Ewing on the deck of Atlantis. Courtesy Woods Hole Oceanographic Institution Archives.

luxury yacht, and headed out on the first of more than 50 missions, circling the globe and spending more than 300 days annually at sea, year after year. Doc's philosophy was simple – get as many cores as possible from as many places as possible, collect as much data as possible, and document everything. By the 1960s, Doc had two ships circling the globe and competing to see who could take the most cores, the most northerly and southerly cores, the deepest-water cores, and the longest cores. Thanks to Doc Ewing, Lamont's "core library" soon became the largest and best in the world, bar none. [14]

THE PACEMAKER OF THE ICE AGES

Scientists now had more cores than they knew what to do with, but the three pivotal ice-age questions remained: where, when, and why? The brainstorming and sleuthing continued, often down false trails. Here and there a spark of interest in the Milankovitch theory still flickered, but the fire was long dead.

A turnaround began in the 1960s when Lamont geochemist Wallace (Wally) Broecker and collaborators examined ancient coral-reef terraces in the Bahamas and Barbados. Broecker developed a way to age-date the coral reefs using uranium-series "clocks" that had been incorporated in the skeletons of corals. When he discovered that the coral reefs had age-dates consistent with times of minimum ice cover and high sea level, predicted by the Milankovitch theory, Broecker jubilantly announced, "The often-discredited hypothesis of Milankovitch must be recognized as the number-one contender in the climatic sweepstakes." [15]

The turnaround continued as the result of a chance encounter in September 1969. John Imbrie, a skilled mathematical geologist from Brown University, was invited by Emiliani to present his findings at an international scientific meeting in Paris. Imbrie had developed a mathematical technique to estimate the history of ocean temperature based on the foram species found in cores. Unavoidably, Imbrie was late and his lecture was rescheduled for four o'clock on Friday. On a beautiful afternoon in Paris at the end of the work-week, even the most dedicated scientists had been lured away from the lecture hall by the city's charming distractions. Imbrie ended up speaking to an audience of two. Half of his audience spoke no English; the other half was Nicholas Shackleton. [8]

Nick Shackleton, great-nephew of the Antarctic explorer Ernest Shackleton, would become a highly regarded scientific explorer in the field of paleoceanography. From his lifelong laboratory at Cambridge, Shackleton made early groundbreaking improvements to techniques to measure oxygen-18 in deep-sea fossils. He would later win the Vetlesen Prize, the Nobel Prize of the earth sciences. Among the other two dozen winners are Wally Broecker, John Imbrie, and the first recipient, Doc Ewing.

When Shackleton met Imbrie in Paris, they discovered they had a lot in common. Working completely independent from each other, and through separate lines of research, both scientists had come to the same tentative conclusion – oxygen-18 variations did not measure the history of ocean temperature (as suggested by Emiliani) but, rather, *ice volume*. They both reasoned that when ice sheets form, they take more of the lighter oxygen-16 isotope from the ocean and store it as ice, leaving oxygen-18 behind. This seemingly subtle distinction had major implications, in that any core in the world would hold the same paleoclimate record and not be affected by local variations in temperature.

The following spring Imbrie met with James Hays, who was studying Antarctic surface-dwelling foraminifera at Lamont. Hays laid out a proposal that would soon bring John Imbrie and Nick Shackleton onboard in making scientific history. Hays was convinced that reconstructing Pleistocene ocean history was too big a job for any single scientist or institution. They needed an organization that would coordinate all the independent laboratory efforts to solve the ice-age problem. Imbrie agreed. Within a year the ambitious CLIMAP project had been launched, eventually involving nearly 100 scientists from eight countries. Five years later, the group published a worldwide map showing the temperatures of the ocean and the distribution of glaciers at the height of the last ice age.

The project's second goal was to measure the ice-age oscillations of Pleistocene climate. Scientists had only to count the oxygen-18 isotope highs and lows in a given core to determine how many ice ages had come and gone. Yet, actually *dating* these ice ages was a much more daunting problem. The breakthrough in carbon-14 dating in the 1950s enabled scientists to date the top 40,000 years of a core, but the highly maddening problem was they had no way of dating the bottom of the core or, for that matter, any section older than 40,000 years.

The first key that eventually unlocked the chronology of ocean cores had been found in 1906, in a French brickyard. Bernard Brunhes, a geophysicist studying the Earth's magnetic field, discovered that as newly baked bricks cool, the iron-rich particles align themselves parallel to the direction of the Earth's magnetic field. Bruhnes then discovered that cooling lava flows also acquire the direction of the Earth's magnetic field. After measuring the direction of magnetization in several ancient lava flows, he discovered, to his utter amazement, that some of the flows were magnetized in the opposite direction from the present magnetic field – the North Pole was transformed into the South Pole, and vice versa. Brunhes concluded that, sometime in the past, the Earth's magnetic field must have been reversed. The idea seemed so preposterous that almost no one believed him. [8]

In the 1920s, Motonori Matuyama, a Japanese geophysicist, found evidence that Brunhes was correct. After studying lava flows in Japan and Korea, Matuyama concluded that the Earth's magnetic field had flipped at least once during the Pleistocene. Subsequent investigation convinced him that magnetic-field reversals had occurred many times during geological epochs older than the Pleistocene. To the scientific community, if one reversal had seemed preposterous, *many* was too ridiculous to even consider.

It wasn't until 1963 that the magnetic-reversal hypothesis was finally confirmed by Allan Cox and Richard Doell of the US Geological Survey, and by G. Brent Dalrymple of the University of California at Berkeley. To honor the memory of their pioneering colleagues, they agreed to name the current epoch of "normal" polarity the "Brunhes Epoch," and the earlier epoch of reversed polarity the "Matuyama Epoch."

This magnetic reversal could be found in ancient iron-rich rocks all over the world. Using potassium–argon age-dating, the Brunhes–Matuyama boundary was dated at around 730,000 years ago (later revised to 780,000 years). But would this signal be recorded in ocean cores? As early as 1956, Doc and a colleague attempted to find out, but after several attempts they abandoned the effort. A decade later, Neil Opdyke at Lamont developed a technique for analyzing the paleomagnetism of soft core sediments.

The next challenge was to find a core long enough that it recorded the Brunhes–Matuyama reversal. A CLIMAP core-reconnaissance team was formed to find this Rosetta Stone. In December 1971, they discovered a core from the western equatorial Pacific Ocean that looked long enough. When Opdyke analyzed the core, he found the Brunhes–Matuyama boundary 12 meters below the top of the core. Hays immediately sent samples of the core to Shackleton for isotopic analysis.

In June 1972, Shackleton presented his results to a breathless group at Lamont. He demonstrated that the Brunhes Epoch could be divided into 19 isotopic stages. With dates firmly fixed at both ends of the core – by a magnetic reversal at the bottom and by radiocarbon dates at the top – and by assuming a steady rate of sedimentation, the age of each isotopic stage could be interpolated. Shackleton also clearly established that the oxygen-18 variations were a climate proxy for ice volume.

With the isotopic record finally dated, Hays, Imbrie, and Shackleton decided to see if the isotopic curve corresponded with the 22,000-year precession and 41,000-year tilt cycles predicted by Milankovitch. But there was yet another problem. A 100,000-year cycle (which may have been eccentricity, but wasn't a major part of the Milankovitch theory) was so dominant that they couldn't find the "higher" frequencies. After much head scratching, Hays suggested that the cores had accumulated too slowly (1–2 cm per 1,000 years) to register the higher-frequency precession and tilt cycles. The search was on for a core that had an accumulation rate of greater than 3 cm per 1,000 years, and went back at least half a million years to cover several glacial cycles.

With virtually thousands of cores available at Lamont and elsewhere, it was like searching for a needle in a haystack. Nearly a year later, Hays found a core from the southern Indian Ocean that seemed to have a higher accumulation rate. Unfortunately, the core extended back only about 300,000 years. After more searching, Hays found a similar core from a nearby site. This core was missing the early part of the record, but the two cores could be analytically spliced together to yield a continuous record spanning 450,000 years. Finally, they were ready to do the analysis.

Imbrie's spectral analysis showed cycles remarkably consistent with the three orbital parameters. The dominant cycle was a period of 100,000 years which they attributed to eccentricity. The second spectral peak of 42,000 years nearly matched the 41,000-year cycles of the tilt of the Earth's axis. The third peak had a period of 23,000 years, close to the precession period. On December 10, 1976, Hays, Imbrie, and Shackleton announced their findings in a 12-page article in *Science*. They boldly summarized, "It is concluded that changes in the earth's orbital geometry are the fundamental cause of the succession of Quaternary ice ages." A half-century later, Milutin Milankovitch had finally arrived. [16]

In spite of the excitement, several key problems remained. The Milankovitch theory predicted that the tilt cycle of 41,000 years should dominate, yet it was only half as strong as the 100,000-year cycle. Another conundrum is that the 100,000-year cycle became dominant only over the past one million years. Prior to this time, the 41,000-year tilt period appears as the dominant cycle, with the 100,000-year cycle either absent or very weak. An even bigger question was still unexplained – why did Earth enter a period of ice ages some 2.6 million years ago?

Hays' group had estimated the dates of the deep-sea core layers by transferring information from the "Rosetta Stone." Yet, a foolproof confirmation of the Milankovitch theory still required a more reliable chronology of the core layers. Just 12 years later, such a chronology would come from a hole in the Earth less than 30 miles (50 km) southeast of Yucca Mountain.

DEVILS HOLE

Devils Hole is an open fissure formed by faulting, located at the discharge end of a regional aquifer in Death Valley National Park. About 50 feet (15 m) below land surface is a warm water pool. To the casual observer the pool appears unremarkable, yet to divers it is the entryway to a deep labyrinth of subterranean chambers and narrow passages. The maximum depth is unknown, but greater than 300 feet (90 m). In 1965, two divers

Figure 16.2 Left to right, USGS scientists Wil Carr, Barney Szabo, and Ike Winograd in Death Valley National Park in 1977. Background shows 1 to 2 million-year-old calcite veins. These veins of groundwater origin led directly to study of younger veins in Devils Hole. Photograph courtesy US Geological Survey.

disappeared and were never found, despite round-the-clock search efforts by 44 divers over a three-day period. [17]

Over time, Devils Hole became recognized as a unique natural laboratory for studies ranging from hydrogeology to evolutionary biology. The "Hole" is home to the endangered Devils Hole pupfish, so named because the small fish dart about like playful puppies, pecking at the substrate or chasing after one another. The fish evolved from Pleistocene-age ancestral stock and their habitat is vigorously protected. In the late 1960s, pumping for irrigation lowered the water level in Devils Hole, threatening to expose the narrow, submerged shelf on which the pupfish feed and breed. The courts ordered pumping to cease and the water levels subsequently recovered.

In the early 1980s, Ike Winograd began geochemical investigations with USGS scientists Alan Riggs and Ray Hoffman. Riggs and Hoffman made dozens of dives to collect samples and to survey the subterranean environment. Ike soon recognized Devils Hole as a potential paleoclimate bonanza. For hundreds of thousands of years, water moving through Devils Hole deposited a small amount of calcite along the walls, eventually building up a layer about a foot thick. In 1982, Riggs and Hoffman painstakingly removed a large block of calcite from the wall for oxygen-18 analysis. The hope was that it would provide a long-term record of global paleoclimate. The results exceeded expectations. The 1982 specimen provided a 250,000-year paleoclimate record. A few years later, a 40 cm-long subsurface core extended back 500,000 years. Using uranium-series dating, Winograd and his team were able to date these cores with unprecedented precision. [18–19]

When their results were published in *Science* in 1988 and in 1992, the Milankovitch theorists were jarred from their complacency. The Devils Hole oxygen-18 record nearly matched the results from deep-sea cores, but the *timing* was significantly different. The Devils Hole record showed the penultimate ice age ending *before* the Milankovitch theory predicted increases in solar radiation that would trigger such mass melting. The Hole also recorded a well-developed glacial–interglacial cycle 450,000 to 350,000 years ago; when the Milankovitch theory indicated none should occur. And if this wasn't enough to upset the apple cart, Devils Hole also showed an unexplained increase in the length of glacial cycles as the record approaches the present day. Finally, Devils Hole calcite records indicated that the duration of the last interglacial was about twice as long as suggested by the deep-sea cores. Suddenly, the Devils Hole record threatened to derail the Milankovitch theory.

Proponents of the Milankovitch theory were not about to take this lying down. Critics suggested that the Devils Hole record was predominantly a regional record, and therefore of limited global significance. Ike and his team were invited to Lamont and Brown University where they were drilled about every detail, making defense of a Ph.D. dissertation seem like a walk in the park.

The Devils Hole record was later shown to be a proxy of Pacific Ocean sea-surface temperature off the coast of California. However, similar sea-surface temperature increases that predate deglaciation also appeared in other records, in both the northern and southern hemispheres. Perhaps the warming energized the global hydrologic cycle, augmenting the transport of moisture to high latitudes and the growth of ice sheets. The bottom line was that the link between the ice ages and solar insolation remained unclear.

When the second *Science* paper by the Winograd team was published in 1992, Wally Broecker (the scientific big-gun who two decades earlier had declared the Milankovitch theory "the number-one contender in the climatic sweepstakes") acknowledged that the Devils Hole chronology was the best we have. With his usual flair for the dramatic, Broecker warned, "just to be safe – climate modelers should start preparing themselves for a world without Milankovitch." [20–21]

AN UNSOLVED MYSTERY

James Croll, Milutin Milankovitch, Doc Ewing, Cesare Emiliani, Wally Broecker, John Imbrie, James Hays, Nick Shackleton, and Ike Winograd all had the willpower and intellectual stamina required to attack a "cosmic problem." Numerous other scientists, many of considerable stature, have approached the ice-age problem from various angles. For months at a time, scientists and technicians have lived at some of the coldest places on Earth to extract ice cores containing paleoclimatic records. The more they looked, the more complicated the explanation became. Today, most scientists believe that orbital variations play a key role in the coming and going of ice ages, yet it is now clear that other factors – such as ocean and wind currents and the natural cycling of carbon dioxide and methane – are of considerable, if not greater, importance. There remain many unknowns. As one prominent paleoclimatologist noted, "the sheer number of explanations for the 100,000-year cycle and for carbon dioxide changes seem to have dulled the scientific community into a semipermanent state of wariness about accepting any particular explanation." [22]

The ice ages are the most fundamental events to have happened on the planet over the past one million years. No other natural phenomenon comes even close. Yet, after 150 years of intense study by some of the most brilliant scientific minds of their time, the cause of the ice ages remains a mystery.

Projecting climate forward is even more difficult than looking into the past. When paleoclimatologists gathered in 1972 to discuss how and when the present warm period would end, a slide into the next glacial seemed imminent (within a thousand years or so). Today, scientists believe we may be in an unusually long interglacial, one that may last another 10,000–50,000 years or more. There are natural reasons to explain this, as well as the effects of burning fossil fuels and other anthropogenic disturbances. [23–24]

Beyond the "short-term" effects of greenhouse-gas induced climate change, the onset of the next ice age is much more than an academic question. The Earth's long-term climate will play a critical role in our ability to contain high-level nuclear waste over the eons. In the final analysis, the inability to adequately explain the ice ages also serves as a reminder of human limitations in predicting the far-off future. Within this context, the regulatory requirement to predict much more subtle phenomena one million years into the future (even on a probabilistic basis) seems like pure folly. Nevertheless, in 2004, that was the court's ruling for Yucca Mountain. Apparently, no one stood up and said, "You have got to be *kidding*?"

Leaving almost no stone unturned

You never feel quite as comfortable about a site

as the day you start to study it.

Wendell Weart [1]

On June 4, 2008, at 8:40 a.m., a moving truck pulled up to the Nuclear Regulatory Commission and dropped off 15 sets of the Department of Energy's license application to construct a nuclear waste repository at Yucca Mountain. Each set weighed 110 pounds (50 kg) and totaled 8,600 pages. It was more than a quarter-century since work began at Yucca Mountain and a decade after the federal government had promised the nuclear industry an operational repository for their wastes. This delivery marked an important milestone, yet it would be at least another decade until the repository opened – perhaps never, if Nevada had its way. [2]

Comprising hundreds of studies, the license application was the most complex application ever completed. Never before had mere mortals been assigned the task of making scientific predictions spanning such geologic timeframes.

TOTAL SYSTEM PERFORMANCE ASSESSMENT

A good part of the case presented to the NRC was based on a *Total System Performance Assessment*, or TSPA. Using a series of linked computer models, the TSPA strove to provide quantitative answers to three basic questions: What might happen in the future to the waste? How likely are different scenarios? What are their consequences? The TSPA was the grail of this difficult and extended quest. The effort was led by Sandia National Laboratories, world leaders in performance assessment.

The first challenge in undertaking the TSPA was to make a list of every conceivable factor that might affect containment of the waste over the next million years. In regulatory lingo, these are called "features, events, and processes," or FEPs for short. *Features* are the physical components of the natural setting and engineered system. *Events* include earthquakes, volcanic eruptions, and human intrusion. *Processes* are the day-to-day phenomena that affect the waste, such as moisture flow and corrosion. While no approach can ensure that all possibilities have been considered, every effort was made to list any and everything that might be relevant. The inventory began with an international database of more than 1,200 FEPs developed from brainstorming by nuclear programs worldwide. This list was supplemented with several hundred FEPs specific to the Yucca Mountain Project. [3]

Not every FEP had to be considered. Those that are sufficiently unlikely or their consequences unimportant can be ignored. On this basis, meteor impacts and erosion of the mountain to the repository level were screened out from the TSPA. Acts of war and sabotage were automatically excluded. These possibilities are impossible to predict and are present for every alternative, including where the waste now resides. Screening out events and processes from inclusion in the TSPA was not without controversy. One event screened out – flooding of the repository by a rising water table – would prove particularly contentious.

Once the FEPs were nailed down, the next step was to analyze how each FEP, or combination thereof, could play out in different scenarios. Developing the scientific basis and computer models for analysis of these scenarios made up the lion's share of the work on the TSPA. In the final assessment, four "scenario classes" were evaluated. A *nominal* scenario class evaluated the events and processes likely to occur. An *early failure* scenario class examined the possibility of waste packages failing from factors such as defective welds. An *igneous* scenario class examined volcanic intrusion or eruption. A *seismic* scenario class examined earthquakes. A separate analysis looked at the possible effects of future human intrusion by an exploratory drill hole through the repository. [4]

The nominal analysis "follows the water." How much water gets into the mountain, where does it go, how fast does it get there, and what are its temperature and chemical composition along the way? The analysis began by estimating the amount and distribution of precipitation, and how this might be affected by climate change. Most of the precipitation runs off or is returned to the atmosphere through evapotranspiration. The rest percolates into the mountain. Eventually, some of this water reaches the repository level, about 1,000 feet (300 m) down.

If the rate of percolation is sufficiently high, water will seep into the repository tunnels. [5]

Seepage into the tunnels would have two important consequences. First, water accelerates waste package degradation through corrosion, particularly if it seeps directly onto the waste packages. Second, water is the vehicle that will transport radionuclides to the water table another 1,000 feet beneath the repository. Once in the saturated zone, the water and contaminants begin to move horizontally as groundwater flows underneath and away from the mountain. Groundwater would dilute the concentration of radionuclides, but also may allow the contaminants access to the biosphere. This can occur naturally at springs and seeps, or by humans pumping wells for water supply or irrigation.

There is no doubt that these things will happen. What matters is how fast they occur. How fast does the water move through the unsaturated zone and into the repository tunnels? How fast do the canisters and spent fuel cladding corrode? How quickly are the exposed radionuclides mobilized by the available water? How fast does the now contaminated water move to the water table below and then beyond Yucca Mountain? How well do the natural and engineered barriers serve to retard the rates of these processes? To answer these questions, the TSPA brought together a staggering amount of information on the physical, chemical, thermal, mechanical, hydrologic, and geologic processes in play at Yucca Mountain. [5]

The TSPA was not for the faint of heart, requiring almost two decades and hundreds of people to complete. It was built by stringing together computer models of widely differing levels of detail. Each model (climate change, unsaturated zone flow, seepage into the tunnels, spent fuel corrosion, etc.) represented a major effort and challenge to represent processes ranging from molecular to regional scales. The output of one model served as input to others. These linkages were not necessarily straightforward, often involving so-called "coupled" processes that act in combination with each other. A good example is the coupling of unsaturated zone flow and temperature near the waste packages: heating the mountain near the waste will alter flow in the unsaturated zone. Concurrently, the amount of unsaturated zone flow will affect the temperature.

The TSPA models are highly simplified descriptions of the real world, yet collectively required thousands of parameters to describe the properties of the modeled systems. Examples of model parameters are the permeability (ease of water movement) in different parts of the mountain, and the properties affecting corrosion of the waste containers.

Incomplete knowledge and natural variability result in considerable uncertainty in the model parameters. Even if the mountain could be disassembled piece-by-piece, the physical, chemical, and biological processes would remain uncertain. This uncertainty does not necessarily translate to significant risks, yet it does mean that a range of results are possible. Performance assessment helps determine which of the uncertainties are of most concern.

To address the uncertainties, scientists on the Yucca Mountain Project used a probabilistic method developed by Stanislaw Ulam and others working at Los Alamos in the 1940s. The method was code named *Monte Carlo simulation* after the famous casino in Monaco – a casino frequented by Ulam's uncle. Instead of specifying a single value for each model parameter, Monte Carlo simulation assigns a probability distribution. Each model parameter thus has a range of possible values from which to select.

Monte Carlo simulation has an interesting origin. Stanislaw Ulam was a brilliant mathematician and an avid solitaire player. In 1946, while convalescing from an illness, Ulam wondered what the chances were that a game of solitaire will end successfully. After spending a lot of time trying to estimate the odds by using complex combinatorial calculations, he decided a more practical method might be to simply play the game 100 times and count the number of successful plays. From there, Ulam recognized the possibilities for solving much more complex problems using random sampling with the help of a computer. Ulam described the idea to fellow mathematician, John von Neumann. The world's first computer had just been built at the University of Pennsylvania, where von Neumann applied the Monte Carlo technique to the problem of estimating neutron diffusion in fissionable material. This complex problem involves following a large number of neutron chains as they undergo scattering, absorption, fission, and escape. At each stage, a sequence of decisions has to be made based on statistical probabilities. [6]

Today, Monte Carlo simulation is commonly used in computer modeling of complex systems in many fields. As applied to the TSPA for Yucca Mountain, Monte Carlo simulation involves a marathon stint at the table. First, a value for each model parameter is obtained by random sampling from the parameter's probability distribution – akin to a computerized "rolling of the dice." The TSPA models are run using these values to simulate repository performance over the next million years. The result is but one plausible outcome given the range of possible model parameter values. The computer rolls the dice again to obtain a second set of values for the model parameters and generates another million-year simulation. The process repeats itself hundreds of times. The final collection of model simulations is then analyzed statistically to obtain a probability distribution for the outcomes.

Monte Carlo simulation provides a way to translate uncertainties in the model parameters to uncertainty in the TSPA predictions. That's the theory, but there are major challenges. The approach assumes the models capture the essential features of the system and that the uncertainty in model parameters can be adequately quantified.

The TSPA models estimated radiation dose rates over the next million years for a hypothetical person living near Yucca Mountain. These dose rates were compared to regulatory limits, thereby assigning the equivalent of a pass/fail grade for the repository. If the calculated dose rate stayed below the regulatory limits, a passing grade was assigned. If not, the repository received a failing grade. The basic idea was simple, but the devil was in the details. Who is this person? Where does he or she live? How much radiation exposure is acceptable? The answers came through a long, convoluted process.

THE EPA STANDARDS

In 1985, the Environmental Protection Agency set regulatory standards applying to any high-level waste repository. Among these standards, radioactivity from a repository should cause no more than 1,000 deaths over the next 10,000 years – on average one death every ten years. This standard was very high, matching the estimated number of deaths by people who would have been exposed to the uranium used for nuclear power *if it had never been mined*. The standards also specified the allowable rate of radionuclide release from the repository, and the minimum travel time for groundwater flow from the repository to the accessible environment. These standards were challenged and overturned by the courts. The EPA was back to base one. [7]

Seven years passed. Frustrated with the delay, Congress took matters into its own hands. The Energy Policy Act of 1992 mandated that standards be set specifically for Yucca Mountain and should follow recommendations by the National Academy of Sciences (NAS). In 1995, an NAS committee concluded that a standard for Yucca Mountain should be based on the risk to the average member of a *critical (high-risk) group*. Given that it's impossible to predict human activities and lifestyles far into the future, the committee recommended that the definition of this group should be based on an extension of current conditions. Highly speculative future scenarios should be avoided. The hypothetical individual representing this group came to be known in the EPA regulations as the *reasonably maximally exposed individual*, or RMEI. [8]

The RMEI lives 12 miles (20 km) from Yucca Mountain in the direction of groundwater flow, the closest downgradient region in which significant farming is practised today. He or she has a diet and lifestyle representative of today's residents of the town of Amargosa Valley, Nevada. The RMEI is potentially exposed to radiation from several sources: drinking water from a contaminated well; eating local crops grown with contaminated well water; consuming meat and milk from farm animals raised on contaminated water and crops; and breathing contaminated air. The RMEI has no idea that any of these bad things are happening, so they take no precautions.

Theoretically, RMEI could have been placed closer to Yucca Mountain. However, the harsh terrain and deep groundwater closer to the mountain make it cost-prohibitive to irrigate crops using today's technology. And so RMEI stayed put. The State of Nevada challenged the placement of RMEI as *arbitrary and capricious* but the courts upheld EPA, who "need not *prove* that humans will never settle [closer]; the agency needs only a reasonable basis for believing that they are unlikely to do so." [9]

With the eventual loss of institutional control it is always possible that, for some unforeseen reason, humans might drill into the repository. To assess the resilience of the repository, the NAS recommended that the consequences of human intrusion be calculated, but advised against setting a quantitative standard for human intrusion – it's simply impossible to determine the odds of such an event.

While the NAS committee recognized that predicting the behavior of human society over extremely long periods is beyond the reach of science, it took a different tack for long-term predictions of repository performance, concluding that compliance with the standard should be evaluated for the time of peak risk. The only limit on the timeframe was the long-term stability of the geologic regime. According to the NAS committee, this was on the order of one million years. The committee acknowledged that any standard, including the timeframe for compliance, should be a policy decision by the EPA, informed by science. [8]

The greatest risk at Yucca Mountain would not come from the shorter-lived radionuclides, like cesium-137 and strontium-90, which initially make the waste hot and highly radioactive. The longer-lived radionuclides, like plutonium and neptunium, would gradually work their way to the accessible environment and begin to peak in RMEI's neighborhood several hundred thousand years in the future.

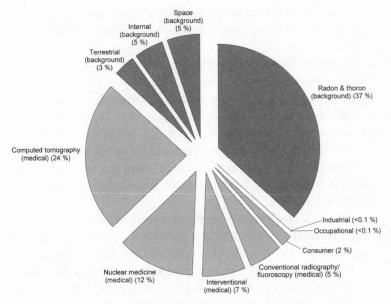

Figure 17.1 The average American is exposed to 620 millirem of radiation each year. About half of this exposure comes from natural sources (primarily radon gas), while the other half comes from human-made sources (primarily medical procedures). Reprinted with permission of the National Council on Radiation Protection and Measurements, http://NCRPpublications.org.

The EPA chose to stick with the 10,000-year timeframe for measuring compliance with standards. The peak dose during one million years would be estimated, but only as an "indicator" of long-term repository performance. The EPA argued that they had used the NAS recommendations as a starting point but were not bound to follow them absolutely. The NAS committee had likewise stressed such policy flexibility. The State of Nevada and environmental groups vehemently disagreed. In 2004, the US Court of Appeals upheld these contentions and ordered EPA to more rigorously address the NAS recommendation for one-million-year compliance. [8–9]

Another controversy developed around how much exposure was acceptable. The effects of radiation on humans depend on many factors, including the type of radiation, length of exposure, distance from the radiation source, and the susceptibility of the exposed cells. To account for these factors, radiation exposure is measured in a unit called "rems" (Roentgen equivalent man). The unit is named after Wilhelm Roentgen, the German physicist who discovered X-rays – a feat that earned him the first Nobel Prize in Physics in 1901. A rem is a large amount of radiation. For long-term chronic exposures, the millirem (one thousandth of a rem) is commonly used.

Based on studies of uranium miners, radium dial painters, and survivors of Hiroshima and Nagasaki, scientists are able to predict with some certainty the consequences of exposure to radiation at very high levels. Doses well over 10,000 millirems (10 rems) can cause radiation sickness – weakness, reddened skin, and reduced blood cell counts. Doses of several hundred thousand millirems are often fatal. However, the consequences of radiation exposure are much more difficult to detect and predict at low levels, particularly with doses below 1,000 millirems.

In a year, the average person in the United States is exposed to approximately 620 millirems, about half from natural background sources such as rocks, cosmic radiation from outer space, and the radioactive carbon and potassium found in most living things. By far, the largest source is radon in rocks. Radiation from outer space increases with land elevation and is about twice as high for a person living in Denver, Colorado as a person living at sea level. Flying across the country exposes a person to about 3 millirems. This amount adds up for airline crews, who are annually exposed to more radiation than workers in nuclear power plants. [10]

The EPA's initial standard for RMEI's first 10,000 years was 15 millirems per year. By comparison, a chest X-ray is about 10 millirems and a mammogram about 300 millirems. The 15-millirem limit corresponds to an annual risk of about 8.5 fatal cancers per million people.

In 2005, after their setback in the courts, EPA proposed a two-tiered standard. The annual limit of 15 millirems would continue to apply for the first 10,000 years, but would be bumped up to 350 millirems for the succeeding 990,000 years. For a benchmark, the agency used the difference between the annual natural background radiation received by residents of the Amargosa Valley (350 millirems) and those of Colorado (700 millirems). In essence, during the next million years people living near Yucca Mountain should not be exposed to any more radiation than Coloradans are today.

The idea of subjecting future Nevadans to Colorado conditions did not sit well with the State's politicians. "This is junk science at its worst," charged Nevada Governor Kenny Guinn. What EPA has proposed is "voodoo science and arbitrary numbers," claimed Senator Harry Reid. Representative Jon Porter called it "irrational and misguided." [11]

When EPA released its final standard in 2008, the second-tier dose limit was reduced from 350 to 100 millirems per year. This revised limit was more consistent with international recommendations for radiation doses received by the general public above background levels and medical exposures. The State of Nevada once again challenged the standard in the Washington, DC Circuit Court.

Many additional controversies arose. EPA established a separate requirement that groundwater meet current drinking-water standards for 10,000 years, thereby pitting the agency against DOE, the NRC, the NAS Board of Radioactive Waste Management, and the nuclear industry – all of whom argued that this additional standard added undue complexity with negligible impact on protection of the public. EPA eventually prevailed. In addition, no one was satisfied with how a "representative volume" of groundwater would be used to calculate radionuclide concentrations in RMEI's well water. Clearly, the debates could go on forever with unavoidable arbitrariness in determining who might be most exposed, where they live, their lifestyle and sophistication, their uses of the local groundwater, and how high or low to set the standards. [12]

Regulating radioactive waste disposal over such an immense timeframe stands in stark contrast to regulations of other hazardous wastes that may be even more persistent in the environment. Examples include mining wastes, deep-well injection of hazardous liquid waste, and solid wastes containing carcinogenic metals at Resource Conservation and Recovery Act (RCRA) sites. The longest compliance time required by EPA for any of these wastes is 10,000 years for deep-well injection of hazardous liquid waste. A typical permit for an RCRA solid waste management facility is for 30 years, and the operator bears responsibility for less than a century. [13]

The short regulatory compliance times for these other wastes does not mean that they don't pose a potential long-term hazard. David Okrent and Leiming Xing, scientists at the University of California Los Angeles, analyzed what would happen over the long term at an RCRA site for disposal of arsenic, chromium, nickel, cadmium, and beryllium. Assuming loss of societal memory a thousand years into the future, residents of a farming community at the RCRA site would face an estimated 30 percent lifetime probability of cancer. [14]

The EPA standards are a societal pledge to limit radiation exposure to very low levels for 40,000 future generations of people living near Yucca Mountain (assuming similar life spans over one million years). In almost every other aspect of environmental protection, society is unwilling to make concessions to even the next two or three generations.

TSPA CREDIBILITY

The TSPA had a central role in evaluating the merits of the Yucca Mountain site; therefore, its credibility was paramount. Early in the Yucca Mountain program, the primary role of the TSPA was to evaluate the *relative* risk of different repository designs and identify data needs. Almost everyone agreed that the TSPA was useful for this purpose. Controversy arose when the TSPA was used in an *absolute* sense to compare model predictions with EPA's regulatory standards.

Many scientists recognize the value of models to develop a simplified understanding of the real world, yet remain highly skeptical of their use to predict the future for even a few decades – let alone for the time periods involved in containing high-level radioactive wastes. Even for the engineered components, there is no fully satisfactory way of validating the results of the TSPA models. In spite of these limitations, using computer models for prediction is irresistible given society's demands for certainty. It is simply not good enough to say that a site *looks* safe. Regulators require quantitative predictions to compare to standards for site performance, with the assumption being that predictions based on our knowledge of physics and chemistry are better than no predictions at all.

The credibility of the Yucca Mountain TSPA was confounded further by its complexity. The completed TSPA required several weeks to run. During their intensive evaluation of the license application, the NRC staff would have the time and resources to acquire an understanding of the models, but others would find it virtually impossible to "look under the hood" in any rigorous way. Even the NRC would be unable to examine the TSPA computer codes in depth. This led to suspicions that the TSPA was like a large black box whose results could be manipulated by adjusting hidden knobs.

The black box metaphor was reinforced by the use of conservative models and assumptions for poorly understood phenomena. For example, early versions of the TSPA assumed that once a crack appeared in a waste package, it would lose all ability to isolate the waste. In theory, conservative models and assumptions provide a margin of safety. The problem is that too many such simplifications cloud the understanding of what is actually going on. Approaching a process conservatively in the TSPA models may demonstrate compliance with the regulations, but it raises questions about whether DOE had adequate knowledge of the critical processes. This concern would be highlighted repeatedly by DOE's most constructive critic – the Nuclear Waste Technical Review Board (NWTRB).

The NWTRB was created by the 1987 Amendments to the Nuclear Waste Policy Act as an independent source of expert advice on the Nation's nuclear waste program. The NWTRB is composed of 11 members who are selected in a similar fashion to the Supreme Court justices, but with term limits. When a member of the NWTRB retires or their term expires, the current President appoints a new member. While no presidential appointee is entirely free of the political slant of the acting administration, the NWTRB functions with a high degree of autonomy. New members must be selected from a slate of candidates nominated by the National Academy of Sciences. The members are eminent scientists and engineers who have distinguished themselves in fields related to nuclear waste disposal.

The Board has the authority to look into any technical issue related to the Nation's nuclear waste program at any time. Although the NWTRB has no authority to require DOE to implement its recommendations, Congress assumed that DOE would heed the Board's views "or clearly state its reasons for disagreeing." As authorized by law, the Board must cease functioning within a year after the first high-level waste repository opens. In spite of the law's intent for a limited existence, the course of events suggests the Board's mission has no end in sight.

Creation of the NWTRB is sometimes credited to science writer Luther Carter, who pointed out the value of the independent Environmental Evaluation Group (EEG) at WIPP. However, the NWTRB and EEG differ in one key aspect – the NWTRB was formed not to represent the locals or the State, but to provide independent expert technical advice to DOE and Congress. [15]

The Board members tend to be true believers in performance assessment, with the phrase "risk-informed, performance-based regulation" a mantra in their daily professional lives. Nevertheless, the NWTRB emphasized that it would never be possible to rely solely on the TSPA to demonstrate repository safety. Other factors, such as the importance of multiple barriers and performance confirmation monitoring, would be important to make the safety case. The NWTRB also repeatedly emphasized that DOE must demonstrate that it had a fundamental understanding of how the repository would perform. That is, mastery of the subject matter was at least equal in importance to passing the TSPA test. The Board challenged DOE not to be satisfied with simply falling back on conservative assumptions in the TSPA as a convenient argument that the safety case had been made. The NWTRB did not win every battle, but had a major impact in pressuring DOE to directly address several important scientific issues.

While the technical community debated the scientific basis of the TSPA, the public viewed the TSPA and repository risks through an entirely different lens. To evaluate risk in daily life, most people rely much more on their gut reaction than risk assessments by experts. These *perceptions of risk* are influenced by a daily barrage from the news media highlighting dangers and disasters occurring throughout the world. At the same time, people are generally willing to tolerate higher risks from activities seen as highly beneficial, and either within their control or managed by an institution they trust. While perhaps poorly informed about true risks of various hazards, the public's "basic conceptualization of risk is much richer than that of the experts and reflects legitimate concerns that are typically omitted from expert risk assessments," writes Paul Slovic, a leading authority in the field of risk analysis. [16]

Views about risks associated with controversial matters, such as nuclear energy, are strongly held and resistant to change. For example, in 1989, three years after the Chernobyl explosion, the government of Taiwan launched an expensive public relations campaign to promote public support for building a new reactor. In spite of presenting the issue in the most reassuring of terms, the elaborate program not only failed in its mission but even appeared to increase anxieties by reminding everyone of the awesome power of nuclear energy. [17]

Another example of the difference between the public and technical view of risk is the analysis of low-probability, high-consequence events – such as an igneous intrusion at Yucca Mountain. The probability of an igneous intrusion that would intersect the repository tunnels and damage the waste packages, or worse yet, bring the waste to the surface in a volcanic eruption is extremely low, having an annual probability of about one chance in 60 million. Yet the impact of such an event could be huge. The TSPA analysis essentially multiplied a very small probability by a huge impact to arrive at a "mean probability-weighted dose" that peaks at about 1 millirem per year – about the same radiation dose that is received in a two-hour airplane flight. While scientists may be comfortable with such an expression, the ordinary citizen knows that either the mountain blows or it doesn't. [4]

The TSPA continued to evolve from its first completion in 1991 to the final license application. While the uncertainties remained great, the results of each generation of TSPA increasingly suggested that the Yucca Mountain repository would meet regulatory standards. In the final TSPA, even the upper end of the estimated dose rate was well below the regulatory limit. *If* the technical basis of the TSPA were accepted by the NRC, it seemed likely that DOE was going to get a passing grade – and might even "ace" the license application test.

Ironically, with so much invested, hardly anyone among the general public had even heard of the TSPA. And if they had, it would not make much difference in winning them over. The public needs more concrete evidence. One place to start is with *natural analogues*, where processes similar to those expected to occur in a nuclear waste repository took place over vast periods of time.

NATURAL ANALOGUES

There are more than 100 caves in southern France with pre-historic art work. The oldest authenticated of these, the Chauvet cave near Marseilles, contains paintings of long extinct mammoths and animals that no longer live in Europe, such as the rhinoceros. The charcoal used in the paintings has been age-dated at around 32,000 years old. [18]

In this sub-humid region that receives about three times the annual rainfall at Yucca Mountain, and in a cave with 99 percent humidity, it seems impossible that such fragile paintings could have survived over such an epochal, and previously much wetter, time span. One explanation is that the paintings were created more recently, with the artist using 32,000 year-old charcoal. Yet, soot deposited on the cave ceiling from oil lamps used by prehistoric humans has been dated at over 26,000 years old.

Chauvet cave shows evidence of water flow down the walls where the paint has deteriorated, yet the murals are remarkably intact. Water would destroy these ancient murals, as demonstrated in the famous cave at Lascaux, discovered in 1940 by five teenage boys when their dog vanished into the ground. All the paint has been dissolved from a block that fell onto the Lascaux cave floor, while the 600 Paleolithic paintings on the walls are remarkably preserved. [19–20]

Painted rock shelters are much more numerous than painted caves. This is not surprising, in that shallow overhangs are more numerous than true caves and artificial light is not required for the artist to work. Surprisingly, in an overhang of even a few yards these paintings have been preserved from percolating water for many thousands of years. Painted rock shelters have been found in over 400 sites in India, and in a variety of climatic zones, with some receiving 50 inches (130 cm) or more of rain each year. Rock art has been found on every continent except Antarctica. Africa has more than any other, with some sites estimated to be 27,000 years old. Experts believe some of these African paintings may have been created 40,000 years ago. [21]

In addition to rock art there are numerous biological remains. Spirit Cave in Nevada is famous for its 9,000-year-old mummy with his scalp completely intact, including a small tuft of hair. Packrat middens composed of twigs, leaves, and everything else "the rat dragged in" provide further evidence of long-term preservation. These large piles, cemented by dried packrat urine, are found in caves and shallow overhangs throughout the Southwestern United States. The middens easily dissolve in water, yet many have been preserved for 20,000 to 40,000 years. [20]

In the 1980s, Ike Winograd became intrigued with this remarkable preservation of artifacts found in caves and rock shelters. Ike was not pursuing his new hobby out of traditional archaeological interest. Could it be that these artifacts provide some answers for how to store high-level radioactive waste? He noted the "amazing preservation of even delicate organic and inorganic objects placed in the unsaturated zone at shallow depths and *generally without intent of preservation* [italics added] on the part of our ancestors." Ike argued that "Certainly modern man should be able to equal and improve upon the practices of his ancestors in his attempt to isolate solidified toxic wastes from the environment for millennia to tens of millennia." Were it not for good preservation of artifacts in the unsaturated zone, there would be far less archaeological evidence of how ancient civilizations lived and functioned. [22]

Archaeological relics provide some compelling evidence that an underground repository could work, yet Ike freely admitted the difficulties. Past climatic conditions at archaeological sites were different from today's climate, and their hydrogeologic settings do not perfectly match any site being studied for a geologic repository. Furthermore, archaeological remains were never subjected to intense radiation or a sustained heat pulse well above the boiling point of water. Finally, the archaeological record may be strongly biased toward successful preservation because the unsuccessful ones are long gone. [22]

John Stuckless, a geochemist at the USGS, devoted years to documenting these archaeological artifacts. While Stuckless admits that "Null evidence cannot be evaluated easily," he cites evidence suggesting that these Paleolithic paintings have all, at least partially, stood the test of time. Ancient cave artists didn't just paint – they also carved into the rock. Stuckless found that cave etchings without paint "seem to be lacking." In addition, if some of the paintings have been completely destroyed, one would expect a range of preservation, from largely destroyed to fully preserved. No such examples have been found. Except where part of a mural is positioned below the high-water mark and the paint in that area is gone, the paintings are remarkably uniform in their preservation. [21]

These prehistoric paintings and biological remains provide realworld evidence, supporting computer predictions of the tendency of percolating water to move *around* openings within the unsaturated zone.

Figure 17.2 Exploratory Study Facility at Yucca Mountain.
Source: US Department of Energy.

This tendency depends partly on the size of the openings – the smaller the diameter of the cave, the more likely water will flow around it. Most caves with highly preserved artwork have openings larger than the tunnels proposed for waste emplacement, suggesting that the tunnels may remain even drier than the caves. [20]

Natural analogues provide an independent line of evidence for timeframes that cannot be studied by laboratory and field-scale experiments. Moreover, the public is more likely to understand and trust natural analogues than the output from complex computer models. For these reasons, scientists sought natural analogues for other features and processes at Yucca Mountain. An example is uranium deposits at Peña Blanca in Chihuahua, Mexico. These deposits are in a very similar climatic, geologic, and geochemical setting and provide a natural analogue for the long-term behavior of uranium in spent fuel in an oxidizing environment like that at Yucca Mountain. [23]

EXPLORATORY STUDIES FACILITY

For many years, scientists studied Yucca Mountain from the surface. They mapped the geology, excavated hundreds of pits and trenches, drilled almost 500 boreholes, and instrumented more than 25 wells. To get the scientists underground where they could actually see and test the rocks near the proposed repository horizon, a tunnel known as the Exploratory Studies Facility, or ESF, was built into the side of the mountain. [24]

Minimizing damage to the rock during ESF tunnel construction was essential, so blasting was out of the question. As a result, DOE custom-built the world's largest and most expensive tunnel boring machine. Looking something like a gigantic caterpillar, this machine slowly munched its way through the rock and carved an opening 25 feet (8 m) in diameter. The resulting "exploratory" tunnel would be large enough to use for the final repository. This fact did not go unnoticed by the State of Nevada, which accused DOE of beginning repository construction before first determining if the site was suitable.

Excavation of the ESF tunnel began in 1994. Three years later, the mountain-eating machine broke through to daylight at the end of its 5-mile (8 km), horseshoe-shaped path. The tunnel enters the mountain from the east side, heads southward parallel to the mountain ridge, and then turns back toward the east, exiting the mountain on the same side it began. The path was "an apt metaphor for the project," quipped David Applegate, Director of Government Affairs for the American Geological Institute. The $13 million tunnel boring machine was eventually painted white and served as a highlight of Yucca Mountain tours. It now sits at the site with no takers willing to pay even its worth as scrap metal. [25]

The ESF is close to, but not actually *in*, the proposed repository section of the mountain. Under pressure from the NWTRB, the Department of Energy excavated a smaller diameter tunnel that crosses over the ESF and provides access to rock units in the proposed repository horizon. Using convoluted DOE terminology, this was called the Enhanced Characterization of the Repository Block Cross Drift – with "drift" being another term for tunnel. The NWTRB emphasized that constructing the Cross Drift conforms to standard engineering practice – you should not embark on a major underground project without seeing firsthand what the actual geology looks like.

A railroad line laid through the 5-mile tunnel transported scientists and equipment to and from their field studies. More than 20 major studies took place. In one, water containing chemical tracers was released in the Cross Drift above the ESF in order to examine the relationship between water percolating through the mountain and seepage into the drifts. In another study, moisture was monitored near the Ghost Dance Fault to evaluate possible preferential pathways provided by faults.

Another series of studies examined how heat affects hydrologic, mechanical, and chemical processes in the proposed repository. Among

Figure 17.3 Tunnel boring machine begins excavation at Yucca Mountain. *Source:* US Department of Energy.

these was the world's largest underground heating experiment. Known as the drift-scale heater test, nine large steel canisters containing heating elements were placed end-to-end in a sealed-off section of drift and measuring about half a football field in length. Fifty boreholes were then drilled horizontally away from the canisters, where smaller "wing" heaters were placed to mimic the effects of heat coming from adjacent drifts. The canisters and wing heaters had a total maximum power output of approximately 280,000 watts. Finally, about 4,000 sensors were installed throughout the heated drift and surrounding rocks to enable detailed recording of temperature, relative humidity, water content, mechanical changes in the rocks, and even microseismic events. [24]

The heaters were cranked up in late 1997 and continued near full blast for over four years. There were a few power failures along the way. The cool-down phase took another four years. During the test, more than 70,000 cubic feet of rock were heated above the boiling temperature for water. Based on typical electric rates, the heating bill was about $9,000 per month. [26]

The drift-scale heater test provided important insights into how intense heat changes the properties of the fractured tuff. Scientists

Figure 17.4 Electrician wiring canister for drift-scale heater test.
Source: US Department of Energy.

measured how water moved away from the heat sources as vapor, con-
densed as it cooled, and then drained down through fractures. The
test provided detailed insights into water and gas compositions, mineral alterations in rock samples, and mechanical effects of heat-induced
expansion of the rocks. Results from the heater test were of particular interest in trying to resolve long-standing debates over how hot the
repository should be allowed to get.

HOW HOT IS HOT?

During the 50 to 300 years before the repository is sealed (during the monitoring and retrievability phase), most of the heat from radioactive decay
would be removed by natural and forced ventilation. Once the repository
was closed, however, surface temperatures of the waste packages would
rapidly increase and remain above the boiling point of water for about
1,000 years – slightly longer than the time from the Norman conquest of
England to the present day. This period is known as the "thermal pulse."
    A fundamental question in the repository design was how the sur-
rounding rocks would handle this intense and sustained heat. There were

two schools of thought. In a *high-temperature* design, the rocks would be allowed to exceed the boiling point of water. In a so-called *low-temperature* design, the temperature of the rocks would be kept below the boiling point. Lower rock temperatures could be achieved in a number of ways – spacing the drifts further apart, greater distances between waste packages, juxtaposing cooler and hotter waste packages, longer aging of waste packages at the surface, and enhanced natural ventilation of the repository after closure.

The Department of Energy favored a high-temperature design that would drive moisture away from the packages during the thermal pulse. A high-temperature design would also result in a smaller repository "footprint." More waste could be packed into a given area, resulting in fewer miles of drifts, less area to study, and reduced construction costs.

The USGS and NWTRB favored a low-temperature design. They argued that high temperatures could induce physical and chemical changes in the rocks that might affect repository safety and would greatly complicate predictions of future repository performance. The NWTRB viewed minimizing uncertainties in projected repository performance as a key safety element that merited consideration – right up there with the TSPA, natural analogues, and other evidence. [27]

The general idea of the DOE high-temperature design was that water in fractures and pores would vaporize and be driven away from the emplacement tunnels. Ideally, much of the vapor would condense and drain harmlessly through the rocks between the tunnels. If the high-temperature design worked as envisioned, it would buy more than a thousand years of dryness around the waste while the radioactivity cools down. Eventually, the dried out zone would start to contract as the heat output from the waste packages declined. After a few thousand years, the fractures and rock matrix would return to near their former moisture levels. In comparison, for a low-temperature design, the rocks would remain below boiling, no water would vaporize, and a dried out zone would not develop. [28]

The problem with the high-temperature design is that it introduces many uncertainties. The water vapor might condense above the repository and "reflux" (drain) directly into the tunnels as the thermal pulse declines. Expansion and contraction of the rocks from the intense temperature changes would cause changes in fracture openings (and possibly more seepage into the tunnels), as well as increased rockfall from tunnel ceilings onto the waste packages. Mineral species like silica and calcite would precipitate in fractures as the water evaporates and some of the water in the mountain could become more corrosive from this

evaporative concentration. These and other changes are extremely difficult to predict.

Among the growing list of uncertainties with a high-temperature
design, even common everyday dust became an issue. Just like in one's
home, dust would accumulate on the waste-package surfaces, particularly while the repository is open. Obviously, no one was going to go
down there to dust them off.

Plenty of dust is generated in the surrounding desert environment,
but some comes, by way of global transport, from dust storms in Asia.
Dust would enter the repository in the air used for ventilation. Most of the
dust would come from the rock walls and construction materials in the
tunnels, including ground-up material from tunnel boring, particulates
from diesel exhaust, abraded neoprene and fiber from conveyor belts,
and all kinds of metal and concrete particles. The result is that while
dust is a common nuisance, it also has a complex chemistry. [29]

In a high-level waste repository, the chemistry of dust matters.
Of particular concern was that soluble salts in the dust could absorb
moisture from air in the tunnels to form spots of corrosive brines on
the surfaces of the waste packages – a process known as deliquescence.
Once deliquescence-induced localized corrosion was initiated in pits
and crevices, propagation rates could be very rapid. Deliquescence only
occurs at higher temperatures, so neither the NWTRB nor DOE considered it to be a problem for the low-temperature design.

Even for the high-temperature design, DOE did not consider deliquescence to be a problem and screened it out from the TSPA. The NWTRB
saw the matter differently. In October 2003, the NWTRB sent a strongly
worded letter to DOE summarizing their concerns about deliquescence
and other forms of waste-package corrosion during the thermal pulse.
The Board was convinced that "the data in hand show that localized
corrosion is likely." [30]

The Department of Energy took immediate exception to the tone
and content of the letter. They argued that the amount of soluble salts
would be small, thereby limiting the potential for localized corrosion
by deliquescence. Moreover, the Department contended, nitrates in the
dust would reduce the potential for corrosion. Unconvinced, the NWTRB
repeated their concerns about deliquescence over the next several years.
[31–32]

In an attempt to resolve this debate, the project went on a major
dust collecting spree. Dust was vacuumed from the walls and equipment
in the ESF and Cross Drift, brushed from the canisters used in the
drift-scale heater test, sucked in from the atmosphere, brushed from

rocks and bushes on the Yucca crest, and collected from nearby areas including a missile silo on the Nevada Test Site. The dust samples were analyzed by high-tech methods, such as X-ray fluorescence and scanning electron microscopy. Among other findings, the dust sampling experiments revealed that heating depleted the nitrates that would protect against deliquescence, further fueling the debates. In contrast, studies sponsored by the Electric Power Research Institute suggested that the deliquescence issue was irrelevant to the safety case. [29, 33–34]

At the time of the license application, an independent review group concluded that the TSPA was "deliquescence ready," but also identified a new issue that needed more investigation – the so-called "tree in the sidewalk problem." In this scenario, expansion that occurs when metal is converted to an oxide could act as a wedge between the inner and outer layers of the waste package. This illustrates the never-ending scientific issues that arose as investigations proceeded. We turn to a few of the most prominent issues in the next three chapters to provide a sample of the scientific challenges, and how they played out with the media, the public, and among politicians. [35]

18

Surprise

There are known knowns. These are things we know that we know. There are known unknowns. That is to say, there are things we know we don't know. But, there are also unknown unknowns. These are things we don't know we don't know.

Donald Rumsfeld [1]

Thomas Chrowder Chamberlin (1843–1928) had a remarkable career. He is noted for his contributions to glaciology and for his early hypothesis that the ice ages might follow a natural cycle driven by feedbacks involving CO_2. Chamberlin also served as president of the University of Wisconsin, president of the American Association for the Advancement of Science (the Nation's leading professional science organization), and founded the *Journal of Geology*. One of his most lasting contributions, however, is his insightful discussion of the method of multiple working hypotheses.

Using the flowery language of the time, T. C. Chamberlin described the dangers of "parental affection" for a favorite explanation or theory. He warned, "The moment one has offered an original explanation for a phenomenon which seems satisfactory, that moment affection for his intellectual child springs into existence." Accordingly, this "intellectual affection" should be replaced with "intellectual rectitude," by considering every rational explanation for new phenomena. In this manner, the investigator becomes "the parent of a family of hypotheses: and, by his parental relation to all, he is forbidden to fasten his affections unduly upon any one." [2]

Although well accepted today, the method of multiple working hypotheses was a novel idea during Chamberlin's time. It was also easier said than done. Upon hearing Alfred Wegener discuss his theory of continental drift, Chamberlin immediately dismissed the idea, saying "If we are to believe this hypothesis, we must forget everything we learned in the last seventy years and start over again." A corollary of the method is to expect surprises. [3]

In 1990, exactly 100 years after Chamberlin published his method of multiple working hypotheses, the National Research Council, the principal operating agency for the National Academy of Sciences, issued a report entitled "Rethinking High-Level Radioactive Waste Disposal." The report did not mention Chamberlin by name, but paralleled many of his ideas. Among the report's conclusions was that the US high-level nuclear waste program was unlikely to succeed. This was a remarkable statement by an assembly of experts in the nuclear field. [4]

The pessimistic view of these experts centered on the failure of DOE managers to acknowledge the inevitable. Changes in concepts and design for high-level waste disposal would be required as unexpected geological features were encountered and scientific understanding progressed. Ignoring this reality would eventually backfire and undermine public trust. The Academy advocated an approach that was more accommodating to new insights, unexpected information, and changing circumstances. The report was largely ignored, but turned out to be a premonition of events to come. One surprise after another was encountered in the decade ahead.

UNCERTAINTY IN THE UZ

There are reasons why a geologist became closely associated with the concept of multiple working hypotheses. Earth scientists rely on geologic data that are sparsely distributed and incomplete. Rocks exposed on the surface of the Earth represent only the tip-of-the-iceberg of the underlying geology, and boreholes provide only a very limited peek below the surface. It is possible to get a broader subsurface view by using geophysical surveys, yet these have limited resolution and provide only indirect evidence. The ESF tunnel at Yucca Mountain provided a unique, yet still limited, view inside the mountain. As a result, Earth scientists rely on their training, previous experience, and intuition to translate the available data into 3-D representations of the geologic system and its properties. Inevitably, multiple interpretations exist.

Geologic interpretations underpin the subsequent *conceptual model*, which is a simplified version of the system upon which the computer models are built. The conceptual model is the most critical part of the modeling process. It controls assumptions about what processes and features should be included in the analysis. While the computer model is built from algorithms and equations, the conceptual model can be expressed by words, pictures, and ideas.

The conceptual model is inherently subjective, building on experiences from similar sites or related problems, and on the background and biases of the scientific investigators. As such, it should be revised and refined as new information is gathered. The conceptual model of unsaturated-zone flow at Yucca Mountain is a good illustration.

Prediction of water flow through the unsaturated zone (UZ) has long been a major challenge in hydrologic science. Work over several decades at Yucca Mountain by scientists at Lawrence Berkeley National Laboratory, Los Alamos National Laboratory, and the US Geological Survey resulted in major scientific advances in the understanding of how water flows through fractured rocks in a dry environment. Among the challenges in modeling UZ flow at Yucca Mountain were how to represent the innumerable fractures, the role of major faults, and the possibilities of lateral flow and perched water. [5]

While the infiltration rates of water into the mountain may be low, the mountain is far from bone dry. Considerable water in the UZ is strongly held by capillary forces. If you take a cubic foot of granite from a potential high-level waste site in Sweden and squeeze all the water out of it, then go to Yucca Mountain and do the same, you may get more water out of the rock at Yucca Mountain. Although the granite in Sweden is below the water table, it has a lot less void space to contain water. Nonetheless, there was a strong expectation that Yucca Mountain would be "dry." [6]

The unsaturated zone of Yucca Mountain consists of a thick sequence of tuffs formed by ash from volcanic eruptions over millions of years. Some of the ash was very hot when deposited and the particles fused together, forming a hard dense rock called *welded* tuff. These welded rocks are brittle and highly fractured, creating potential water pathways through otherwise almost impenetrable rocks. In addition, some of the ash had cooled when deposited, forming *nonwelded* tuffs that have much less fracturing and are less dense, with greater pore space for flow between grains.

The first detailed conceptual model of the UZ at Yucca Mountain was published in 1983 by USGS geologist Bob Scott and co-workers. Scott estimated that three percent of the approximately 200 millimeters (8 inches) per year of precipitation that falls on Yucca Mountain enters the unsaturated zone as net infiltration (infiltration of water below the root zone of plants). This works out to be about 6 millimeters per year, or the thickness of a stack of 3 nickels. The majority of precipitation runs off or returns to the atmosphere. Scott envisioned the infiltrating water to move vertically as fracture flow in the welded units, and as flow through pores between grains in the nonwelded units. Although major faults passing through the mountain could be either sealed or open to flow, Scott cautioned that faults should be assumed to be open in the absence of evidence to the contrary. [5, 7]

The same year another USGS scientist, Gene Roseboom, published a summary report on the potential role of the UZ in arid regions for a nuclear waste repository. Expanding on the ideas of Ike Winograd, the report was highly influential in convincing DOE and the NRC on the merits of the thick unsaturated zone for waste disposal. Roseboom's conceptual model was fairly consistent with that of Bob Scott, with downward flow of water through fractured zones averaging about 4 mm per year. [8]

Although the estimates of percolation through the mountain by Scott and Roseboom may seem very small, the estimated rate was about to get much smaller. In 1985, two USGS scientists, Parviz Montazer and William Wilson, published a report that became a basic reference on the hydrogeology of Yucca Mountain. They, and others to follow, suggested that water flowing through a fracture under unsaturated conditions would be quickly pulled into the adjacent rock matrix by capillary forces, in the same manner that water is pulled into a thin straw when it touches the surface. This idea led to major changes in the conceptual model for two key geologic units – the *Topopah Spring tuff*, a thick welded unit which was to be the host rock for the repository, and the overlying *Paintbrush nonwelded tuff* (commonly called the *PTn*). [9]

According to this revised conceptual model, capillary forces serve as a barrier to flow between the PTn and the Topopah Spring tuff. As a result, water in the PTn is diverted laterally until it reaches major faults that cut through the Topopah Spring tuff. In this scenario, most of the water flow is restricted to the major faults. In effect, the PTn acts like an umbrella, or "tin roof," over the proposed repository horizon. What little water passes through the tin roof was presumed to flow mostly through the pores between grains. This slow lane would require several hundred thousand years to reach the water table. In contrast to the conceptual models of Scott and Roseboom, fracture flow through the Topopah Spring tuff was considered insignificant. At most, 1 mm per year was transmitted through the Topopah Spring tuff – and perhaps much less. These views persisted for more than a decade, and were included in the first TSPA model for Yucca Mountain. [10]

Although all of the estimates were low, the difference between less than 1 mm per year and the earlier estimate of 4–6 mm per year was more than academic. Water could seep into the repository drifts only if the percolation rate exceeded a certain threshold, and the larger estimates might do just that. As more data were collected, the extremely optimistic view of flow through the Topopah Spring tuff changed.

An early clue came from neutron logging. In this technique, a radioactive source that emits neutrons is lowered into boreholes drilled into the mountain. The neutrons bounce off the atoms of most elements in the surrounding geologic media. In contrast, collision with a hydrogen atom – which has about the same mass as the neutron – reduces the neutron's speed, like in a game of billiards. When the neutron is slowed sufficiently, it becomes absorbed into one of the heavier nuclei. The absorption causes emission of a gamma ray, which can be detected by a scintillation counter – another neat little tool placed in the borehole. In this way, neutron logging can estimate the density of hydrogen atoms and, in turn, water molecules. By making multiple measurements over time, the rate of water movement can be estimated. Infiltration rates estimated by USGS scientists using this technique suggested that the PTn would have to be an extraordinarily effective "tin roof" to limit seepage into the Topopah Spring tuff to the prevailing estimates. [5]

There was further evidence for water percolation through the fractures of the Topopah Spring tuff. Calcite deposits, lining many of the fractures, had formed from evaporation of percolating water. In addition, temperatures below land surface increase with depth in a systematic way, as heat flows from the Earth's hot core to its surface. This "geothermal gradient" can be offset by cooler water percolating downward. After taking the mountain's temperature down boreholes, estimates of percolating water from temperature profiles were more in line with those of Bob Scott and Gene Roseboom. [11]

In 1996, events took a dramatic turn. As the tunnel boring machine carved its way through the mountain, a team of scientists from Los Alamos National Laboratory, led by June Fabryka-Martin, collected rock samples from the walls of the ESF. The samples were analyzed for chlorine-36, a radioactive isotope formed naturally in the upper atmosphere by cosmic rays. Surprisingly, the chlorine-36 levels were higher than could be explained by natural processes. [12]

An alternative explanation had its roots in the Atomic Age. During the 1950s and 1960s, atmospheric testing of nuclear weapons greatly magnified concentrations of chlorine-36 in the atmosphere. It now

Figure 18.1 Robot equipped with video cameras in Exploratory Study Facility.
Source: US Department of Energy.

appeared that these *bomb-pulse* levels of chlorine-36 were showing up in the tunnel samples. If this was truly the case, water was taking *less than 50 years* to travel 200 to 300 meters from land surface to the exploratory tunnel. The bomb-pulse levels mostly showed up in samples collected near faults and zones of concentrated fractures, implicating those features for fast pathways. [12]

The results appeared to be a final nail in the coffin for the presumed lack of fracture flow in the Topopah Spring tuff. The PTn "tin roof" now appeared to be more of a "torn wet blanket" with two distinct flow systems: a fast track with travel times of less than 50 years where faults or concentrations of fractures cut through the PTn, and a much slower track elsewhere. However, the chlorine-36 findings gave no indication of *how much* water flowed through each of these two pathways. This subtlety was largely lost in the ensuing controversy, as was the question of whether the fracture flow was an asset or a liability. [10]

The State of Nevada pointed to the chlorine-36 data as proof-positive of a fundamental design flaw, and quoted DOE that the mountain contained "billions of water conducting fractures." However, from the outset, USGS scientists and others had argued that a well-connected fracture network would keep water draining through the mountain rather than ponding around the waste canisters. In addition, the repository would be located away from major fault zones.

The association of the higher chlorine-36 levels with faults and fracture zones was conceptually plausible, yet the possibility of erroneous results remained. Chlorine-36 is measured using highly sophisticated accelerator mass spectrometry and reported as a ratio of the radioactive isotope to total chloride in units of 10^{-15}. That's one part of chlorine-36 for every *million billion* parts of chloride. Sampling requires extreme care to avoid contamination by outside sources of chlorine-36. This was particularly true for Yucca Mountain, given its proximity to nuclear testing at the Nevada Test Site.

In 1999, DOE initiated confirmatory studies led by scientists at the USGS. The laboratory analyses would be done at Lawrence Livermore National Laboratory. Core samples were collected from 50 boreholes drilled near to where the bomb-pulse chlorine-36 concentrations had been found by the Los Alamos scientists. After this extensive sampling and testing, the USGS/Lawrence Livermore team failed to detect *any* bomb-pulse chlorine-36 values. In contrast, analyses of some of the samples by Los Alamos were consistent with their earlier results. Possibilities for the differences included sample contamination and micro-environmental controls on occurrence. After failing to achieve the same results, the two groups documented their work in a single report in which, essentially, they agreed to disagree. [13]

Scientists at the University of Nevada-Las Vegas (UNLV) also initiated confirmatory studies, attempting to replicate the Los Alamos sampling along the ESF walls. In the end, one of their samples yielded a possible bomb-pulse value. [14]

In spite of the lack of consensus about the chlorine-36 findings, studies suggest that if fast flow paths do exist, they represent only a small portion (about 1 percent) of the total water movement through the UZ at Yucca Mountain. Nevertheless, the chlorine-36 findings shook the very foundation of DOE's confidence in Yucca Mountain as a natural barrier to radionuclide movement. [15]

NATURE VERSUS ENGINEERING

Limitations in the ability to predict the long-term behavior of a geologic repository can be offset, in part, by adopting a multiple-barrier or *defense-in-depth* philosophy for radionuclide containment. Defense-in-depth means that the safety of the repository does not depend on the performance of any single barrier. These multiple barriers comprise both natural and engineered barriers. Natural barriers include the geologic system's capability to dilute, retard, and even retain radionuclides during transport. The engineered-barrier system includes the waste form (reactor fuel assemblies with zirconium cladding for spent fuel and glass logs for defense wastes), the canister or waste package, and any backfill between the canister and adjacent host rock.

In 1986, as the investigations of Yucca Mountain were ramping up, the Department of Energy was sufficiently confident in the natural system that it claimed the NRC standards could be met without *any* additional engineered barriers. This view persisted for some time. After the chlorine-36 findings, DOE did an abrupt about-face, claiming that almost all the waste isolation would be dependent on the engineered barriers – with the geologic system now being relegated to defense-in-depth. Critics charged that "the original concept of geologic disposal had been turned on its ear." [16–17]

In 2001, the State of Nevada filed suit in the US Court of Appeals, arguing that basing repository performance almost solely on engineered barriers meant that DOE could approve permanent storage "at virtually any physical site in the United States." Washington attorney Joseph R. Egan, who led Nevada's legal campaign to block the Yucca Mountain site, suggested that an ideal candidate for permanent storage was the basement of DOE headquarters in Washington, DC. A nuclear engineer-turned-lawyer, Egan was so opposed to the site that, before he passed away in 2008, he requested that his ashes be spread at Yucca Mountain with the words "radwaste buried here only over my dead body." [18–19]

While the principal challenge with natural barriers is characterizing the local geology, the principal difficulty with engineered barriers is the lack of data on their long-term performance. In deciding to make the engineered barriers the first line of defense, the long-term performance of the waste packages took on a whole new level of importance. As such, the waste packages would be specially designed with two protective layers. A stainless steel inner layer would give the package structural rigidity, while an outer layer would provide resistance to corrosion. The outer layer would be 1 inch (25 mm) thick and made of Alloy 22 – a highly corrosion-resistant mixture of nickel, chromium, and molybdenum.

Alloy 22 was developed in 1981, but nickel–chromium alloys date back more than a century with each succeeding generation providing improved resistance to corrosion. In 2001, Roger Newman, a prominent corrosion expert, confidently told the NWTRB: "I'd have one of these [waste packages] in my back yard, and if anybody wants to pay me to put it there, I'd be happy to discuss it with them later." Not everyone was so confident. Short-term tests of Alloy 22 in anticipated repository environments yielded extremely low general corrosion rates, suggesting waste package lifetimes on the order of 100,000 years may be achievable. Yet, these were only short-term tests that were extrapolated out 100 millennia. [20–21]

By the late 1990s, uncertainties about the long-term behavior of the waste packages, coupled with concerns about the natural system, caused DOE to add a new feature to the engineered barriers. The proverbial "tin roof" of the PTn would be replaced by a titanium drip shield covering the entire length of the waste packages, protecting them from rock falls and dripping water. Named after the Titans of Greek mythology, titanium is called the "space-age metal," having light weight, high strength, high resistance to corrosion, and extraordinary resistance to fatigue. Bicycle enthusiasts love it. As one woman put it, a titanium bicycle is "the one God rides on Sunday." [22]

Yet, even a titanium drip shield would not be impregnable. DOE predicted that it would begin to fail after about 40,000 years. Nor would it solve every problem. For example, a drip shield would not prevent accumulation of dust on the waste packages. And there was the much more basic question of whether the drip shield would ever actually be installed.

To cover every waste package, the drip shield would require more than 10,000 sections, *each* about 20 feet long (6 m) and weighing more than 4 tons. The technology for remote installation had yet to be developed, or even prototyped. Construction of the drip shield would place a huge demand on the world market for titanium. To top it off, the drip shield would not be emplaced until just before the repository is closed. That could be up to 300 years from now – imposing a multi-billion dollar burden on future generations at a time perhaps longer than the present age of the United States. [23]

Given the uncertainties over whether the drip shield would be installed, the State of Nevada charged that it would "make a mockery of the TSPA" to include it in the calculations. The State had a point, although the drip shield might not be needed if some of the TSPA conservatisms could be removed through further study. DOE also recognized that the drip shield may be less important, or even unnecessary, for a low-temperature repository design. This admission raised yet another question about DOE's preference for a high-temperature design. [24–25]

PLUTONIUM TRANSPORT AT THE NTS

No matter how clever and robust the engineered barrier system might be, one thing was virtually guaranteed – it would eventually fail. At this juncture, waste containment would rely solely on the natural system. Sorption (attachment) of radionuclides to the rocks through which the water flows would then play a critical role at Yucca Mountain. Determining this role proved to be elusive.

Sorption is a complex process that varies with the chemistry of the water and the mineral surfaces through which the water moves. It is usually a temporary stop-over, with the contaminant later released back into solution to be sorbed further down the water pathway to another mineral surface. This delay buys time for radioactive decay, while also slowing down contaminants before they reach the accessible environment.

In 1997, Annie B. Kersting of Lawrence Livermore National Laboratory and her associates reported finding plutonium and other radionuclides hitching a ride on tiny particles known as *colloids* in groundwater at the Nevada Test Site (NTS). These submicron-sized particles can remain suspended and navigate their way with the groundwater as it flows through small pores and fractures. Scientists call this *colloid-facilitated transport*. [26]

Kersting's findings challenged the conventional wisdom about the role of sorption in containing plutonium. Under most natural conditions, plutonium has extremely low solubility in water and readily sorbs onto a variety of minerals. For these reasons, it was assumed that the plutonium would basically stay put.

Kersting's plutonium findings were not a total surprise. Plutonium had been found in groundwater throughout the NTS as a consequence of the large number of nuclear blasts set off below the water table. Nor was it a surprise to find plutonium hitching rides on colloids. Pioneering field studies examining colloids and radionuclides had been carried out years earlier at the NTS by Robert Buddemeier of the Kansas Geological Survey and James Hunt of the University of California Berkeley. Buddemeier and Hunt found that many of the low-solubility radionuclides in groundwater were associated with colloids. Additional studies at Rocky Flats in Colorado and at other weapons sites had confirmed the association of plutonium with colloids. The big surprise with Annie Kersting's findings was just how far the plutonium had traveled. [27–28]

Kersting discovered that the source of the plutonium was not the closest nuclear blast site, but a more distant one. Through careful measurement of the isotopic ratio of plutonium-240 to plutonium-239, Kersting and her collaborators had the equivalent of the nuclear blast's fingerprints. The plutonium isotopic ratio in their samples exactly matched only one underground test – the Benham test, detonated 28 years earlier and slightly under a mile away. Here was compelling evidence that at least small amounts of colloid-hitching plutonium had been transported relatively rapidly by groundwater over significant distances through fractured tuff – the same volcanic rock as at Yucca Mountain. [29]

Kersting's surprising results originated from work outside the Yucca Mountain project, but led to one of its more significant and unanticipated uncertainties. Colloids are ubiquitous. They form from broken down rocks, plants, and soil. At the Yucca Mountain repository, they would also form from degradation of spent nuclear fuel and the glass used to solidify defense high-level waste. Because of the long half-lives of many of its isotopes, plutonium would comprise much of the waste when the engineered barriers fail. In 10,000 years, plutonium would represent about 90 percent of the total waste inventory.

Field studies have shown that colloid-facilitated transport is the dominant way in which groundwater transports plutonium, yet many questions remain about the nature and extent of this transport. How reversible is sorption on colloids? How does plutonium "partition" between colloids and the fixed rock matrix? To what degree are colloids filtered out during transport? How would changing chemical and thermal conditions over the lifetime of the Yucca Mountain repository affect the stability of colloids? And finally, how would plutonium travel through the thick UZ beneath the repository to the saturated zone where colloid-facilitated transport would more likely occur? The answers to these and other questions now needed to be factored into the TSPA based on the best available, but limited, information.

GROUND ZERO AT YUCCA MOUNTAIN

On March 5, 1995, an article prominently placed on the front page of the Sunday edition of the New York Times announced to the world, "Scientists Fear Atomic Explosion of Buried Waste." The article reported that two Los Alamos scientists had theorized that leaching of plutonium from the glass logs used for defense wastes could release a sufficient amount of fissile material into a small area to reach criticality. In a worst case scenario, this would cause a large nuclear explosion at Yucca Mountain, scattering radioactivity to the winds, into groundwater, or both. About two weeks later, the Times reported that the theory had gained support from a team of scientists at the Savannah River nuclear site in South Carolina – when, in fact, the Savannah River group had only looked at the plausibility of part of the hypothesis. [30–31]

Politicians and the media were quick to react with their own chain reaction. The next day, Nevada Senator Richard Bryan was on the floor of the Senate displaying an enlarged copy of the *Times* article and accusing DOE of a cover-up. It didn't help that four days earlier, DOE Secretary Hazel O'Leary and the head of the Yucca Mountain Project, Dan Dreyfus, had testified to Congress that no scientific issues were holding back progress at Yucca Mountain. Meanwhile, the Nevada newspapers were full of doomsday scenarios, including cartoons with huge mushroom clouds coming out of Yucca Mountain. [32]

The basic tenet of the two Los Alamos scientists was that dispersal of plutonium from the leaking waste might cause the surrounding rocks to act as a moderator of neutrons in place of water. Moderation – a seeming oxymoron for a nuclear reaction – is necessary to slow the neutrons and make them more likely to trigger fission. Otherwise they are more likely to bounce off fissionable nuclei, like skipping rocks across a stream. Previous researchers had evaluated the possibility of the waste going critical and concluded that, if a critical mass did form, any chain reaction would inevitably shut itself down. Nature had even provided a precedent in Gabon, a small country on the west coast of Africa.

In the early 1970s, routine analyses of high-grade uranium ore from Oklo, Gabon turned up some very strange results – the ore was abnormally low in uranium-235, the fissile isotope used to sustain a nuclear chain reaction. The uranium-235 content in the Oklo samples came in at just 0.717 percent. This was odd. Everywhere else on Earth, the moon, and even in meteorites, uranium-235 atoms made up the same exact 0.720 percent of the total uranium. Although this discrepancy may seem insignificant, it was enough to alert a worker at a French nuclear fuel plant. [33–34]

French physicist Francis Perrin and a team of scientists were called to investigate. After reviewing the facts, Perrin concluded that the ore deposits at Oklo contain the remnants of a natural nuclear reactor – the missing uranium-235 had been consumed in a fission reaction set off by Mother Nature. Intensive follow-up study of the ore showed that the expected radioactive byproducts of a nuclear reaction were present, confirming the French scientists' hypothesis. Existence of a natural reactor requires unique circumstances – uranium ore of the proper grade and distribution in the rocks, no elements that would absorb neutrons, and water present part of the time in just the right amounts to serve as a moderator.

Nearly 20 such natural nuclear reactors have since been found in Gabon, all of which burned some 2 billion years ago. The nuclear fission reactions turned on and off over time. Each heating cycle boiled the water away. When water returned, the reactors would go critical again.

This was a nuclear version of geysers, which build up heat and pressure, boil off their supply of groundwater in a spectacular display, refill with groundwater, and repeat the cycle. The energy release at Oklo was slow – not the rapid energy release required for an explosion. Scientific evidence suggests that the nuclear reactions at Oklo must have been repeated over hundreds of thousands of years. The average power output was probably less than 100 kilowatts – enough to run a few dozen toasters. [34]

Calculating back about 2 billion years, there was sufficient uranium-235 to permit nuclear fissions to occur, providing other conditions were right. This is no longer the case. Uranium-235 decays more than six times faster than the abundant uranium-238, so the fissile fraction, which was about 3 percent two billion years ago, is now 0.7 percent.

The Oklo site is regarded as a natural analogue for containment of radioactive wastes – the radioactive products of these natural fission reactors have not migrated even after such an extremely long time. The plutonium has moved less than 10 feet (3 m) from where it was formed almost 2 billion years ago. However, the analogy between Oklo and Yucca Mountain is weak. The highly reducing conditions (lacking oxygen) and abundant organic matter at Oklo are quite different from conditions at Yucca Mountain. Nevertheless, the Oklo results provide some compelling evidence for the potential containment of radioactive wastes over thousands of millenia.

Not everyone was completely surprised by the Oklo findings. In 1956 a scientist, Paul Kuroda, had described conditions under which a natural nuclear reaction could occur. Four decades later, Charles Bowman and Francesco Venneri, the two Los Alamos scientists featured in the *Times* article, were describing how wastes placed in the natural system might cause criticality.

Bowman and Venneri had been working on an advanced technology known as accelerator transmutation of waste (ATW). The ATW technology works by firing a beam of protons to burn off long-lived radioactive isotopes, while also generating electric power as part of the bargain. The ATW technology was to some extent in competition with geologic disposal of the long-lived isotopes. Nonetheless, the two scientists' thesis had to be addressed. The approach taken by the press and the scientific community offer stark contrasts in how this took place.

Bowman and Venneri first broached their findings internally at Los Alamos in September, 1994. In spite of serious doubts about the validity of the thesis, the lab administration realized they had a potential bombshell. The two scientists were asked to write a paper explaining their reasoning. John Browne, head of energy research at Los Alamos, organized an extensive review of the paper with "red," "blue," and "white" teams. The red team, known as the "murder board," was assigned to tear apart the paper and find everything wrong with it. The blue team would try to support the findings by making as many positive assumptions as they could. The white team, composed of senior members of the lab, would serve as neutral referees and make a recommendation to the lab administration. Each team had experts covering a broad range of nuclear energy and waste disposal topics. [32]

Upon review, all three teams soundly rejected the hypothesis. Even the blue team found it impossible to defend the work. The reviewers reasoned that even if the physically implausible were to happen, the amount of energy released would not lead to an explosion. Not to be dissuaded, Bowman and Venneri insisted that the internal debate left their work "honed and strengthened." Bowman reasoned that the whole weapons program at Los Alamos is "devoted to concentrating nuclear material to make it explode. We show that dispersion can make an explosion. Ours is the inverse." [32]

The two scientists went about revising their paper. By February 1995, they had a new draft, leaving the lab management wondering what to do next. Los Alamos Director Sig Hecker later remarked, "If one of our scientists has some idea that has potentially large political implications, and quite clearly this one did ... we've got to allow him to go ahead and develop the idea and then really subject it to the scientific process of peer review and publication." Yet, Hecker was torn by the idea of seeming to give a Los Alamos National Laboratory imprimatur to such a wild hypothesis. He and John Browne decided the solution was to have the white team write up its review as a scientific paper, so that both it and the Bowman–Venneri paper could be submitted simultaneously to journals. [32, 35–36]

News of the controversy reached *New York Times* reporter William J. Broad. Efforts were made to convince Broad to allow the debate to be published first in a peer-reviewed journal or presented at a scientific meeting, before alarming the public. But Broad had his scoop and was not dissuaded. When the article appeared on the front page of the Sunday *Times*, the head of the red team at Los Alamos angrily called the article "a sensationalistic story ... the sole purpose of which is not to inform but to inflame and sell papers." The *Times* science editor defended the paper's actions by explaining that newspapers have different criteria than scientific journals. Scientists caustically call this "publishing in the *New York Times*," referring to the newspaper as a pseudo journal. [32]

The following year, a review by virtually the entire nuclear engineering faculty at the University of California Berkeley, along with outside experts, concluded that the potential for such a nuclear explosion was not credible. The *New York Times* did not report on these findings. [37]

The criticality issue continued to be studied up to the submittal of the license application. Later studies focused less on plutonium from defense wastes and more on the possibilities for restarting the nuclear reaction in the spent nuclear fuel. This might occur if the repository was flooded after the canisters were breached. Yet, estimates of the likelihood of a criticality event remained low enough that it was screened out of the TSPA. If a criticality event should occur, most estimates of the consequences were marginal increases in temperature and fission products – in other words, a short-term reactor, not an explosion. [38]

The controversies over criticality, chlorine-36 in the ESF, and plutonium hitching a ride on colloids at the Nevada Test Site were but three of many surprises that confronted the Yucca Mountain Project. Some of these surprises were minor, others highly controversial. All of them reinforced Chamberlin's warnings, more than a century earlier, about the pitfalls of parental affection in any scientific endeavor.

19

Shake & bake

Civilization exists by geological consent subject to change without notice.
Generally attributed to Will Durant

Little Skull Mountain, 12 miles (20 km) from Yucca Mountain, was suddenly alive and rocking during the week of July 4, 1992. The proximity of the 5.6 magnitude earthquake to the proposed nuclear repository caught considerable media attention. A State of Nevada spokesman called it a "wake-up call." Many residents in the region, accustomed as they were to past nuclear detonations, slept right through the 3 a.m. event. The Department of Energy reassured everyone that the quake was well within the magnitude for which they were designing the repository. The DOE spokesman even suggested that the event was a big plus, in that it would help characterize the region's seismic hazards. [1]

The Little Skull Mountain earthquake occurred less than 24 hours after the powerful magnitude 7.3 Landers earthquake shook the Mojave desert in California, about 180 miles (300 km) to the south. Clearly, there was a connection. Almost immediately after the Landers quake, a series of micro-earthquakes began in the vicinity of Little Skull Mountain. The Landers earthquake, recorded as the second largest in southern California in the twentieth century, definitely woke a few people up. [2]

While the media made hay and the State raised the red flag, the Little Skull Mountain quake came as no surprise to geologists. Yucca Mountain lies in the tectonically active Basin and Range. This huge geographic province consists of small, north–south trending mountain ranges separated by nearly flat desert basins. The basins are filled with sediments eroded from the mountains over the eons. The Basin and Range includes almost all of Nevada, the western half of Utah, southeastern California, and the southern part of Arizona. G.K. Gilbert, the nineteenth-century

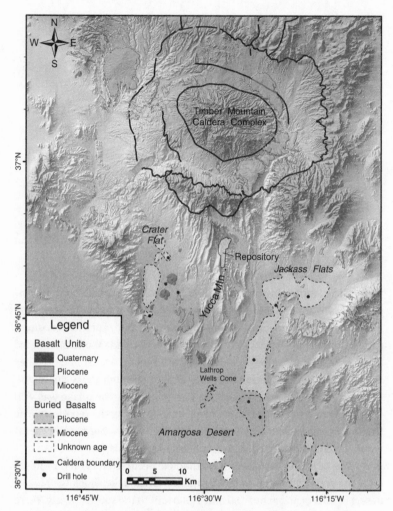

Figure 19.1 Shaded relief map illustrates the southwest Nevada volcanic field in the vicinity of Yucca Mountain. Miocene basalts are greater than 5.3 million years old. Pliocene basalts are 5.3 to 2.6 million years old. Modified from reference [5].

geologist who coined the term *Basin and Range*, likened the ranges to an "army of caterpillars marching north out of Mexico." [3]

The Earth's crust in this region has been stretching in an east–west direction for about 20 million years, continuing to the present day. During this time the sites of today's Reno and Salt Lake City, on opposite sides of the province, have moved 50 miles (80 km) apart. The crustal extension has produced a series of tilted blocks separated by faults, like a row of books leaning on a shelf. The downthrown blocks are the basins and the upthrown blocks form the ranges. The present-day geology of Yucca Mountain is the product of this crustal extension, faulting, and erosion – along with a history of spectacular volcanism. [4]

VOLCANIC HAZARDS

A series of huge volcanic eruptions began 16 million years ago throughout southwestern Nevada. A few million years later, the volcanic tuffs of Yucca Mountain formed and were later uplifted by the block faulting. Fortunately for Las Vegas, these *large-volume* eruptions ceased long ago. Scientists agree that such catastrophic episodes are not a threat. Concerns about future volcanism in the region center on *small-volume* basaltic volcanoes, known as cinder cones.

A dozen small cinder cones lie within 12 miles (20 km) of Yucca Mountain. Eleven of these are in Crater Flat, a broad alluvium-filled basin west of Yucca Mountain. The other is the Lathrop Wells cone located to the south of Yucca Mountain. Six of the cinder cones have erupted in the past million years; the most recent being the Lathrop Wells cone about 77,000 years ago. [5]

The presence of these geologically young volcanoes sparked intense study and debate about the potential for future volcanism at Yucca Mountain. Southwestern Nevada constitutes one of the least active, but longest lived, basaltic volcanic fields in the western United States. The Department of Energy notes that 99.9 percent of the volcanic deposits were produced more than 7 million years ago. Critics point out that it would not take much magma in the wrong place to cause major problems at the proposed repository. During the first few thousand years after waste emplacement, volcanism dominates the risk at Yucca Mountain, as it would be the only way to quickly transport radionuclides from the mountain to the biosphere. [5–6]

The debate over volcanism eluded consensus for many years. Traditional volcanic hazard studies look at the possibility of eruption at an existing volcano. The threat at Yucca Mountain is from a *new* volcano (or new igneous activity) at an *unknown* location. The small number of past eruptions suggests that the probability of a future volcanic event is low, yet paradoxically, the rareness of such volcanic events also leads to large uncertainty.

Most basaltic eruptions begin as upward movement of vertical sheet-like dikes of magma. Typical basaltic dikes are several feet wide

Figure 19.2 Lathrop Wells cinder cone. Photograph by Greg Valentine.

and a few kilometers long. The dikes, which may or may not reach land surface, lead to two basic scenarios. In the *eruption* scenario, magma pierces a section of the repository and continues to the surface through one or more volcanic vents. Any waste packages in the path of the volcano's conduit are destroyed and radioactive waste is spewed into the atmosphere. The effects would be experienced immediately. In the *intrusion* scenario, an igneous dike spills into the repository but does not erupt at the surface. The effects would be delayed by the time required for water to transport the released radionuclides from damaged waste packages to the biosphere.

Significant disagreement exists among scientists regarding how magma would flow into the repository tunnels and how many waste packages would be destroyed. Would the magma enter the repository like slow moving molasses, cooling quickly and affecting relatively few containers? Or would it burst into the tunnels with shock waves powerful enough to break open repository tunnels? Lacking sufficient real-world examples, scientists debated the possibilities using computer models of magma physics. [7–8]

In addition to the questions about possible igneous activity and its consequences, there was the underlying question of whether magma would ever intersect the repository in the first place. A straightforward way to address this question is to draw a circle around Yucca Mountain, count the volcanoes within it, and determine their ages. The frequency of volcanic events estimated from this information is then modified to account for the footprint of the repository. Experts applied this basic approach in considerable detail using various geological and geophysical data along with detailed models for the behavior of a volcanic field. [5]

In 1980, the first estimates of volcanic activity at the repository site put the annual probability at about 1 in 100 million. This was right at the Nuclear Regulatory Commission's cutoff point for inclusion in the TSPA, but not below it. By some accounts, 1 in 100 million is also roughly the same possibility as the ultimate low-probability, high-consequence event – global mass extinction from the impact of an asteroid or comet. [9]

In the mid 1990s, DOE convened a panel of ten experts, mostly volcanologists, to conduct a formalized "ask-the-experts" approach to estimating the probability of volcanism and its uncertainty. The method, called *expert elicitation*, brings together a panel of experts and mathematically combines their individual estimates. The goal is to obtain a probability distribution and range of uncertainty representative of the larger scientific community. Of course, the end result is affected by who serves on the panel, and the pool of qualified participants is not very large.

Using a formal nomination process, ten panel members were selected from a group of 70 scientists. Expertise mattered, but equally important were strong communication and interpersonal skills, as well as flexibility and impartiality. The experts were asked to act as objective evaluators of the various theories. Their job was to listen to proponents of different positions and then weigh each of these theories in making their estimates. [10]

After workshops and field trips to bring everyone up to speed, professional interviewers spent two days with each panel member extracting key information. Each of the ten experts independently arrived at an annual probability distribution for a volcanic event intersecting the repository. The average of these estimates was about 1 in 70 million, later revised to about 1 in 60 million. In the scheme of things, this was not far from the original 1 in 100 million estimate. [10]

The expert elicitation did not end the debate. The results were challenged not only by the State of Nevada but also by scientists working for the NRC. Arguing that conservatism was needed, the NRC used an estimate of 1 event in 10 million in their assessments. Still portending a rare event, the higher probability by NRC scientists came about, in part, from their assumption that faults and deep tectonic structures may provide pathways for the ascent of magma directly into Yucca Mountain. There was also disagreement about whether the time between eruptions was increasing or decreasing. Volcanism is known to be episodic. While most geoscientists consider the volcanism in the Yucca Mountain region to be waning, a few argued that we could be in the middle or end of a quiescent period. [11–13]

A further difficulty in evaluating past volcanic activity was that evidence of basaltic volcanoes, particularly those older than 2 million years, may be hidden by sediments. *Magnetic data* would be the key to additional insights. The basic idea is that buried volcanoes are more magnetic than the surrounding sediments, and can be detected by a sensitive magnetometer placed on the ground or towed behind an aircraft. The magnetometer records minuscule variations in the intensity of the Earth's magnetic field. After factoring out solar winds and regional variations, local magnetic anomalies provide evidence of possible buried volcanoes. This was not a slam-dunk test. There might be other causes of the anomalies. The only way to know for sure was by exploratory drilling at the anomaly locations.

In 1999, three years after the expert elicitation, an aeromagnetic survey conducted by the USGS revealed multiple magnetic anomalies in Crater Flat to the west and the Amargosa Desert to the south of Yucca Mountain. The Department of Energy contended that this new information had only a minor effect on their previous estimates of volcanic risks. The Nuclear Regulatory Commission was unconvinced and insisted more information was needed. The Energy Department could hardly ignore this request. As Kevin Crowley at the National Academy of Sciences put it, "When your regulator says you've got a problem – you've got a problem." [12, 14–15]

The Department of Energy made a commitment to gather additional geophysical data, drill into several of the anomalies, and age date samples of basalt collected from them. The ages were critical. If the magnetic anomalies represented recent volcanism, this would suggest the volcanic hazard was greater than estimated.

A high resolution magnetic survey flown from a helicopter identified about 30 anomalies. Of the seven anomalies drilled, four turned out to be buried basalts. One basalt was about 4 million years old; the other three were older than 9 million years. The results reinforced the expert elicitation estimates of igneous activity. Significantly, the results confirmed that the dikes feeding the volcanoes intruded along faults parallel to Yucca Mountain – not cutting across it. [5]

By now, a fair consensus was reached on the probability of an igneous event intersecting the repository, but DOE was unable to resolve the differences of opinion about the impacts of an igneous intrusion. In their final TSPA, the Energy Department took a conservative approach, assuming that an igneous intrusion intersecting the repository would damage *all* waste packages and drip shields. In other words, radionuclides from almost 12,000 waste packages would be exposed and ready for transport. Largely because of this assumption, igneous intrusion became the major risk factor according to the TSPA, although it remained well below the regulatory standard. The second largest TSPA risk came from the other scenario that was assumed to impact all of the waste packages – ground motion (shaking) from a seismic event. [16]

SEISMIC HAZARDS

Although the Basin and Range is considered to be a region of active tectonics, the Yucca Mountain environs has been "a surprisingly inactive place," according to the USGS, "at least for the past 500,000 years or so." Nevertheless, the region has more than 30 mapped faults. The proposed repository at Yucca Mountain is bounded by faults, but it is highly unlikely that a fault rupture would shear the waste packages. The only fault within the area proposed for waste emplacement – the Sundance fault – created a stir upon discovery, but shows no evidence of activity during the past 2 million years. [17–18]

The principal seismic hazard comes from vibratory ground motion, potentially causing rock falls or tunnel collapse, that would damage the waste packages or drip shields and alter water flow through the repository. In addition, surface facilities for receiving and handling the waste prior to emplacement in the repository must be built to withstand earthquakes.

While the time scales for seismic hazards at the surface facilities are those normally considered in engineering design, prediction of seismic risk in the repository for one million years after closure is a whole different ballgame. Yet, the situation is not as risky as it might seem. During an earthquake, most of the shaking is due to seismic waves traveling along the surface. Underground tunnels are less likely to be damaged than aboveground structures. Shaking is also more intense in unconsolidated sediments than it is in a mountain's solid rock. The Little Skull Mountain earthquake broke windows and caused minor damage to surface facilities at the Yucca Mountain site, but there was no damage in two tunnels drilled into Little Skull Mountain a couple of miles from the epicenter. In addition, mine tunnels at the Nevada Test Site have withstood ground motions from underground nuclear explosions that exceed any ground motion anticipated at Yucca Mountain. [19–20]

Estimating the frequency and magnitude of future earthquakes suffers many of the same difficulties as estimating volcanic hazards. As a result, a similar process of expert elicitation was used. Referred to as probabilistic seismic hazard analysis (PSHA), the approach had three steps. First, scientists identified the location of potential earthquake sources and defined their characteristics. Next, they estimated how frequently earthquakes of various magnitudes might occur at each source location. Finally, they estimated the amount of fault displacement and associated ground motions that will occur given earthquakes of a particular magnitude. The uncertainty in each of these estimates was also evaluated. The process took three years and involved several dozen scientists and engineers. [21]

While the PSHA was widely considered state-of-the-art at the time, it had a major limitation. At very low probabilities, the results predicted extreme ground motions that have *never* been documented in actual earthquakes. To place more realistic bounds on these extreme events, follow-on studies looked at the physical limits of what theoretically *could* happen (based on rock strength) as well as studies of what *had* happened at the site. The most creative studies for the latter case took a close look at precariously balanced rocks scattered around the mountain. [22]

Precariously balanced rocks develop on the land surface as water seeps into cracks in fractured bedrock, causing it to break down. As the loose material erodes away, the rocks settle on one another leaving some in a bizarre balancing act. James Brune, Director of the University of Nevada Seismological Laboratory, came up with the novel idea of using these rocks for seismicity studies. By testing or estimating the force required to topple precariously balanced rocks around Yucca Mountain, Brune realized he could estimate an upper bound on ground motions that have occurred during the time the rocks have been in their present balancing act. If a larger earthquake had occurred, the rocks would have toppled. [23]

While the basic idea is simple, the calculations are not. Brune perfected the technique by examining rocks within areas of recent earthquakes of known magnitude, looking at the effects of underground

Figure 19.3 Marginally stable, precariously balanced rocks on the west face of Yucca Mountain. Photograph by Thomas C. Hanks, US Geological Survey.

nuclear explosions at the Nevada Test Site, pushing and pulling on rocks in the field with specialized equipment, and setting up *shake-table* tests in the laboratory.

In addition to knowing what it would take to knock the rocks over, Brune needed to know how long the rocks had been balanced. There was an ingenious solution to this part of the problem. After erosion exposes buried rocks, bombardment by cosmic rays creates traces of rare isotopes like beryllium-10, aluminum-26, and chlorine-36 on the rock surfaces. By determining the amount of these isotopes, geologists can estimate exposure ages. When the balanced rocks near Yucca Mountain were sampled, many proved to be more than 10,000 years old. Several had maintained their balancing act for more than 30,000 years. This valuable long-term information supplemented the much shorter historical record of earthquakes.

The seismic hazard studies at Yucca Mountain have been called "by far the most complete paleoseismic history for any place on the planet." Even so, surprises are part and parcel of estimating earthquake hazards. This was vividly demonstrated by Japan's devastating 2011 Tohoku earthquake and tsunami, which defied expectations of the maximum event possible at that location. [22, 24]

In 2007, drilling to obtain geotechnical information revealed that a fault ran right under the planned location of aboveground concrete pads for temporarily storing spent fuel casks. After DOE moved the location of the planned facilities, Bob Loux, Nevada's chief anti-Yucca spokesman, called it "just-in-time engineering." Politicians and the media proclaimed the project had unwittingly stumbled upon a previously unmapped fault. In reality, the fault was well known but could now be located more precisely beneath the sediment. Such findings were a reason the drilling was done in the first place. [25–26]

Surprises come in various packages. In May 1986, Alan Riggs and Ray Hoffman were diving in Brown's Room, an underground chamber in Devils Hole. The two USGS scientists were suddenly startled by low, moaning sounds in the normally silent chamber, followed by noises that resembled draining a bathtub. Their initial surprise was heightened when the pool in Brown's Room began to drain, followed by oscillating water levels that sloshed back and forth for over an hour. Riggs and Hoffman were experiencing first-hand the effects of a magnitude 7.7 earthquake nearly 3,000 miles (5,000 km) away in the Aleutian Islands! It was about this same time that the question of what happens to the underground plumbing at Yucca Mountain during an earthquake was becoming one of the most controversial issues in the entire project. We turn to this matter in the next chapter. [27]

The project gets into hot water

The whole aim of practical politics is to keep the populace alarmed (and hence clamorous to be led to safety) by menacing it with an endless series of hobgoblins, all of them imaginary.

H. L. Mencken [1]

In 2002, Jared L. Cohon, outgoing chairman of the Nuclear Waste Technical Review Board (NWTRB) and President of Carnegie-Mellon University, looked back at major accomplishments of the Board. Among these, he observed, "Proving something not to be true is the hardest thing to do in science, and a decidedly unglamorous undertaking. Yet, the Board did not shy away from the challenge presented by the hypothesis of geothermal upwelling. I think we did a very effective job in marshalling limited resources and helping to spawn reviews of what was a very complicated and controversial issue." [2]

The issue referred to by Jared Cohon – the question of hot water upwelling into the repository – was among the most complicated and controversial issues faced by the Yucca Mountain Project. While nearly every scientist even remotely associated with the project considered the possibility so unlikely it had been screened out of the TSPA, a few persistent individuals relentlessly stoked the issue to keep it alive. As it turned out, proving upwelling *not* to be true required scientific detective work beyond Sherlock Holmes' wildest dreams.

UPWELLING WATER

Under today's conditions, the proposed repository would sit about 1,000 feet (300 m) above the water table at Yucca Mountain. During past glacial epochs, the climate was much cooler and wetter with large freshwater lakes in the now arid Death Valley. Yet, evidence of previous water-level rises indicated that the repository would remain well above the water table even during a much wetter, cooler climate. [3]

Aside from climate change, another possibility for a rising water table would be if a volcanic intrusion formed a barrier downstream from Yucca Mountain. In such an event, the water table would rise in the same way a dam impounds water. However, such a barrier to flow seemed out of the realms of possibility. In the unlikely event volcanism did occur, geologic investigations found that it would form narrow dikes along faults that were largely parallel to the direction of groundwater flow. In other words, no barrier and no rising water. A third possibility was that earthquakes could change the stresses in the rocks, causing rapid upwelling of water along faults by *seismic pumping*. Although there was no evidence of an earthquake in the area ever having shifted the water table by more than tens of feet, this possibility became the focus of a drawn-out and highly contentious debate.

The controversy began in the mid 1980s, when Jerry Szymanski, a DOE geologist working on the Yucca Mountain Project, became convinced that earthquakes near Yucca Mountain had repeatedly caused pulses of thermal waters to well up from deep under the mountain. Szymanski warned that future upwelling would flood the repository with hot corrosive fluids and initiate a calamity of vast proportions, possibly spreading radioactivity throughout California and Nevada. To his mind, this scenario of hot water shooting up into the proposed repository was a virtual "certainty." [4]

Szymanski developed his theory after observing nearly vertical mineral veins containing calcium carbonate and silica (opal) exposed in a trench excavated across a fault bordering Yucca Mountain. The trench, designated Trench 14, was one of about two dozen that had been excavated by bulldozer and backhoe to examine past fault movement. Szymanski immediately concluded that the veins in Trench 14 could only be explained by water upwelling from great depths.

There was another much more plausible explanation. Most scientists believed the veins were just one more example of soil development in desert regions, where limited precipitation and huge evaporation forms carbonate and silica deposits. Regardless of the fact that there was nothing uncommon about his discovery, Szymanski spread his message with missionary zeal. In numerous meetings and field trips he proselytized about his discovery – that every other scientist had *missed*. At first DOE ignored him. When he built up a small cadre of followers, Szymanski was told to document his findings in a report.

After completing a draft of his report, Szymanski sent copies to DOE management and to a colleague working for the State of Nevada. The 1987 Amendments to the Nuclear Waste Policy Act had just passed, and here was one of DOE's own saying the site was a disaster waiting to happen. The State employee passed the report along to Nevada Governor Richard Bryan, who immediately went to the press. [4]

Szymanski's report had not gone through the peer-review process, and when it finally did, it failed twice. A panel of 40 federal scientists chaired by William Dudley of the USGS reviewed the report, as well as a group of experts working for the State of Nevada. Both review teams saw nothing remarkable about finding calcium carbonate and silica veins in the desert. When asked if he had read these comments, Szymanski told *Science* magazine he hadn't bothered. He dismissed the reviews as the work of contractors. Szymanski's final 911-page report was released to the public in July 1989. [4–5]

In 1990, science reporter William J. Broad picked up on the story and penned a feature article in the *New York Times*. Broad wrote that Yucca Mountain may one day hold the "most dangerous nuclear facility in the world." He painted Szymanski as a modern-day folk hero – the lone little guy fighting the big, dumb bureaucracy. Broad chronicled Szymanski's life from his birth to Polish nobility, the end of his family's prosperity during World War II, their resistance to invading Germans and Russians, his immigration to the US, penniless, with his family in tow, and his rise to positions of responsibility and influence. In contrast, according to Broad, Szymanski characterized his scientific colleagues as "decent men who were blinded to the obvious by bureaucratic inertia, a desire for career and financial security and a fear that the nation and the nuclear industry would be thrown into turmoil if Yucca Mountain were to be abandoned." The minor space given to other scientists' views was presented in such a biased way that most, if not all, readers would assume that Yucca Mountain was a disaster waiting to happen, with a government conspiracy pushing it through. [6]

When the *New York Times* story came out, scientists at the USGS were so aghast they hardly knew where to begin. Broad's blatantly pro-Szymanski feature story seriously crossed the line of fair and objective journalism. Twenty of the Survey's senior scientists wrote a lengthy letter to the *Times* expressing their concern about the inflammatory nature of the article, the biased presentation of the "scientific evidence," and the implication that scientists at the USGS and the DOE National Laboratories were either incompetent or had compromised their integrity because of fear of losing their jobs. The letter concluded, "Szymanski has dismissed honest criticisms of his ideas as 'banality of thought,' instead seeking scientific legitimacy from the press." The *New York Times* published two short paragraphs of the scientists' letter, along with a mailing address for readers to obtain the full copy. [7]

The pot was now on high boil. To counter the negative internal reviews he had received, Szymanski requested an external peer review. By agreement with DOE, Szymanski picked two university professors sympathetic to his views and DOE selected three scientists familiar with the Yucca Mountain Project. While the protocol for the review called for a single report, the differences of opinion could not be resolved. The three DOE-picked reviewers found no merit in Szymanski's theory, while the two university professors not only supported Szymanski, they thought he had underestimated the risk. [8–10]

In an effort to bring closure to the matter, the National Research Council of the National Academy of Sciences (NAS) appointed a panel to review Szymanski's theory. The panel reviewed all the evidence, interviewed previous reviewers and other experts, and spent several days in the field visiting sites deemed critical to the cases made by scientists on both sides of the issue. In 1992, in its final report, the NAS panel concluded that Szymanski's "proof of upwelling water" was related to the volcanic eruptions that formed the rocks more than 10 million years ago, *or* were "classic examples of arid soil characteristics recognized worldwide." [11]

The panel cited a host of physical and chemical evidence. The texture, chemistry, and mineralogy of the fracture fillings resembled carbonates in local soils, while features typical of hydrothermal springs were nowhere to be found. Oxygen isotopes indicated the temperature of the calcite veins at the time of their formation was too low to be from upwelling water. Isotopes of strontium and uranium in the fracture fillings were within the range of values for local soils but different from any known groundwater in the vicinity of Yucca Mountain. The 17-member NAS panel was unanimous – there was no evidence whatsoever to support Szymanski's theory. The NWTRB agreed with the Academy panel's assessment, but left the door open for additional evaluation "if further significant data or modifications" came to light. [12–14]

By this time, Szymanski had pretty much burned his departmental bridges and had been relegated to a small windowless office. He resigned his position with the Department of Energy in May 1992 and went to work as a contractor for the State of Nevada. Soon after, the Little Skull Mountain quake caused the water table at Yucca Mountain to rise, not by 1,000 feet to the proposed repository level, but by about a foot. Nevertheless the upwelling water debate was far from resolved. [8, 14]

In January 1997, Szymanski petitioned the NWTRB to consider new evidence of his upwelling hypothesis. Over the past five years he had not been idle. Szymanski submitted 11 reports to the Board. The Nevada Attorney General's office added three more. The NWTRB agreed to take another look. The Board asked four scientists, who had not been involved in the previous debate, to review the new evidence.

The reviewers were unswayed. Among other problems, they cited the "selective use of information," "tenuous fits of lines to scattered small data sets," "lack of mention of important dissenting material," "reliance on dubious conclusions reported in earlier reports," and "assertions presented as proofs." With these reviews in hand, the NWTRB once again concluded that the case for upwelling water was not credible. However, there remained one perplexing unresolved issue – tiny pockets of fluids found in minerals chipped from the Exploratory Studies Facility (ESF). This became the focus of the next phase of the controversy. [15–17]

FLUID INCLUSIONS

At the time of the 1992 NAS report, samples were available only from surface trenches and drill cores. Beginning in 1994, excavation of the ESF tunnel provided more direct observations and the opportunity to collect much better samples. As the ESF tunnel was bored, scientists discovered that the rocks are riddled with cavities ranging from thumb size to the diameter of a basketball and larger. Called *lithophysal* cavities after the Latin word for "rock bubble," they were formed by gases as the volcanic rocks cooled and solidified. Some of the lithophysal cavities and fractures are lined with minerals like calcite, opal, and chalcedony. These *secondary minerals*, several millimeters to a few centimeters in thickness, were formed sometime after the volcanic tuffs were laid down. Trapped within some of these thin secondary minerals are microscopic pockets known as *fluid inclusions*.

Interest in fluid inclusions goes back more than a thousand years to Abu Raihon Al-Beruni (973–1048), a prolific scientist and philosopher in Central Asia. Al-Beruni hypothesized that fluid inclusions in gems were remnants of liquids originating in the womb of the Earth. He called them "juices of Earth." [18]

Al-Beruni was correct in the sense that fluid inclusions provide samples of waters that were present when the mineral formed. Remarkably, they serve as tiny thermometers of the temperature at the time the mineral crystallized. If the fluid inclusion is *all liquid*, then the fluid was probably trapped near today's temperature of the rocks. On the other hand, if the fluid inclusion consists of two phases – a liquid *and a vapor* bubble – then the fluid was trapped at elevated temperatures. As the mineral cooled, the fluid contracted and the vapor bubble filled the void. To determine the approximate temperature at which the fluid was trapped, the cooling process is reversed by heating the inclusions. As the temperature rises, the fluid expands and the vapor bubble shrinks. The temperature at which the bubble disappears is assumed to be the temperature at which the mineral originally formed. [19]

Initially, Yucca Mountain Project scientists found only *all liquid* fluid inclusions, supporting the hypothesis that the calcite formed at low temperature. Then along came Dr. Yuri Dublyansky, a member of the Siberian branch of the Russian Academy of Scientists and a scientific expert for the State of Nevada. Dublyansky identified two-phase (liquid *and vapor*) fluid inclusions in samples of secondary minerals he collected from the ESF. The fluid inclusions indicated depositional temperatures ranging from 35 to 85 °C (95 to 185 °F). Dublyansky heralded his findings as unambiguous evidence for upwelling hydrothermal fluids. [20]

Dr. Robert Bodnar, a fluid inclusions expert on the previous four-scientist review team, was intrigued by Dublyansky's findings. After Dublyansky visited his laboratory at Virginia Tech, Bodnar became convinced that the fluid inclusion data were real. In Bodnar's view, this was strong evidence that at least some of the secondary minerals in the vicinity of the proposed repository had been formed at elevated temperatures. But there was still the critical question of the age of the fluids. If the fluid inclusions were remnants of the mountain's volcanic history millions of years ago, then they had little bearing on future possibilities for hydrothermal upwelling. If, on the other hand, some of the fluid inclusions that formed at high temperatures were young (say a few hundred thousand years, as asserted by Dublyansky), that was a different matter. Without the timing information, the data could not be interpreted in any meaningful way. [21]

In an effort to fully settle the matter, the age-dating was done through a joint program between federal and State of Nevada scientists. The effort was led by scientists at the University of Nevada-Las Vegas (UNLV) and the USGS. Participants met on a regular basis to establish a common methodology for sample collection and handling, and to share the results of their investigations. Dublyansky was included as an observer but, to his dismay, was never a full member of the team.

Meanwhile, further evidence accumulated that supported downward water percolation, rather than upwelling. As the ESF was explored, scientists discovered that only a small portion of the fractures and lithophysal cavities contain calcite and opal coatings. This suggested that the unsaturated zone was never inundated with water. If it had been, all rock voids would, or at least should, contain calcite and opal. Furthermore, mineral coatings were found only on the footwalls (i.e. upward facing surfaces) of faults and fractures and on the lower half of lithophysal cavities. This was consistent with thin films of water flowing *downward* through the open spaces. [22]

In May 2001, after years of detailed study, scientists from UNLV and the USGS, along with Yuri Dublyansky and Robert Bodnar from Virginia Tech, presented their findings to the Nuclear Waste Technical Review Board. Dr. Jean Cline, principal investigator for UNLV and coordinator of the overall study, confirmed that a record of hot fluids at Yucca Mountain could be observed in fluid inclusions distributed across the repository site. But there was much more to the story. The UNLV scientists had mapped the chemistry and age of the secondary minerals in microscopic detail, using an electron microprobe and uranium–lead dating. Uranium–lead dating provides ages by comparing the relative amounts of uranium and its final radioactive decay product, lead. This common method of dating geologic events has been used to estimate the age of the Earth. While it was not possible to date the calcite with this technique (there was too little uranium), some of the opal and chalcedony could be dated. A minimum age for a fluid inclusion in the calcite was obtained by dating the overlying opal or chalcedony. [23]

Dr. Cline concluded that hot fluids were present more than 5 million years ago, but there was no record of hot fluids in younger rocks. Dr. Cline also noted that the secondary minerals at Yucca Mountain do not contain characteristics typical of hydrothermal mineralization. The two-phase fluid inclusions are sparse. The vein style is very simple. The mineralogy is indicative of formation at low temperatures. And the minerals were always cooling – they were never cool and then heated up, as would be the case with periodic upwelling. The USGS findings paralleled those of Dr. Cline, while Yuri Dublyansky argued that the crystal structure of the secondary minerals was not possible by downward percolating water. [23]

Robert Bodnar, now serving as an independent reviewer, confirmed the quality of the fluid inclusion data, and also discussed additional evidence against the hydrothermal upwelling hypothesis. Quartz would be expected as a primary mineral in hydrothermal systems but was rare at Yucca Mountain except for the early-stage minerals. Echoing Cline's observation, Bodnar noted that temperature reversals are expected in hydrothermal systems but there was no evidence for them at Yucca Mountain. He explained that Dublyansky's use of crystal morphology to infer the mineral-forming environment was commonly used in the former Soviet Union, but long ago abandoned as unreliable by scientists in Western countries. [23]

During the question and answer period, Szymanski complimented the UNLV researchers on their impressive database. It was a problem of interpretation, he said, and the choice was up to the Board. Handing in his last report, Jerry Szymanski announced that he would now terminate his involvement. He thanked the Board members for their "indulgence" in listening to him over the years. [23]

The extensive investigations by the UNLV and USGS scientists had found no evidence of hydrothermal activity at Yucca Mountain during the past 5 million years. The UNLV team concluded that "the hypothesis of geologically recent upwelling hot water is not valid and should not disqualify Yucca Mountain as a nuclear waste repository." The USGS scientists had come to a similar conclusion. Even the State of Nevada agreed that it would pursue "other scientific questions about the mountain's stability" than the upwelling issue. In January 2002, J. Russell Dyer, project manager for the DOE Yucca Mountain Project, notified NWTRB Chairman Jared Cohon that DOE considered the upwelling waters hypotheses had been "adequately addressed and may be discounted." Having spent many millions of dollars on the issue over the course of 15 to 20 years, DOE was more than ready to move on. [8, 22, 24–25]

Dublyansky remained recalcitrant. Over the next decade, he published numerous journal articles promoting the upwelling hypothesis and submitted comments challenging virtually every article written by the USGS and UNLV researchers. He continued to emphasize the limitations of the USGS thermal modeling to explain the unusual extended period of cooling from 13 million years ago, at the time of eruption, to 5 million years ago. The Nuclear Regulatory Commission also questioned the validity of the thermal modeling scenarios for this initial cooling period, but accepted that hydrothermal upwelling could be excluded from the TSPA. [8]

In 2006, the NWTRB received yet another request to review the upwelling theory because of uncertainties in the thermal modeling. The request came from Harry W. Swainston, former Nevada Deputy Attorney General, now a private attorney associated with a group of "concerned scientists." The NWTRB rejected the request, stating in no uncertain terms, "the Board does not believe that an understanding of the thermal history of Yucca Mountain during periods between the deposition of Yucca Mountain rocks and about 5 million years ago is relevant to predicting repository performance over the next one million years." The case was finally closed. [26]

A COMPACT PLEISTOCENE RECORD

The secondary minerals at Yucca Mountain represent the only physical record of water flow in the unsaturated zone over geologic time. USGS geochemist James Paces and his co-workers decided to examine the youngest layers of opal for evidence of Pleistocene climate signals. The challenges were great. If such a record existed, it would be exceedingly compact – one million years compressed into less than a millimeter of mineral deposition.

Paces and his team analyzed the microscopic opal layers with state-of-the art instrumentation at Stanford University. Using cathodoluminescence, growth stages in the opal were identified as alternating bright and dark bands. Then by using SHRIMP (Sensitive High-Resolution Ion Micro-Probe) – a remarkable instrument that despite its name fills a large room – they were able to determine radiometric age dates at selected spots on the growth bands with resolutions around 20 microns (approaching the size of bacteria). [27–28]

The age-calibrated luminescence in the opal growth bands correlated remarkably well with the Devils Hole calcite records, and also displayed the 100,000- and 41,000-year cycles associated with paleoclimate records. The strong association of the opal growth bands with climate variations was compelling evidence that the secondary minerals at Yucca Mountain formed in response to climate-driven processes, not upwelling water. Even more amazing, the study also revealed fairly uniform rates of secondary mineral growth during a period spanning several glacial–interglacial climate cycles. This strongly suggested that extreme climatic shifts at the land surface were greatly attenuated at the level of the proposed repository. This remarkable story was preserved in less than a millimeter of rock. [28]

THE SCIENTIFIC LEGACY OF YUCCA MOUNTAIN

Refuting the upwelling hypothesis was one of many pioneering scientific achievements at Yucca Mountain. The technical scope of the Yucca Mountain Project, breadth and depth of interdisciplinary studies, intense public review, and the level of quality assurance and regulatory oversight put the project in a class of its own. [29]

The scientific challenges arose in many ways. Some, such as the volcanic and seismic hazards at Yucca Mountain, were identified as significant issues from the outset. Others were surprises – sometimes complete surprises – as the studies moved underground. The findings of chlorine-36, the multiyear drift-scale heater test, and the fluid inclusion studies are only a few examples of the hundreds of studies that led to major scientific advances and insights. Regardless of its ultimate fate, the Yucca Mountain Project created a vast scientific legacy:

- The project made major advances in understanding the flow of water and transport of contaminants through unsaturated fractured rocks. The models and concepts developed at Yucca Mountain can be applied to all kinds of hazardous waste disposal and hydrologic evaluations. As one researcher noted, "Largely as a result of the Yucca Mountain site assessment, fracture modeling has been elevated from a knotty problem of mostly academic interest to an issue with national, if not international attention." [30]

- The project included the largest and most comprehensive seismic hazard evaluation ever performed for a site. Many of the approaches to probabilistic hazard analyses for earthquakes that were developed at Yucca Mountain are being applied to other nuclear facilities, including re-evaluations of the earthquake hazards at the Nation's nuclear power plants. [31]

- The probabilistic volcanic hazard analysis at Yucca Mountain is now considered the standard for conducting such analyses. [32]

- The project made innovative use of natural analogues as qualitative evidence of processes that might occur over the immense timeframes required for waste containment. [33]

- The integrated examination of physical, chemical, thermal, mechanical, hydrologic, and geologic processes at play at Yucca Mountain have provided insights into how to address other complex geologic problems, such as carbon sequestration in deep underground geologic formations. [34]

In addition to these scientific advances, and when all is said and done, Yucca Mountain is the closest the United States has ever come to taking responsibility for its high-level nuclear waste.

Part III   No solution in sight

21

A new President, new policies

A commission. You know, that's Washington-speak for "we'll get back to you later."

> Barack Obama, 2008 Presidential campaign, Grand Junction,
> Colorado, September 17, 2008 [1]

Solving the nuclear waste dilemma requires staying the course over decades with a technically complex and politically sensitive program. In spite of setbacks over the years, the Yucca Mountain Project was surprisingly resilient, surviving four administrations from Ronald Reagan to George W. Bush. Things were looking up when the Department of Energy submitted its license application to the Nuclear Regulatory Commission in June 2008. This optimism proved to be short-lived.

"Yucca Mountain is not an option," Steven Chu told a Senate hearing in March 2009. Chu had six weeks under his belt as the newly appointed Secretary of Energy for the Obama Administration. After more than 25 years and a $10 billion investment at Yucca Mountain, Secretary Chu's sole explanation was, "I think we can do a better job." He did not cite technical or safety issues, nor did he identify alternatives. The new Energy Secretary subsequently announced that a Blue Ribbon Commission would be created to obtain advice on the path forward. Among insiders, this became known as the "anything but Yucca Mountain" commission. [2]

That same month, President Obama spoke at the annual meeting of the National Academy of Sciences where he emphasized his administration's strong support for science. The previous Bush Administration had repeatedly been accused of manipulating scientific findings to deny human-induced climate change and support its anti-regulation agenda. "Under my administration," Obama announced, "the days of science taking a back seat to ideology are over." The President added, "To undermine scientific integrity is to undermine our democracy. It is contrary to our way of life." [3]

In spite of these lofty statements, politics was clearly trumping science when it came to Obama's actions on Yucca Mountain. It was still anyone's guess if Yucca Mountain would meet the stringent tests that would allow it to become a fully licensed and operating repository. One thing, however, was clear – there was no compelling reason to circumvent the ongoing evaluation by the Nuclear Regulatory Commission.

Virtually all observers attributed the decision to pull the plug on Yucca Mountain as political payoff to Senate Majority Leader Harry Reid (D-NV). Nevada was a swing State in the election and Obama had pledged to kill Yucca Mountain, if elected. Obama also needed Harry Reid to help usher through Congress his ambitious political agenda on health care, climate change, and regulation of the financial industry. Moreover, Reid was up for re-election in 2010, and a dead Yucca Mountain would boost his struggling campaign.

A month after Secretary Chu's announcement, 17 Republican Senators sent him a few pointed questions that were on quite a few people's minds. "Have you discovered in a few short weeks, research that discredits the scientific work produced by the National Academy of Sciences, the Nuclear Waste Technical Review Board or any of the National Labs?" "Are you aware of any conclusions by the Nuclear Regulatory Commission that would preclude completion of the license review?" "Did you consult with the Secretary of the Navy regarding possible disruption to spent nuclear fuel defueling operations and storage plans?" "Your decision may cause delays in the clean-up of DOE former weapons complex sites. Did you consult with the relevant governors regarding DOE's potential non-compliance with its commitments under state agreements?" [4]

Initially, Energy Secretary Chu did not kill the project. He would let the licensing process play out as a "learning experience." The Nuclear Regulatory Commission would continue to review the license application and the Department of Energy would answer inquiries about the application, but little else. Yet, Senator Reid was still not satisfied. The NRC learning experience could drag on for years, and Reid worried that a future administration might resurrect the project. [5]

Finally, on July 31, 2009, Reid jubilantly announced his long sought victory – the Obama Administration had agreed to eliminate all money for pursuing a repository license at Yucca Mountain. Seven months later, the Department of Energy filed a motion with the NRC to withdraw the Yucca Mountain license application "with prejudice," essentially making it impossible to ever resurrect Yucca Mountain as an option. The withdrawal motion set off a firestorm of bipartisan protests in Congress. In addition, South Carolina, Washington State, and others filed lawsuits to halt shutdown activities. [6–7]

All lawsuits aside, DOE could not simply take its football and go home. The license withdrawal motion had to go through a lengthy administrative process before the NRC would decide whether DOE could take back its application. In June 2010, the NRC's semi-autonomous Atomic Safety and Licensing Board dealt a blow to Chu's motion by unanimously ruling that only Congress could withdraw the license application. In their view, the Secretary of Energy should not be allowed to make such a momentous decision at his whim. [8]

The following month, 91 Members of Congress sent a harshly worded letter to Secretary Chu demanding that he halt all efforts to terminate the Yucca Mountain Project. Undeterred, the Department of Energy continued at full throttle toward complete closure of the project. According to DOE, it ceased to exist on September 30, 2010. Officials later admitted to the Government Accountability Office (GAO) that "they had never seen such a large program with so much pressure to close down quickly." Staff was shoved out the door, and equipment was abandoned or declared excess and trucked away. The US scientific capacity to study any kind of geologic repository was, as one member of the Blue Ribbon Commission put it, "almost completely dismantled." Top-notch scientists would likely think twice about working on a nuclear waste repository in the future. [9–12]

While the Nuclear Regulatory Commission continued to review the license application, Senator Reid held yet another ace in his hand. Gregory B. Jaczko, Chairman of the Nuclear Regulatory Commission, was formerly Reid's senior aide and close associate. Jaczko had been appointed NRC Chairman by President Obama in 2009. On October 4, 2010, Jaczko announced that the Nuclear Regulatory Commission would end its review of the Yucca Mountain license application. He cited the lack of a budget as justification. In so doing, Jaczko ignored the Atomic Safety and Licensing Board ruling that the NRC lacked such authority, the ongoing challenge in the courts, and the will of Congress. [13]

This was but one of many controversies that swirled around Chairman Jaczko. A report by the NRC Inspector General accused Jaczko of intimidating NRC staff members who disagreed with him and withholding information from the other Commissioners to gain their support. Members of Jaczko's staff called his actions "bizarre," "unorthodox," and "illegal." The other four Commissioners – two Republicans and two

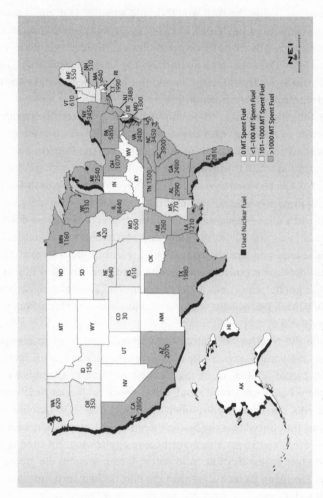

Figure 21.1 Map of metric tons of spent (used) fuel in each State, at the end of 2010. Courtesy Nuclear Energy Institute.

Democrats – wrote to the White House Chief of Staff that Jaczko's behavior was "causing serious damage" to the agency and "creating a chilled work environment." [14–15]

The NRC Commissioners could uphold or reverse the Atomic Safety and Licensing Board's ruling. For more than a year, Jaczko refused to call a formal meeting to vote on the case. The Chairman did not publically state his reason, but the foot-dragging kept the lawsuits at bay. The courts ruled that legal challenges to the withdrawal of the license application were premature until the government (i.e. the NRC Commissioners) made a final decision. When the Commissioners finally did vote in September 2011, the vote was a 2 to 2 tie. Only four out of five Commissioners voted, as one Commissioner had to recuse himself because of previous involvement with the Yucca Mountain Project. With this vote complete, the Court of Appeals is currently (April 2012) considering the legal challenges to the Yucca Mountain closure. [16]

By now, any semblance of a functioning US nuclear waste program had completely broken down. It was up to the Blue Ribbon Commission to try to put Humpty-Dumpty back together again.

THE BLUE RIBBON COMMISSION

President Obama's Blue Ribbon Commission on America's Nuclear Future was a classic government cop-out. Nevertheless, the 15-member BRC took its job seriously. The Commission was co-chaired by Lee Hamilton and Brent Scowcroft. Hamilton, a 17-term Congressman from Indiana, previously served as vice chairman of the 9/11 Commission. General Scowcroft was National Security Advisor under Presidents Gerald Ford and George H. W. Bush.

As the BRC prepared its report, the worst nuclear disaster since Chernobyl illustrated how problems that are completely unexpected can arise suddenly out of the blue. At 2:46 p.m. on March 11, 2011, a magnitude 9.0 earthquake occurred 110 miles (180 km) off the coast of Japan. Moments after the quake, Fukushima's operating units began to shutdown, but it was too late. At 3:23 p.m., a tsunami almost 50 feet (15 m) high overwhelmed the Fukushima Dai-ichi nuclear power plant, causing a loss of power and core meltdowns, fires, and radiation leaks.

Among the many serious problems at Fukushima were concerns about the release of radiation from the spent fuel stored in pools. Operators were unable to run the pumps that circulate water to keep the spent fuel cool, or even to monitor water levels in the pools. The Fukushima pools contained many fewer assemblies than what is typically stored in US spent fuel pools. A similar unforeseen catastrophe in the United States could be even more disastrous.

The BRC report, released in January 2012, acknowledged that the elements of its strategy "will not be new to those who have followed the U.S. nuclear waste program over the years." In other words, they didn't come up with any new ideas because there are none. The Commission affirmed the long-standing consensus that disposal in a geologic repository is the only feasible, permanent solution to the high-level waste problem. With no apparent sense of irony, the Commission called for "prompt efforts to develop one or more geologic disposal facilities." The report carefully side-stepped taking a position on Yucca Mountain, nor did it identify any scientific advances that would make the Yucca Mountain repository obsolete. Left unsaid was that, with Yucca Mountain *off* the table, the entire country was again back *on* the table for possible sites. The BRC advised greater US international leadership to address safety, nonproliferation, and security concerns in nuclear waste management, but chastened that "the United States cannot exercise effective leadership . . . so long as its own program is in disarray." [17]

The BRC report breathed new life into interim storage, calling for prompt efforts to develop one or more consolidated interim storage facilities. The Commission voiced concern about nuclear waste stranded at decommissioned reactor sites and in packed spent fuel pools at operating reactors. In the event of a Fukushima-type emergency, they warned, the US nuclear industry has no place to move the fuel.

Nuclear power plants are not the only places with excess spent nuclear fuel. By January 1, 2035, the US Navy must remove its spent fuel from the Idaho National Laboratory or the government could face a penalty of $60,000 each day the waste remains in Idaho. Along with a similar agreement between DOE and Colorado, the penalties could total about $27 million annually. Yet, money is only part of the problem. If the milestone is not met, Idaho can refuse to receive any more spent fuel from the Navy's fleet of nuclear submarines and aircraft carriers. [18]

The BRC recognized that, in the past, tribal support among the Goshutes and local support at Knoxville, Tennessee had not been sufficient to overcome State-level opposition (NIMS) to interim storage. To be successful, State concerns must somehow be ameliorated while capitalizing on local support. The BRC also recognized that *real* progress must be demonstrated in developing a geologic repository to assure that an interim site would not become a de facto long-term solution. In spite of these impediments, the BRC expressed confidence that satisfactory interim sites could be found.

A crucial part of this optimism, and a central theme of the BRC report, is based on a tactical change. Instead of trying to enforce their will on a State and local community (the ever recurrent NIMS and NIMBY problems), the BRC recommended an open, transparent, consent-based approach. Of course, 25 years earlier this approach had gone down in flames with interim storage.

LOOKING FURTHER AFIELD

As part of their study, the Blue Ribbon Commission sent delegations to Sweden, Finland, France, Japan, and the United Kingdom to investigate their repository programs. They also heard a presentation about Canada's nuclear waste program. Much was learned.

While a volunteer approach seems to be working in Sweden, Finland, and France, the jury is out in other countries. After decades of failure with top-down, technically centered efforts, the UK and Canada are seeking volunteer communities through phased, open processes. Thus far, communities in the UK near Sellafield are discussing the *possibility* of participating in the process. In Canada, several communities are in the earliest stages of discussion and information gathering. Canada bears watching, given that their strong provincial governments are somewhat comparable to US States. [19]

In Japan, the situation is murkier. In 2002, the government launched a voluntary repository siting process. When the mayor of Toyo-cho, southwest of Tokyo, announced that he would respond positively to the solicitation, opposition immediately arose with the local community and from governors of nearby prefectures. The mayor was soundly defeated in an election that served as a referendum on his actions. Since then, no other community has stepped forward. Needless to say, the meltdown of the three reactor cores at Fukushima has seriously diminished any possibility of finding a volunteer community to host a geologic repository. [19]

For three decades, Sweden and the United States had the most advanced spent fuel repository programs in the world. Suddenly, the roads diverged in 2009 when President Obama shut down Yucca Mountain while the Swedish repository program forged full steam ahead. This was a busy year for the Swedish Nuclear Fuel and Waste Management Company (SKB) which recommended a site, complete with locals on board, for repository construction. Two years later, SKB submitted their repository permit application and related environmental impact statement. As things now stand, Sweden expects to have a fully operational nuclear waste management and disposal program in place by 2025.

Leif G. Eriksson, a self-described "nuclear waste undertaker," has had extensive involvement with both the Swedish and US repository programs since 1978. Eriksson minces no words in diagnosing the two primary "root causes" of the now all-but-defunct US repository program, citing the "pertinacious lack of will by Congress" and the "pestilent performance of the implementing organization, the U.S. Department of Energy's Office of Civilian Radioactive Waste Management (OCRWM)." [20]

In Eriksson's view, the proof is in the pudding. The SKB is owned and controlled by Sweden's nuclear industry, with upper management positions filled by nuclear energy and waste management career professionals. In the United States, the positions of director of the OCRWM all the way up to the Secretary of Energy are filled by politically appointed individuals who often have limited expertise and experience in nuclear waste management and disposal, and usually come and go with the administration at whose pleasure they serve. By way of circumventing NIMBY, a cornerstone of the Swedish program has been to seek out strictly volunteer communities. (NIMS is not a problem in Sweden or Finland – they don't have States.) In an unusual move, the community selected for the repository received only one-quarter of the "benefits package." The remainder went to the runner-up to compensate for the lost future benefits resulting from not hosting a repository. In contrast, the US siting approach has been largely driven by finding a suitable host rock and engineering design, with little attention given to solving the human components of the problem. [19–20]

The Blue Ribbon Commission recommended learning a few things from the Swedes, by way of developing an approach to siting and developing nuclear waste management and disposal facilities in the United States that is "consent-based, transparent, and standards- and science-based." While all this looks great on paper, the difficulties of finding such a happily-ever-after triad among a suitable site, consenting State, and volunteer community, cannot be lightly dismissed. In addition, the US has the dubious distinction of being the most litigious country in the world, whereas Sweden's socialist economy sets the stage for enhanced cooperation and teamwork. [17]

Finland also appears to be in the repository home-stretch, with a projected opening date of 2020. Like Sweden, the Finnish program enjoys local cooperation and adequate funding for their site, Onkalo (meaning *hidden* or *cave*). As with the US Nuclear Waste Fund, the

Finnish government created a cleanup fund for their nuclear waste. However, unlike in the United States, the fund is actually funding the project.

Onkalo lies about 185 miles (300 km) northwest of Helsinki, and is being hewn out of solid bedrock that will reach 1,600 feet (500 m) below land surface. At completion, a spiraling track around the repository will measure three miles (5 km). The spent fuel will be secured through a system of multiple barriers – bedrock of the Fennoscandian Shield, iron canisters encased by corrosion-resistant copper, and a bed of bentonite clay to cushion the canisters and keep water at bay.

The BRC report discussed at length the underlying reasons why the US nuclear waste program is in complete disarray, while Sweden and Finland seem to be getting the job done. At the same time, the report fails to acknowledge the extent to which all three countries share a serious problem in making an airtight case for containment. Getting the waste in the ground is one thing – *keeping it there* over geologic timeframes is an entirely different matter.

The major potential disruptive events at the Yucca Mountain site are earthquakes and volcanoes, while Sweden's and Finland's far northerly latitude means they must design their repositories for the next ice age. When this happens (and with no reason to think it won't), somewhere around *two vertical miles of ice* (3 km) will again descend on the northerly latitudes. This weight will deform the bedrock and compress the canisters and fuel rods. Assuming no engineering or construction defects or completely unexpected problems, the canisters have been designed to withstand these intense pressures. [21]

However, even thornier problems loom during the "rapid" retreats of future ice sheets. As the great weight of ice is lifted, the resulting increased seismic activity could fracture the rocks and bentonite clay surrounding the canisters. The Fennoscandian Shield bedrock is full of faults, fractures, and bedrock caves (described as "explosive fracturing") resulting from seismic glacial unloading. During the last deglacial some 10,000 years ago, paleoseismic evidence reveals events larger than the 1964 Alaska earthquake. Compounding this problem, water from the rapidly melting massive ice sheets will percolate deep into the subsurface, carrying oxygen with it. Exposure of the copper canisters to this oxygenated water would jeopardize their corrosion resistance. [22]

The Swedes and Finns have accepted these far distant uncertainties in moving ahead with their waste problem. The two countries analyze conditions far into the future, but require a much less prescriptive and quantitative analysis than in the US. "It boils down basically to trust," comments Timo Aikas, vice-president in charge of engineering for the Onkalo facility. "When you make a decision concerning this kind of thing... there will always be uncertainty." [21]

France is also moving forward with a repository program. After reprocessing their spent fuel to squeeze out another 20 percent of energy, France ends up with spent MOX fuel and high-level reprocessing wastes. Their planned repository site for vitrified high-level waste (not spent fuel) lies in northeast France, near the small town of Bure. Scientists are convinced that thick clay beds (argillite) with their extremely slow water movement and high sorption capacity can safely house this high-level waste. France has no plans for burial of their spent MOX fuel, which is being stored indefinitely at the reprocessing plants in de facto centralized interim storage.

The French take a pragmatic view to the uncertainty problem. Patrick Landais, the scientific director of the French National Radioactive Waste Management Agency, explains that a few thousand years after the repository is sealed, "the stainless steel would corrode away until it was ruptured by the pressure of the rock, leaving the vitrified waste, and the rock itself, to provide containment." He adds that rock (argillite) is not an absolute barrier, and therefore "The idea of a geological safe does not exist." On the other hand, Landais points out that it should take hundreds of thousands of years for the radionuclides to reach the surface. By that time, their low concentrations and lower levels of radioactivity would render any environmental contamination negligible. [23]

These European countries have been able to elicit local and government support while also viewing uncertainty over geologic time spans as part of the nature of the beast. They are, therefore, moving ahead. On the other hand, the United States has been unsuccessful in coordinating local and State support with the federal government's will. On top of this deadlock, although the regulations require demonstrating "reasonable expectation" of repository performance, the US has become fixated with absolute certainty in finding a guaranteed-safe solution for up to a million years out. The price is a stymied repository program.

While failing to fully address the sociological and safety-case differences between the US and Sweden, Finland, and France, the Blue Ribbon Commission report recommended looking for a strictly volunteer community for hosting an interim storage facility or geologic repository. While this was tried, unsuccessfully 25 years ago for interim storage, at that time the US had a stronger economy and lower unemployment. It's possible that in today's more competitive world, monetary incentives and jobs might finally make the elusive volunteer proposal a reality.

Ironically, Nye County, Nevada, which includes Yucca Mountain within its boundaries, has long supported a geologic repository at Yucca Mountain because of the jobs and economic development the facility would bring with it. Six weeks after the BRC released its report, officials in Nye County sent a letter to Steven Chu. "Nye County, Nevada," the letter read, "hereby provides notice to you, the Secretary of Energy, that we consent to host the proposed repository at Yucca Mountain." A week later, Nevada Governor Brian Sandoval wrote to Secretary Chu, informing him that, no matter what Nye County says, the State is not going to change its official position – Yucca Mountain is not going to happen. [24–25]

The BRC also recommended creating a new organization dedicated solely to implementing the nuclear repository program. This recommendation came as a surprise to almost no one. Beyond the truism that reorganization is a staple of committee recommendations, the current DOE organization for nuclear waste management has virtually no supporters. Even former DOE managers called for removal of this responsibility from the Department.

The Department of Energy has long been a troubled agency. Pointing to several GAO reports, Richard and Jane Stewart in their book, *Fuel Cycle to Nowhere*, summarize many of the problems. Among these are difficulty recruiting and retaining qualified personnel to discharge its highly demanding tasks, excessive power handed over to contractors, and an organizational culture of secrecy and isolation inherited from the AEC. The latter problem has been exacerbated by government lawyers who clamp down on disclosure of any records or views that might prejudice the government's case. [26]

In a review of lessons learned from Yucca Mountain, the Nuclear Waste Technical Review Board pointed to the lack of continuity of management, personnel, and funding as a major deficiency. "Contractors came and went, and managers cycled in and out, while the amount of money available in the next fiscal year was always in doubt, and seldom under the control of the management of the program." This was no way to run a railroad, let alone a program facing such long-term challenges. [19]

According to the BRC, the Department of Energy might retain responsibility for defense waste and other DOE-owned waste (this required further evaluation), but responsibility for civilian spent fuel should be vested in a new single-purpose, congressionally chartered federal corporation, not unlike the Tennessee Valley Authority. Such an organization, supposedly, would be above politics and focus instead on getting the job done. The BRC recognized that this new entity could only succeed if it were empowered with sufficient authority and resources. A key was access to funding that did not depend on the whims of Congress. [12, 17]

Since the Nuclear Waste Policy Act of 1982, utilities have been paying into a Nuclear Waste Fund intended to finance waste disposal. Amounting to more than $750 million each year, these costs are passed along to the customers of nuclear power. Except for a few special cases, such as the damaged core from the Three Mile Island accident, no waste has been picked up.

Worse yet, the Nuclear Waste Fund is a fund in name only. The revenues have not been set aside in an account earmarked for waste disposal. Instead, Congress has used the money for general government spending, while congressional opponents of Yucca Mountain worked hard to keep funding for Yucca Mountain as low as possible. Including interest, the current unspent balance in the fund is nearly $27 billion. By a sleight of hand, this $27 billion is used to reduce the national debt, but (theoretically) must eventually be paid out for its intended purpose. [17]

The United States stands alone as the only country where the annual expenditure of funds collected for nuclear waste management is controlled by the national legislature. The Blue Ribbon Commission "spent considerable time on this issue" and recommended legislative changes to extricate the nuclear waste fund from the web of annual congressional appropriations. Meanwhile, it is not lost on anyone that the *only* promise the federal government has fulfilled in spent fuel disposal is to collect the fees. To date, no one has gotten a good return on their investment. [17]

Nuclear waste and our energy future

We have conquered Mother Nature;

now we have only to conquer human nature.

D. R. Knowlton [1]

Imagine that starting tomorrow the world permanently shutters all of its nuclear reactors – in power plants, submarines, aircraft carriers, universities, and medical research laboratories. Even in such an event, massive amounts of spent nuclear fuel, high-level reprocessing wastes, and excess plutonium from dismantled nuclear warheads would still be with us. This huge backlog of wastes remains, whether or not we ever produce another kilowatt of electricity from a nuclear power plant. Regardless of one's stance on nuclear energy, everyone has a stake in finding a long-term solution for the waste.

Adding urgency to solving the waste problem is that nuclear energy is one of the proposed solutions for addressing global warming. In a widely cited article in 2004, Stephen Pacala and Robert Socolow of Princeton University argued that humanity can solve the climate problem by scaling up what we already know how to do. Although technically feasible, the magnitude of this scale-up is monumental. Meeting the world's projected energy needs over the next 50 years, while also stabilizing atmospheric carbon dioxide (CO_2), would require the equivalent of *all* the following actions: (1) increase fuel economy for 2 billion cars from 30 to 60 miles per gallon, (2) cut carbon emissions by one-fourth in buildings and appliances, (3) replace 1,400 GW (gigawatts) of coal plants with natural gas plants amounting to four times the current production of gas-based power, (4) capture and store 80 percent of the CO_2 from today's coal energy production, (5) increase biofuels production to a level that would require one-sixth of the world's cropland, (6) install 2 million 1-megawatt wind turbines that would occupy an area equal to three percent of the United States (some on land and some offshore) and, finally, (7) replace 700 GW of coal-based power generation with nuclear energy, an increase equal to about twice the current worldwide nuclear capacity. [2]

The future of nuclear energy is impossible to predict. In its September 8, 2007 edition, *The Economist* heralded "Nuclear power's new age" on the magazine's cover; less than five years later, the March 10, 2012 cover declared "Nuclear energy: The dream that failed." In spite of the uncertainties, nuclear energy appears to be here to stay for some time to come, even if used as a bridge to a world much more reliant on renewable sources of energy, such as wind and solar. More than 60 nuclear power plants in 15 countries are currently under construction, many of them in Asia. As a result, the nuclear waste problem is a growing worldwide dilemma, leaving us with two fundamental choices – finding a responsible solution to the nuclear waste problem or shifting the burden to future generations. [3]

The technical characteristics of nuclear waste make the disposal problem difficult, yet it is the human factors that have made it intractable. These include a lack of interest in solving the problem, unrealistic demands for earth-science predictions far into the future, eroding confidence in government and institutions, confusion about which "experts" to trust, and the ever present NIMS and NIMBY. A better understanding of these human elements is imperative to avoid past failings.

LACK OF INTEREST

In the early years of the nuclear age, the problem of waste disposal was easily postponed to another day in the mistaken belief that it would readily yield to some simple technical solution. Experts assured the public that perpetual care was neither difficult nor costly. As nuclear energy became controversial, more and more people became opposed to doing anything with the waste. "They don't have any position, except to not deal with it," observed Paul Slovic, an expert on risk analysis who served on the NAS Board on Radioactive Waste Management in the mid 1990s. "Any suggestion you could make would be considered unacceptable," Slovic added. In addition, environmental groups viewed stopping the search for a repository as a way to stop nuclear power, so doing nothing was a good option in their view. [4]

In recent decades, apathy towards solving the problem is spawned by the new holding pattern of dry cask storage. The utilities, of course, would prefer to get the waste off their sites, but on-site storage in dry casks will do as long as the government pays for the storage and the utilities can continue with business as usual. Neither the nuclear industry nor the electric utilities want to rock the boat by voicing concerns too loudly about the need for a repository. If these plans vaporize – as they did for Yucca Mountain – then so does their credibility that the waste problem is well under control.

As a result, the Yucca Mountain Project had few staunch defenders among environmentalists or industry. Its cancellation by the Obama Administration was greeted with a great big yawn. Even in Las Vegas, the editorial pages of local newspapers bemoaned the lack of celebration. Why, they asked, are people not "dancing in the streets?" [5]

A PREOCCUPATION WITH PREDICTIONS INTO THE FAR-DISTANT FUTURE

Disposal of high-level waste is a first-of-a-kind endeavor, further saddled by the ambitious goal to achieve nearly escape-proof containment for time periods beyond our wildest imagination. The scientific imperative in the US has become one of knowing *everything* about *all* phenomena and possibilities before *any* action can be taken. The potential consequences ten thousand, one hundred thousand, even a million years into the future carry almost equal weight to what might happen during the next few hundred years. It is no wonder that paralysis is the result.

This long-term view is unique. Humans continue to deplete the world's resources, cause mass extinction of species, destroy the ocean's fisheries, destabilize the world's climate, and poison the environment with persistent toxic chemicals – many of which will outlive the radioactive ones. In these matters, humans have given little thought for the next few generations, let alone many thousands of years in the future. Serious concerns about the long-term consequences of nuclear waste disposal are appropriate, but only in the context of a suitable realism about scientists' predictive capability. It's also easy to underestimate the immense unknowns associated with future societal values and patterns of human settlement – not to mention natural disasters. Even when the standard was established for 10,000 years, the NAS Board on Radioactive Waste Management warned of the "scientific trap" set by encouraging the public to expect absolute certainty about repository safety – and DOE program managers acting like they could provide it. Paradoxically, this emphasis on predicting repository behavior *thousands of centuries* into the future, stands in stark contrast to the almost complete lack of risk assessment of indefinite on-site storage. [6]

There is another short-term risk not typically considered. While nuclear power plants can be retrofitted with new parts reflecting technical advances, the current US fleet of nuclear reactors is based on designs and technology from the mid 1950s (a time, for example, when only rudimentary digital computers existed). More than half of the Nation's nuclear power reactors have received license renewals beyond their original 40-year license, with many others soon up for review. The decision to renew old plants rather than replace them with newer safer ones is, in part, another holding pattern resulting from the waste dilemma. A number of States continue to uphold moratoriums on building any new nuclear power plants until a repository is available. As a result, these States must rely on an aging nuclear power infrastructure or find large alternative energy sources. With nearly 10 percent of the population of the USA, no State illustrates this dilemma more than California.

California has two nuclear power plants – Diablo Canyon north of Los Angeles and San Onofre on the oceanfront in northwest San Diego County. The licenses for the reactors at these plants expire in approximately a decade (much less than the time required from conception to a fully operating new power plant). Construction at Diablo Canyon began in the late 1960s, but the reactors at the plant were not commissioned until the mid 1980s as a result of regulatory and legal hurdles revolving around seismic hazards. San Onofre is likewise dated and currently shut down due to technical problems. The California moratorium on building any new nuclear power plants until a final resting place for spent fuel is guaranteed means that replacing these plants with newer, safer designs is not on the table, nor likely to be for decades.

The current approach to nuclear waste disposal exemplifies how an overcautious attitude can paralyze action and possibly end up leaving us less safe. As political scientist Aaron Wildavsky cautioned in the late 1970s, "by trying to make ourselves super-safe, for all we know, we may end up super-sorry." [7]

CRISIS OF CONFIDENCE

The current US political system seems incapable of producing long-range answers to big problems or opportunities of almost any sort. The Department of Energy also inherited a legacy of public mistrust and has made little progress in dispelling these concerns. All the same, the task is not an easy one.

President Obama's Blue Ribbon Commission (and many others) has advised transferring the nuclear waste program to a quasi-independent organization that could be more efficient and less affected by politics. Yet, even if such a program's management were improved, it is far from assured that a new organization could avoid the political controversy that has dogged the DOE program. Without fundamental changes to assure continuity across administrations and avoid self-interested interference by politicians, a new organization could be just another version of rearranging the deck chairs on the Titanic. [8]

A realistic starting point for nuclear waste disposal would be to acknowledge that any course chosen will be an imperfect solution. The problem is just too big, too complex, and too long. As investigations proceed, surprises should be expected and this expectation acknowledged from the outset. To recognize these uncertainties, a stepwise decision-making process and phased implementation over decades are required. The public also deserves a realistic and understandable safety case for a repository – as well as an appreciation of the possible consequences of continuing to dodge the problem.

Credibility is a crucial part of solving this dilemma. Schedules and mileposts are an unavoidable part of any nuclear waste program, yet there needs to be enough flexibility that the scientific and technical work can be conducted in a deliberate, careful, and unhurried manner. The United States is the only country in the world to have set a deadline for opening a high-level waste repository, and is among the few countries to rely so heavily on a single candidate site. Both of these characteristics inevitably led to the perception that meeting deadlines is more important than thoughtful deliberation.

One of the biggest stumbling blocks in developing an interim storage facility or a geologic repository is the public's fear and mistrust. Science and technology issues involving nuclear waste are incomprehensible, and therefore frightening, to most nonscientists. Here is where we could learn a lesson from WIPP. The Environmental Evaluation Group (EEG), which was established as an independent technical oversight group at the WIPP site, looked out for the public's concerns. In contrast, the Nuclear Waste Technical Review Board has proved advantageous in challenging DOE in key technical areas, yet its purpose is to advise the President and Congress, not to represent the public's concerns.

Many years ago, science writer Luther Carter argued that, "trust will be gained by building a record of sure, competent, open performance that gets good marks from independent technical peer reviewers and that shows decent respect for the public's sensibilities and common sense." These ingredients do not ensure success, but in their absence, failure is guaranteed. [9]

Figure 22.1 The 100th shipment of Rocky Flats waste to WIPP.
Source: US Department of Energy.

RECOGNIZING SOUND SCIENCE AND ITS LIMITATIONS

In the early days of atomic physics, the word of scientists went unquestioned. David E. Lilienthal, chairman of the AEC from 1946–1950, described this heady time:

> As "master of the Atom," the scientist had transformed the world. His
> views on all subjects were sought by newspapermen, by Congressional
> committees, by organizations of all kinds; he was asked in effect to
> transfer his scientific mastery to the analysis of the very different
> questions of human affairs: peace, world government, social organization,
> population control, military strategy, and so forth. And his authority in
> these nonscientific areas was, at least at first, not strongly questioned. [10]

Scientists are no longer viewed as the fountain of wisdom, or even, necessarily, a trusted source. To a point, this skepticism is healthy. Psychologists like Paul Slovic have found that experts' judgments "appear to be prone to many of the same biases as those of the general public, particularly when experts are forced to go beyond the limits of available data and rely on intuition." To earn public trust, experts need to be competent, truthful, *and* aware of their personal and professional limitations. [11]

Yet, these personal and professional limitations are compounded by a troubling recent development. It is extraordinarily difficult (if not downright impossible) to address the complex problem of high-level nuclear waste in a society where a large percentage of the public places little or no value on *facts*. Today's culture of infotainment, sound bites, fundamentalist religion, ideological extremism and rigidity, and the politics of fear and hate impairs reasoning and thoughtful debate. As an astounding case in point, contemporary Americans are as likely to believe in flying saucers as in evolution. Depending on how the questions are worded, roughly 30 to 40 percent of Americans believe in each. When asked about evolution, President George W. Bush hedged his bets, saying the "jury is still out." Nicholas Kristof of the *New York Times*, quipped "no word on whether he believed in little green men." [12]

This brings us to the problem previously referred to as "publishing in the *New York Times*." Thoughtful differences of opinion among scientists and engineers always have and will exist. Yet, the general public is at a huge disadvantage in distinguishing between legitimate scientific debates and those who use the media to twist science in order to promote their pet theory or personal agenda. These opponents can easily cry "inadequate science" when, in fact, their true concerns are related to other factors.

G. Brent Dalrymple, one of the originators of the theory of plate tectonics and a member of the Ward Valley NAS Committee, summed up the problem: "Skepticism and challenge to the status quo are the stuff of scientific breakthrough and new insight; that is when they are coupled with the arduous, unglamorous, often tedious process of gathering evidence and analyzing data to test a hypothesis. When an individual comes along who short-circuits the process, dodges the scrutiny of the peer-reviewed scientific literature, and takes his hypotheses directly to the press, the scientific community quite rightly views him as suspect, at best over-zealous, and at worst a charlatan. Too often, however, the press depicts such individuals as misunderstood and persecuted mavericks bucking an intolerant and self-protective establishment. That may make for a better story, but the truth is usually something quite different." [13]

Dalrymple was alluding in part to the Szymanski controversy, but also to an individual who had predicted that the alignment of the Sun and Moon with the Earth would trigger a major earthquake on December 3, 1990 along the New Madrid, Missouri seismic zone. A panel of the Nation's leading seismologists reviewed the methodology and concluded it was without merit. Yet, the press continued to report the findings as though the lone prediction had equal standing with the scientific community's view. As a result, many parents kept their kids home from school or sent them donning helmets on the fateful day. Some schools and businesses closed. Needless to say, the earthquake never happened.

Science literacy is much more than number crunching and memorizing facts. It requires a basic understanding of the scientific process and an appreciation for the fact that the more scientists learn, the more questions there are to ask. Without an understanding and respect for this process, the public is vulnerable to self-proclaimed experts who sidetrack efforts through unsubstantiated claims; resort to personal attacks on the integrity of scientists whose findings disagree with their agenda; and point to minor errors or inconsistencies as proof that the whole system is a conspiracy to deceive. All too often, the media have exacerbated this problem. [14]

NIMS AND NIMBY

In his book on nuclear waste, Luther Carter recounted an interview with John O'Leary, Deputy Secretary of Energy under President Carter. Although the interview focused on WIPP, O'Leary unknowingly predicted the fate of Yucca Mountain. "When you think of all the things a determined State can do, it's no contest." O'Leary cited the power a State has to regulate its lands, highways, employment codes, and so forth. The federal courts might strike down each of the State's blocking actions, but meanwhile years can roll by. [9]

Without their cooperation, State and local interests often prevail over national needs. This lesson has been demonstrated repeatedly throughout the history of nuclear waste disposal. To move this mountain of instinctive opposition, citizens of the State and local jurisdiction must see a clear benefit, feel empowered to voice their concerns and have them seriously addressed, and have a basic sense of trust and fair play as the process moves along. People also need to know why it matters. This involves more than an information campaign whereby the operating agency tries to educate the public. What is long overdue is a mature dialogue, as equal partners, between an informed public and the operating agency. Short of this, it is doubtful whether the public will ever come to appreciate why it matters for society to take responsibility for its high-level nuclear waste.

Our story began with President Dwight D. Eisenhower waving a magic wand to start construction of the Nation's first commercial nuclear power plant. It is fitting that we end with Eisenhower's farewell speech as President. Delivered on January 17, 1961, the speech sparked little interest at first. Ultimately, however, it became the most memorable farewell address by a Chief Executive since George Washington urged the fledgling Nation to stick together in the cause of its founding principles – a message that seems equally relevant today. [15]

In his now famous words, Eisenhower coined the term *military–industrial complex*. He cautioned, "In the councils of government, we must guard against the acquisition of unwarranted influence, whether sought or unsought, by the military–industrial complex." Eisenhower then turned to the role of the populace: "Only an alert and knowledgeable citizenry can compel the proper meshing of the huge industrial and military machinery of defense with our peaceful methods and goals, so that security and liberty may prosper together." This same admonition applies to the high-level waste dilemma. After billions of dollars of taxpayers' money and untold hours of study, the manifestation of that goal is no closer today than it was more than a half-century ago.

Discussion questions

SCIENCE AND RISK

1. (Part I) The long-lived radioactivity and intense heat generated by radioactive decay make high-level nuclear waste unique among society's wastes. How do these characteristics affect how the problem is viewed and the difficulty of resolving it?

2. (Part I) Many Americans today have a diminished trust in Congress and the federal government. How do you think most people view scientists?

3. (Part I) For decades, the international scientific community has viewed a geologic repository as the only viable "solution" to dealing with the waste. Nonetheless, many people continue to think that we should recycle the waste. What are the arguments for and against reprocessing? Given these complexities, based on your reading of Chapter 7, should reprocessing be part of the nuclear fuel cycle?

4. (Part I/Part II) While the "uncertainty factor" is a problem with developing a geologic repository, many Americans appear to be more comfortable with the uncertainties associated with spent-fuel pools and open-ended dry cask storage. How would you explain this paradox?

5. (Part II) One of the characteristics of scientific research is that as new information is obtained toward resolving one question, new questions arise. Give examples of this from the book. How do you make any policy decisions regarding high-level waste when the scientific studies continue to raise new questions?

6. (Part II) From a scientific viewpoint, what were the strengths of a proposed geologic repository at Yucca Mountain? What were the weaknesses? How did the scientific surprises and nuances

discussed in the book complicate these strengths and weaknesses? In your opinion, based purely on the science, should Yucca Mountain have continued through the licensing application process or should it have been shut down? What about when we take into consideration competing stakeholder views?

7.   (Part II) Determining the regulatory standards for Yucca Mountain was a highly contentious issue (Chapter 17). Among the controversies was the ruling by the US Court of Appeals in 2004 that the timeframe for measuring compliance with the standards should be one million years rather than the 10,000-year period proposed by the Environmental Protection Agency. Summarize the contributing factors to this ruling. If you had been the judge in this appellate court, how would you have ruled? How would you have supported your ruling?

COMMUNICATION AND THE PUBLIC

8.   (Part I/Part II) Perceptions about the difficulty of managing high-level waste range from it being a trivial problem to a complex, perhaps insoluble problem. Why do such extremely different views exist, and which do you think it is?

9.   (Part I/Part II) How do the public and media distinguish between scientists challenging the status quo as "honest brokers" versus those who use science to promote a personal or political agenda?

10.   (Part II) Government scientists (USGS, Lawrence Berkeley National Lab, etc.) have a limited ability to respond to professional or character attack that comes through the media. Is this lack of a level playing field a problem in solving controversial scientific problems or a necessary check on the government?

11.   (Part II) Good engineering practice usually involves prototyping first-of-a-kind systems. The Yucca Mountain Project placed heaters in the mountain as a multiyear test of the effects of prolonged heat on the repository, but did not undertake any prototyping with high-level nuclear waste. Describe the pros and cons of emplacing a sample of nuclear waste in the mountain as a test. How would you communicate with the public about such a test?

12.   (Part II) Taiwan's public information campaign to build support for a new nuclear power plant seriously backfired (Chapter 17). On the other hand, the Atomic Energy Commission's decision to withhold critical information to create public acceptance for the Nevada Test Site succeeded beyond their wildest dreams. In this Information

Age, how do governments balance the need to fully disclose the risks without creating a knee-jerk oppositional reaction?

POLICY AND POLITICS

13. (Part I) In Chapter 5, the book draws an analogy between the current default "solution" of dry cask storage and shooting an arrow into a wall, drawing a bull's eye around it, and declaring yourself an excellent marksman. Is this an appropriate analogy? Based on your reading, what are the pros and cons of extended dry cask storage?

14. (Part I/Part II) NIMBY (not in my backyard) understates the problem. It is not just communities, but States (NIMS) that don't want a geologic repository within their borders. The Carlsbad, New Mexico community was generally receptive to the idea of WIPP, but the State leaders fought long and hard to keep it out (Chapter 11). Even more complicated is when a sovereign tribe, such as the Mescalero Apaches and the Goshutes, are willing to host a repository and the State legally outmaneuvers them (Chapter 9). How do you bring a suitable site, local community (or tribe) and State together?

15. (Part II) A phased decision-making approach to developing a nuclear waste repository was the key to many of the arguments for the presidential recommendation to move forward with the Yucca Mountain license application in 2002. Do you agree with this approach, or should the project not have gone forward until after all relevant scientific issues had been *fully* addressed?

16. (Part II) The policy process for Yucca Mountain played out over decades, with many ups and downs. Summarize this political process. What were the major policy roadblocks along the way? Could any of these roadblocks have been circumvented, and if so, how?

17. (Part II) The approach of the federal government at WIPP and Yucca Mountain differed. Compare the similarities and differences between the approaches at the two sites. Why do you think WIPP succeeded while Yucca Mountain did not? Could WIPP be a template for how to deal with high-level waste, or could lessons be drawn from both cases?

18. (Part III) The Blue Ribbon Commission recommended developing one or more interim storage facilities. In light of NIMBY and States' refusal to even consider exploratory studies, how can this be accomplished? Is it a legitimate fear that an interim storage site would become a de facto repository? Is it any easier to find an interim storage site than a site for a geologic repository?

19.   (Part III) The Blue Ribbon Commission also recommended look-
       ing for a willing community to host a geologic repository. Is this
       thinking naive?
20.   (Part III) The Blue Ribbon Commission recommended taking
       responsibility for spent fuel away from the Department of Energy
       and creating a new quasi-government agency. What are the pros
       and cons of this recommendation?

GENERAL CONCLUDING QUESTIONS

21.   For over half a century, the federal government's inability to solve
       the nuclear waste dilemma has resulted in a legacy of failure. The
       implications in failing to protect public health and the environ-
       ment are readily obvious, raising the question of the viability of
       building more nuclear power plants. How do you view nuclear
       energy as a viable and responsible energy source? Have your views
       changed, or been reinforced, after reading this book?
22.   How do the scientific, political, and social lessons learned from
       Yucca Mountain translate to other energy related problems, such
       as geologic carbon sequestration and development of shale gas by
       fracking?
23.   Extra credit: how do you solve a problem like this?

References

INTRODUCTION

[1] Rhodes, R. (1981) A demonstration at Shippingport. *American Heritage* 32(4): 66–73.

[2] Shippingport Atomic Power Station: A National Historic Mechanical Engineering Landmark. (http://files.asme.org/ASMEORG/Communities/History/Landmarks/5643.pdf)

[3] *New York Times* (September 17, 1954) Abundant power from atom seen; it will be too cheap for our children to meter, Strauss tells science writers.

[4] Ford, D. (1984) *The Cult of the Atom*. Simon and Schuster, p. 46.

[5] US Congress, Office of Technology Assessment (1985) Managing the nation's commercial high-level radioactive waste. Report OTA-O-171. Washington, DC, p. 4.

[6] Blue Ribbon Commission on America's Nuclear Future (2012) Report to the Secretary of Energy. Washington, DC.

CHAPTER 1. THE AWAKENING

[1] *Gone with the Wind* (1939) Film directed by Victor Fleming and adapted by Sidney Howard from the novel by Margaret Mitchell. (1937) *Gone with the Wind*. New York: Macmillan.

[2] Hamblin, J. D. (2008) *Poison in the Well: Radioactive Waste in the Oceans at the Dawn of the Nuclear Age*. Rutgers University Press.

[3] Walker, J. S. (2009) *The Road to Yucca Mountain*. University of California Press.

[4] Laurence, W. L. (December 17, 1955) Waste held peril in atomic power. *New York Times*.

[5] National Research Council (1957) Report on disposal of radioactive waste on land. Washington, DC: National Academies Press.

[6] Kious, W. J. and R. I. Tilling (1996) This dynamic earth: the story of plate tectonics. US Geological Survey General Interest Publication.

[7] www.mkinghubbert.com.

[8] Hubbert, M. K. (1956) Nuclear energy and the fossil fuels. In: *Drilling and Production Practice*. Washington, DC: American Petroleum Institute, pp. 7–25.

[9] *New York Times* (January 13, 1950) Close check urged on atomic waste.

[10] Boffey, P. M. (1975) *The Brain Bank of America*. McGraw-Hill.

[11] Letter from H. H. Hess to J. A. McCone (June 21, 1960) reprinted in Hubbert, M. K. (1962) Energy resources. National Academy of Sciences-National Research Council, Publication 1000-D, Washington, DC.

[12] Brown, H. (1985) Fallout and falsehoods. *Bulletin of the Atomic Scientists* December: 59–65.

[13] National Research Council (1966) Report to the Division of Reactor Development and Technology, United States Atomic Energy Commission. Washington, DC: National Academy Press.

[14] Smith, R. M. (March 7, 1970) A. E. C. scored on storing waste. *New York Times*.

[15] Blomeke, J. O. *et al.* (1973) Managing radioactive wastes. *Physics Today* 26(8): 36–42.

[16] Metzger, P. H. (1972) *The Atomic Establishment*. Simon and Schuster.

[17] Stacy, S. M. (2000) Proving the principle: a history of the Idaho National Engineering and Environmental Laboratory, 1949–1999. Department of Energy, Idaho Operations Office, Idaho Falls, DOE/ID-10799.

[18] Stewart, R. B. and J. B. Stewart (2011) *Fuel Cycle to Nowhere: U.S. Law and Policy on Nuclear Waste*. Vanderbilt University Press.

[19] *Idaho-Falls Post-Register* (May 26, 1970) Science Academy doubts safety of waste disposal at NRTS.

[20] US Congress, Office of Technology Assessment (1985) Managing the nation's commercial high-level radioactive waste. Report OTA-O-171. Washington, DC, p. 85.

[21] National Research Council (1970) Disposal of solid radioactive wastes in bedded salt deposits. Washington, DC: National Academy Press.

[22] Carter, L. J. (1987) *Nuclear Imperatives and Public Trust*. Washington, DC: Resources for the Future.

[23] Conant, J. B. (1951) Our future in the atomic age. In: *Headline Series, Foreign Policy Association* 90 (November–December).

[24] Ford, D. (1984) *The Cult of the Atom*. Simon and Schuster, pp. 41–42.

[25] www.mkinghubbert.com/tribute/udall/transcript.

[26] Green, H. P. (1982) The peculiar politics of nuclear power. *Bulletin of the Atomic Scientists* April: 3.

[27] Metlay, D. (1985) Radioactive waste management policymaking. In US Congress, Office of Technology Assessment, *Managing the Nation's Commercial High-Level Radioactive Waste*. Report OTA-O-171. Washington, DC, p. 225.

[28] Royal Commission on Environmental Pollution (1976) Sixth report – nuclear power and the environment. London: Her Majesty's Stationery Office.

[29] Mathieson, J. (October 26, 2010) Presentation to the Nuclear Waste Technical Review Board. (http://www.nwtrb.gov/meetings/2010/oct/mathieson.pdf)

[30] Hocke, P. and O. Renn (2009) Concerned public and the paralysis of decision-making: nuclear waste management policy in Germany. *Journal of Risk Research* 7–8: 921–940.

[31] *Spiegel Online International* (April 15, 2011) Merkel takes first steps toward a future of renewables.

[32] Galbraith, K. (November 27, 2011) A new urgency to the problem of storing nuclear waste. *New York Times*.

[33] Nuclear Waste Technical Review Board (2009) Survey of national programs for managing high-level waste and spent nuclear fuel. A Report to Congress and the Secretary of Energy, October. (www.nwtrb.gov)

CHAPTER 2. BRAINSTORMING

[1] *New Scientist* (July 5, 1973) The lady gets her way. 59(853): 14–16.

[2] *New York Times* (April 17, 1951) Atomic death belt urged for Korea.

[3] US Atomic Energy Commission (1950) Eighth semiannual report. Washington, DC: Government Printing Office.

[4] *New York Times* (April 18, 1951) Atomic belt plan held not feasible.

[5] Udall, S. L. (1994) *The Myths of August*. Pantheon Books, p. 32.

[6] Lanouette, W. (2009) Civilian control of nuclear weapons. *Arms Control Today* 39(4): 45–48.

[7] de la Bruheze, A. (1992) Radiological weapons and radioactive waste in the United States: insiders' and outsiders' views, 1941–55. *British Journal for the History of Science* 25: 207–227.

[8] Hewlett, R. G. and O. E. Anderson (1962) *A History of the United States Atomic Energy Commission, Vol. I: The New World, 1939–1946*. Pennsylvania State University Press, pp. 37–38.

[9] *Los Angeles Times* (November 2, 1947) Guided missile may use radioactive waste.

[10] *New York Times* (August 1, 1950) Weapons using radioactive poison pushed by atomic energy board.

[11] *New York Times* (May 20, 1954) An atomic Maginot Line.

[12] *New York Times* (April 18, 1954) Atomic ashes? Dump them on Mars.

[13] US Department of Energy (1980) Final environmental impact statement: management of commercially generated radioactive waste. DOE/EIS-0046F, Volume 1 of 3, pp. 6.142–6.145.

[14] McFarlane, H. F. (2006) The paradox of nuclear waste. *Radwaste Solutions* September/October: 32–36.

[15] Coopersmith, J. (August 22, 2005) Nuclear waste in space. (www.thespacereview.com/article/437/1).

[16] Boffey, P. M. (1975) *The Brain Bank of America*. McGraw-Hill, pp. 106–109.

[17] Broad, W. J. (November 21, 1994) Nuclear roulette for Russia: burying uncontained waste. *New York Times*.

[18] National Research Council (2001) Disposition of high-level waste and spent nuclear fuel: The continuing societal and technical challenges. Washington, DC: National Academy Press.

[19] Dickson, D. (December 23, 1989) Kyshtym "almost as bad as Chernobyl." *New Scientist*.

[20] Brady, P. V. *et al.* (2009) Deep borehole disposal of high-level radioactive waste. Sandia National Laboratories, Sandia Report SAND2009-4401.

[21] Arnold, B. and Brady, P. (March 7, 2012) Presentation to Nuclear Waste Technical Review Board. (http://www.nwtrb.gov/meetings/2012/march/arnold.pdf)

[22] Milnes, A. G. (1985) *Geology and Radwaste*. Academic Press.

[23] Cohen, J. J. *et al.* (1972) In-situ incorporation of nuclear waste in deep molten silicate rock. *Nuclear Technology* 13: 76–88.

[24] Edwards, R. (March 27, 1999) It's got to go. *New Scientist*.

[25] Royal Commission on Environmental Pollution (1976) Sixth report – nuclear power and the environment. London: Her Majesty's Stationery Office.

[26] Chapman, N. and T. McEwen (August 28, 1986) Geological solutions for nuclear wastes. *New Scientist*.

[27] Edwards, R. (June 10, 2005) Secret nuclear waste disposal sites revealed. *New Scientist*.

CHAPTER 3. THE OCEAN AS A DUMPING GROUND

[1] *Childe Harold's Pilgrimage*, Canto IV, Stanza 179.

[2] Ebbesmeyer, C. and E. Scigliano (2009) *Flotsametrics and the Floating World*. HarperCollins.

[3] Gramling, C. (2007) Voyage of the ducks. *Geotimes* (September).

[4] Revelle, R. *et al.* (1956) Oceanography, fisheries, and atomic radiation. *Science* 124: 13–16.

[5] Revelle, R. and H. E. Suess (1957) Carbon dioxide exchange between atmosphere and ocean and the question of an increase of atmospheric CO_2 during the past decades. *Tellus* 9: 18–27.

[6] Keeling, R. F. (2008) Recording Earth's vital signs. *Science* 319: 1771–1772.

[7] Hamblin, J. D. (2008) *Poison in the Well: Radioactive Waste in the Oceans at the Dawn of the Nuclear Age*. Rutgers University Press.

[8] *New York Times* (June 29, 1958) Current is found far down in sea.

[9] *New York Times* (February 5, 1960) Bathyscaph's crew honored by the president for their achievement.

[10] *Science* (1959) Radioactive waste disposal discussed in Monaco. 130: 1700.

[11] Carter, L. J. (1980) Navy considers scuttling old nuclear subs. *Science* 209: 1495–1497.

[12] Sjöblom, K.-L. and G. Linsley (1994) Sea disposal of radioactive wastes: the London Convention 1972. *International Atomic Energy Agency Bulletin* 2: 12–16.

[13] Smith, G. (September 5, 1957) "Hot cargo" gets one-way sea trip. *New York Times*.

[14] *Los Angeles Times* (July 17, 1957) Drifting atomic waste can sunk.

[15] Reston, J. (July 16, 1957) A peril of the atom age. *New York Times*.

[16] *New York Times* (June 21, 1959) Sea areas picked for atom wastes.

[17] *New York Times* (November 16, 1962) A. E. C. rebuts Soviet on pollution of sea.

[18] *New York Times* (October 13, 1960) France to delay atomic disposal.

[19] Carter, L. J. (1987) *Nuclear Imperatives and Public Trust*. Washington, DC: Resources for the Future.

[20] Dickson, D. (1986) Nuclear reprocessing and "the world's most radioactive sea." *Science* 231: 1500.

[21] MacKenzie, D. (April 17, 1993) Russia owns up to sea burial for nuclear waste. *New Scientist*.

[22] Broad, W. J. (April 27, 1993) Russians describe extensive dumping of nuclear waste. *New York Times*.

[23] Pearce, F. (November 20, 1993) Nuclear club offers to help Russia clean up. *New Scientist*.

[24] MacKenzie, D. (April 20, 1996) Russian secrecy could sink nuclear aid. *New Scientist*.

[25] MacKenzie, D. (October 19, 1996) Treason case dropped as Russia signs nuclear deal. *New Scientist*.

[26] Hadfield, P. (July 31, 1999) Nuclear captain is free at last. *New Scientist*.

[27] *USA Today* (August 2, 2008).

[28] Hadfield, P. and D. MacKenzie (November 6, 1993) Nuclear dumping at sea goads Japan into action. *New Scientist*.

[29] Hollister, C. D. *et al.* (1981) Subseabed disposal of nuclear wastes. *Science* 213: 1321–1326.

[30] Krauskopf, K. B. (1988) *Radioactive Waste Disposal and Geology*. Chapman and Hall.

[31] Woods Hole Oceanographic Institution (1999) WHOI Waypoints: Charles Davis Hollister. *Woods Hole Currents* 8(2).

[32] Hollister, C. D. and S. Nadis (1998) Burial of radioactive waste under the seabed. *Scientific American* January: 60–65.

[33] Nadis, S. (1996) The sub-seabed solution. *The Atlantic Monthly* 278: 28–39.

CHAPTER 4. RADIOACTIVITY AND ATOMIC ENERGY

[1] Compton, A. H. (1956) *Atomic Quest.* Oxford University Press, p. 144.

[2] Landa, E. R. and T. B. Councell (1992) Leaching of uranium from glass and ceramic foodware and decorative items. *Health Physics* 63(3): 343–348.

[3] Glasstone, S. and W. H. Jordan (1980) *Nuclear Power and Its Environmental Effects.* La Grange Park, Illinois: American Nuclear Society.

[4] Conn, G. K. T. and H. D. Turner (1965) *The Evolution of the Nuclear Atom.* American Elsevier, p. 136.

[5] Zumdahl, S. S. (1993) *Chemistry.* Third edition. D. C. Heath and Co.

[6] Gephart, R. E. (2003) *Hanford: A Conversation about Nuclear Waste and Cleanup.* Columbus, Ohio: Battelle Press.

[7] http://www.hss.energy.gov/healthsafety/ohre/new/findingaids/epidemiologic/oakridge2/intro.ht.

[8] *Consumer Reports* (1960) Fallout in our milk . . . a follow-up report. February: 64–70.

[9] Washington University School of Dental Medicine (2006–2009) St. Louis Baby Tooth Survey, 1959–1970. (http://beckerexhibits.wustl.edu/dental/articles/babytooth.html).

[10] McGuire, K. (October 21, 2009) St. Louis baby teeth yield new findings on nuclear fallout. *St Louis Post-Dispatch.*

[11] Milnes, A. G. (1985) *Geology and Radwaste.* Academic Press, pp. 16–17.

[12] Bradley, D. J. *et al.* (1996) Nuclear contamination from weapons complexes in the former Soviet Union and the United States. *Physics Today* April: 40–45.

[13] Sjöblom, K.-L. and G. Linsley (1994) Sea disposal of radioactive wastes: the London Convention 1972. *International Atomic Energy Agency Bulletin* 2: 12–16.

[14] Ewing, R. C. (2004) Environmental impact of the nuclear fuel cycle. In: R. Gieré and P. Stille (eds.), *Energy, Waste, and the Environment: A Geochemical Perspective.* Geological Society of London, Special Publication 236.

[15] Ahearne, J. F. (1987) Nuclear power after Chernobyl. *Science* 236: 673–679.

[16] Heeb, C. M. and D. J. Bates (1994) Radionuclide releases to the Columbia River from Hanford operations, 1944–1971. Report PNWD-2223 HEDR. Richland, Washington: Pacific Northwest Laboratory.

[17] National Research Council (2006) Tank waste retrieval, processing, and on-site disposal at three Department of Energy sites. Washington, DC: National Academies Press.

[18] Ahearne, J. F. (1997) Radioactive waste: the size of the problem. *Physics Today* June: 24–29.

[19] Rhodes, R. (1986) *The Making of the Atomic Bomb.* Simon and Schuster.

[20] *New York Times* (February 3, 1939) Revolution in physics.

[21] Seife, C. (2008) *Sun in a Bottle.* Viking.

CHAPTER 5. THE COLD WAR LEGACY

[1] Brinkley, D. (2001) Eisenhower the dove. *American Heritage* September: 58–65.

[2] US Department of Energy (1998) Accelerating cleanup: paths to closure. DOE/EM-0362. Washington, DC.

[3] Rhodes, R. (1986) *The Making of the Atomic Bomb*. Simon and Schuster, pp. 294, 500.

[4] Crowley, K. D. (1997) Nuclear waste disposal: the technical challenges. *Physics Today* June: 32–39.

[5] Gephart, R. E. (2003) *Hanford: A Conversation about Nuclear Waste and Cleanup*. Columbus, Ohio: Battelle Press.

[6] Probst, K. N. and A. I. Lowe (2000) Cleaning up the nuclear weapons complex: does anyone care? Center for Risk Assessment. Washington, DC: Resources for the Future, p. 2.

[7] Williams, S. J. (chairman) (1948) Report of the Safety and Industrial Health Advisory Board. US Atomic Energy Commission, Washington, DC, p. 10.

[8] Crowley, K. D. and J. F. Ahearne (2002) Managing the environmental legacy of U.S. nuclear-weapons production. *American Scientist* 90: 514–523.

[9] National Research Council (2000) Research needs in subsurface science: US Department of Energy's Environmental Management Science Program. Washington, DC: National Academy Press.

[10] National Research Council (2006) Tank waste retrieval, processing, and on-site disposal at three Department of Energy sites. Washington, DC: National Academies Press.

[11] Groves, L. R. (1962) *Now It Can Be Told: The Story of the Manhattan Project*. Harper and Brothers, p. 70.

[12] *New York Times* (August 1, 1950) Weapons using radioactive poison pushed by atomic energy board.

[13] Gillette, R. (1973) Radiation spill at Hanford: the anatomy of an accident. *Science* 181: 728–730.

[14] Levi, B. G. (1992) Hanford seeks short- and long-term solutions to its legacy of waste. *Physics Today* March: 17–21.

[15] Zorpette, G. (1996) Hanford's nuclear wasteland. *Scientific American* May: 88–97.

[16] Gephart, R. E. and R. E. Lundgren (1998) *Hanford Tank Cleanup: A Guide to Understanding the Technical Issues*. Columbus, Ohio: Battelle Press.

[17] National Research Council (2001) Research needs for high-level waste stored in tanks and bins at US Department of Energy sites. Washington, DC: National Academy Press.

[18] *Radwaste Solutions* (2009) New Hanford cleanup deadlines; other D&D updates. September/October: 9.

[19] *Radwaste Solutions* (2011) D&D updates. January–April: 12.

[20] Savannah River site news release (March 3, 2009) SRS begins cleaning outside of underground waste tanks.

[21] Wald, M. L. (March 13, 1996) Factory is set to process dangerous nuclear waste. *New York Times*.

[22] Blue Ribbon Commission on America's Nuclear Future (2012) Report to the Secretary of Energy. Washington, DC.

[23] Stacy, S. M. (2000) Proving the principle: a history of the Idaho National Engineering and Environmental Laboratory, 1949–1999. Department of Energy, Idaho Operations Office, Idaho Falls, DOE/ID-10799.

[24] National Research Council (2011) Waste forms technology and performance: final report. Washington, DC: National Academies Press.

[25] Wald, M. L. (October 10, 2004) Bill allows waste to remain in tanks. *New York Times*.

[26] Hewlett, R. G. and F. Duncan (1969) *A History of the United States Atomic Energy Commission, Vol. II: Atomic Shield, 1947–1952*. Pennsylvania State University Press, p. 47.

[27] Taubes, G. (1994) No easy way to shackle the nuclear demon. *Science* 263: 629–631.

[28] Panofsky discussion from Jackson, J. D. (2009) Panofsky agonistes: the 1950 loyalty oath at Berkeley. *Physics Today* January: 41–47; Panofsky, W. K. H. (1997) A physical heritage of the Cold War: excess weapons plutonium. *Physics Today* April: 61–62; and http://www2.slac.stanford.edu/panofsky_fellow/career.html "A brief biography of Wolfgang K. H. Panofsky."

[29] National Academy of Sciences, Committee on International Security and Arms Control (1995) Management and disposition of excess weapons plutonium. Washington, DC: National Academies Press.

[30] von Hippel, F. N. (2001) Plutonium and reprocessing of spent nuclear fuel. *Science* 293: 2397–2398.

[31] Panofsky, W. K. H. (1997) Disposing of excess plutonium. *Science* 275: 11–12.

[32] Panofsky, W. K. H. (1997) A physical heritage of the Cold War: excess weapons plutonium. *Physics Today* April: 61–62.

[33] Wald, M. L. (December 10, 1996) Agency to pursue two plans to shrink plutonium supply. *New York Times*.

[34] Savannah River Site Facts. (http://www.srs.gov/general/news/factsheets/srs.pdf)

[35] Brooks, L. F. (July 26, 2006) Plutonium disposition and the U.S. mixed oxide fuel facility. Testimony before the House Armed Services Subcommittee.

[36] Becker, J. and W. J. Broad (April 10, 2011) New doubts about turning plutonium into a fuel. *New York Times*.

CHAPTER 6. THE PEACEFUL ATOM AND ITS WASTES

[1] Smith, K. R. (1988) *Energy Environment Monitor* 4(1): 61–70.

[2] Hewlett, R. G. and F. Duncan (1969) *A History of the United States Atomic Energy Commission, Vol. II: Atomic Shield, 1947–1952.* Pennsylvania State University Press.

[3] Gillette, R. (1972) Nuclear safety (I): the roots of dissent. *Science* 177: 771–776.

[4] Stacy, S. M. (2000) Proving the principle: a history of the Idaho National Engineering and Environmental Laboratory, 1949–1999. Department of Energy, Idaho Operations Office, Idaho Falls, DOE/ID-10799.

[5] Bruno, J. and R. C. Ewing (2006) Spent nuclear fuel. *Elements* 2: 343–349.

[6] Cravens, G. (2007) *Power to Save the World.* Alfred A. Knopf, p. 9.

[7] Wood, H. G. *et al.* (2008) The gas centrifuge and nuclear weapons proliferation. *Physics Today* September: 40–45.

[8] US Government Accountability Office (2008) Nuclear material: DOE has several potential options for dealing with depleted uranium tails, each of which could benefit the government. GAO-08-606R.

[9] Bohr, N. and J. A. Wheeler (1939) The mechanism of nuclear fission. *Physical Review* 56: 426–450.

[10] Wheeler, J. A. (1967) Mechanism of fission. *Physics Today* 20(11): 49–52.

[11] Gephart, R. E. (2003) *Hanford: A Conversation about Nuclear Waste and Cleanup.* Columbus, Ohio: Battelle Press.

[12] Hewlett, R. G. and J. M. Holl (1989) *Atoms for Peace and War, 1953–1961.* University of California Press, pp. 37–41.

CHAPTER 7. RECYCLING

[1] Starr, C. and R. P. Hammond (1972) Nuclear waste storage. *Science* 177: 744–745.

[2] Rochlin, G. I. *et al.* (1978) West Valley: remnant of the AEC. *Bulletin of the Atomic Scientists* January: 17–26.

[3] Lester, R. K. and D. J. Rose (1977) The nuclear wastes at West Valley, New York. *Technology Review* 79(6): 20–29.

[4] Gillette, R. (1974) "Transient" nuclear workers: a special case for standards. *Science* 186: 125–129.

[5] Carter, L. J. (1987) *Nuclear Imperatives and Public Trust.* Washington, DC: Resources for the Future.

[6] National Research Council (2006) Tank waste retrieval, processing, and on-site disposal at three Department of Energy sites. Washington, DC: National Academies Press.

[7] New York State Energy Research and Development Authority, http://www.nyserda.ny.gov/Programs/West-Valley/West-Valley-Demonstration-Project.aspx

[8] Metz, W. D. (1977) Reprocessing: how necessary is it for the near term? *Science* 196: 43–45.

[9] American Physical Society (1978) Report to the American Physical Society by the study group on nuclear fuel cycles and waste management. *Reviews of Modern Physics* 50(1), Part II: S8.

[10] von Hippel, F. N. (2001) Plutonium and reprocessing of spent nuclear fuel. *Science* 293: 2397–2398.

[11] Fetter, S. (2009) How long will the world's uranium supplies last? *Scientific American* March: 84.

[12] MIT (2010) The future of the nuclear fuel cycle – an interdisciplinary MIT Study. Cambridge, MA: Massachusetts Institute of Technology.

[13] Kramer, A. E. (November 10, 2009) Power for U.S. from Russia's old nuclear weapons. *New York Times.*

[14] Weinberg, A. M. (1994) *The First Nuclear Era – The Life and Times of a Technological Fixer.* AIP Press.

[15] Van Gosen, B. S. *et al.* (2009) Thorium deposits of the United States – energy resources for the future? US Geological Survey Circular 1336.

[16] Gephart, R. E. (2003) *Hanford: A Conversation about Nuclear Waste and Cleanup.* Columbus, Ohio: Battelle Press, p. 1.9.

[17] *New York Times* (August 23, 1950) Jailed in a theft of plutonium, scientist says he took "souvenir."

[18] *New York Times* (August 25, 1950) Plutonium guards tight.

[19] Shapley, D. (1971) Plutonium: reactor proliferation threatens a nuclear black market. *Science* 172: 143–146.

[20] McPhee, J. (1974) *The Curve of Binding Energy.* Ballantine.

[21] Ford, D. (1984) *The Cult of the Atom.* Simon and Schuster, p. 23.

[22] Hammond, A. (1972) Fission: the pro's and con's of nuclear power. *Science* 178: 147–149.

[23] Hammond, A. L. (1974) Complications indicated for the breeder. *Science* 185: 768.

[24] Gillette, R. (1975) Recycling plutonium: the NRC proposes a second look. *Science* 188: 818, 874.

[25] Patterson, W. C. (1984) *The Plutonium Business and the Spread of the Bomb.* Wildwood House, pp. 83, 152.

[26] Nuclear Energy Policy Study Group. (1977) *Nuclear Power Issues and Choices.* Ballinger.

[27] von Hippel, F. N. (2007) Managing spent fuel in the United States: the illogic of reprocessing. International Panel on Fissile Materials, Research Report No. 3. (http://www.fissilematerials.org/ipfm/site_down/rr03.pdf)

[28] Brown, P. (May 9, 2005) Huge radioactive leak closes Thorp nuclear plant. *The Guardian*.

[29] International Panel on Fissile Materials (2012) Global fissile material report 2011: nuclear weapon and fissile material stockpiles and production. Sixth annual report. (www.fissilematerials.org)

[30] von Hippel, F. N. (2008) Rethinking nuclear fuel recycling. *Scientific American* 298(5): 88–93.

[31] Taylor, A. (November 28, 2011) Protestors disrupt German nuclear waste shipment. *The Atlantic*.

[32] Darst, R. G. and J. I. Dawson (2008) "Baptists and bootleggers, once removed": the politics of radioactive waste internalization in the European Union. *Global Environmental Politics* 8: 17–38.

[33] Strandberg, U. and M. Andrén (2009) Nuclear waste management in a globalized world. *Journal of Risk Research* 12: 879–895.

[34] Blue Ribbon Commission on America's Nuclear Future (2012) Report to the Secretary of Energy. Washington, DC.

[35] Hannum, W. H. *et al.* (2005) Smarter use of nuclear waste. *Scientific American* December: 84–91.

[36] Hewlett, R. G. and F. Duncan (1974) *Nuclear Navy 1946–1962*. University of Chicago Press, p. 274.

[37] Weart, S. R. (1988) *Nuclear Fear: A History of Images*. Harvard University Press, pp. 5–6.

[38] Rhodes, R. (1986) *The Making of the Atomic Bomb*. Simon and Schuster, p. 66.

[39] Sherr, R. *et al.* (1941) Transmutation of mercury by fast neutrons. *The Physical Review* 60(7): 473–479.

[40] Nikitin, M. B. *et al.* (2008) Managing the nuclear fuel cycle: policy implications of expanding global access to nuclear power. Congressional Research Service Report RL34234, Washington, DC.

[41] National Academy of Sciences and National Research Council (2009) Internationalization of the nuclear fuel cycle: goals, strategies, and challenges. Washington, DC: National Academies Press.

[42] Johnson, J. (2007) Reprocessing key to nuclear plan. *Chemical & Engineering News* 85: 48–54.

[43] Lester, R. K. (2006) Energy conundrums: new nukes. *Issues in Science and Technology* Summer.

[44] National Research Council (2008) Review of DOE's nuclear energy research and development program. Washington, DC: National Academies Press.

[45] Marris, E. (2006) Nuclear reincarnation. *Nature* 441(15): 796–797.

[46] National Research Council (2008) The National Academies Summit on America's Energy Future. Washington, DC: National Academies Press, p. 46.

CHAPTER 8. DRY CASK STORAGE

[1] Keyes, R. (2006) *The Quote Verifier: Who Said What, Where, and When*. New York: St. Martin's Press.

[2] Maine Yankee: a brief history of operation, decommissioning, and the interim storage of spent nuclear fuel. (www.maineyankee.com/public/MaineYankee.pdf)

[3] National Research Council (2006) Safety and security of commercial spent nuclear fuel storage: public report. Washington, DC: National Academies Press, p. 8.

[4] Federal Register (October 9, 2008) 73(197): 59551–59570.

[5] Fahey, J. and R. Henry (March 22, 2011) U.S. spent-fuel storage sites are packed. *The Associated Press.*

[6] Alvarez, R. (2012) Improving spent fuel storage at nuclear reactors. *Issues in Science and Technology* Winter.

[7] Metz, W. D. (1977) Reprocessing alternatives: the options multiply. *Science* 196: 284–287.

[8] Sullivan, J. (June 10, 2001) Environment: nuclear plants face storing toxic waste. *New York Times.*

[9] McCullum, R. (June 11, 2009) Presentation to Nuclear Waste Technical Review Board. (http://www.nwtrb.gov/meetings/2009/june/mccullum.pdf)

[10] US Congress, Office of Technology Assessment (1985) Managing the nation's commercial high-level radioactive waste. Report OTA-O-171. Washington, DC, pp. 90–91.

[11] Federal Register (August 31, 1984) 49: 34658.

[12] Federal Register (September 18, 1990) 55: 38472.

[13] Federal Register (December 6, 1999) 64: 68005.

[14] Carter, L. J. *et al.* (2010) Nuclear waste disposal: showdown at Yucca Mountain. *Issues in Science and Technology* Fall: 80–84.

[15] Federal Register (December 23, 2010) 75: 81037.

[16] Electric Power Research Institute (2002) Dry cask storage characterization project. EPRI Report 1002882. Palo Alto, CA.

[17] Electric Power Research Institute (2002) Technical bases for extended dry storage of spent nuclear fuel. EPRI Report 1003416. Palo Alto, CA.

[18] http://www.nirs.org/radwaste/atreactorstorage/drycaskfactsheet07152004.pdf

[19] US Nuclear Regulatory Commission (2007) A pilot probabilistic risk assessment of a dry cask storage system at a nuclear power plant. NUREG-1864. Washington, DC.

[20] Neider, T. (June 11, 2009) Presentation to Nuclear Waste Technical Review Board. (http://www.nwtrb.gov/meetings/2009/june/neider.pdf)

[21] Nuclear Waste Technical Review Board (June 11, 2009) Meeting transcript, pp. 194, 246. (http://www.nwtrb.gov/meetings/2009/june/09june11.pdf)

[22] Rogers, K. (July 31, 2009) Reid declares Yucca Victory. *Las Vegas Review-Journal.*

[23] Blue Ribbon Commission on America's Nuclear Future (2012) Report to the Secretary of Energy. Washington, DC.

CHAPTER 9. INTERIM STORAGE

[1] Arnold, E. (1947) *Blood Brother*. University of Nebraska Press, p. 178.

[2] House Budget Committee (July 27, 2010) Hearing on Budget Implications of Closing Yucca Mountain. (www.agiweb.org/gap/legis111/nuclear_hearings.html)

[3] Johnson, L. and D. Schaffer (1994) *Oak Ridge National Laboratory: The First Fifty Years*. University of Tennessee Press.

[4] Isherwood, D. (1986) The view from Capitol Hill: Congressional Science Fellow's midyear report. *Eos, Transactions, American Geophysical Union* 67(8): 99.

[5] Marshall, E. (1987) Thirty ways to temporize on waste. *Science* 237: 591–592.

[6] Gowda, M. V. R. and D. Easterling (1998) Nuclear waste and Native America: the MRS siting exercise. *Risk: Health, Safety & Environment* Summer: 229–258.

[7] Wald, M. L. (February 13, 1991) Leroy was hired to be a negotiator, but treated like a pariah. *New York Times.*

[8] Satchell, M. (January 8, 1996) Dances with nuclear waste. *US News and World Report*.

[9] Wald, M. L. (November 11, 1993) Nuclear storage divides Apaches and neighbors. *New York Times*.

[10] Carter, L. J. (1994) The Mescalero option. *Bulletin of the Atomic Scientists* September: 11–13.

[11] Hebert, H. J. (June 26, 2006) Store nuclear waste on reservation? Tribe split. *MSNBC*, http://www.msnbc.msn.com/id/13458867

[12] Wald, M. L. (April 18, 1999) Tribe in Utah fights for nuclear waste dump. *New York Times*.

[13] *New York Times* (March 11, 2003) U.S. withholds approval for nuclear waste storage on Indian Reservation.

[14] *New York Times* (September 10, 2005) U.S. panel backs nuclear dump on Indian Reservation in Utah.

[15] Stolz, M. and M. L. Wald (September 9, 2006) Interior Department rejects interim plan for nuclear waste. *New York Times*.

[16] Nuclear Waste Technical Review Board (1996) Disposal and storage of spent nuclear fuel – Finding the right balance. A Report to Congress and the Secretary of Energy, March. (www.nwtrb.gov)

[17] US Department of Energy (2008) Report to Congress on the demonstration of the interim storage of spent nuclear fuel from decommissioned nuclear power reactor sites. DOE/RW-0596. Washington, DC.

[18] *Radwaste Solutions* (2010) Will court ruling put PFS back on track? September/October: 6.

CHAPTER 10. A CAN OF WORMS

[1] Original quote appears to be "If our newspapers knew what is really happening in the world, or could discriminate between the news value of a bicycle accident in Clapham and that of the collapse of civilization," Shaw, G.B. (1934) *Too True to Be Good*, Preface. London: Constable and Company, p. 17.

[2] McFarlane, H. F. (2006) The paradox of nuclear waste. *Radwaste Solutions* September/October: 32–36.

[3] Norman, C. (1984) High-level politics over low-level waste. *Science* 223: 258–260.

[4] Marshall, E. (1979) Radioactive waste backup threatens research. *Science* 206: 431–433.

[5] Reinhold, R. (December 28, 1992) States, failing to cooperate, face a nuclear-waste crisis. *New York Times*.

[6] Cohen, J. (1994) California's disposal plan goes nowhere fast. *Science* 263: 912.

[7] Abelson, P. H. (1995) Low-level radioactive waste. *Science* 268: 1547.

[8] Wilshire, H. (1995) Letters: Radioactive waste at Ward Valley. *Science* 269: 1654–1655.

[9] National Research Council (1995) Ward Valley: an examination of seven issues in Earth sciences and ecology. Washington, DC: National Academy Press.

[10] Reinhold, R. (January 24, 1994) A test case for nuclear disposal. *New York Times*.

[11] Hirsch, R. M. (February 1, 1994) Before nuclear wastes seep into water supply; not a federal report. *New York Times*.

[12] Boxer, B. (February 22, 1994). Deceptions hinder waste-dump debate. *New York Times*.

[13] Hendricks, D. W. and C. W. Fort (1976) Radiation survey in Beatty, Nevada, and surrounding area (March 1976). US Environmental Protection Agency, Office of Radiation Programs, Las Vegas, Nevada, Technical Note ORP/LV-76-1.

[14] Saleska, S. (1970) Low-level radioactive waste: gamma rays in the garbage. *Bulletin of the Atomic Scientists* 46: 18–25.

[15] Walker, J. S. (2009) *The Road to Yucca Mountain.* University of California Press, p. 132.

[16] Prudic, D. E. (1994) Estimates of percolation rates and ages in unsaturated sediments at two Mojave Desert sites, California – Nevada. US Geological Survey Water-Resources Investigations Report 94–4160.

[17] Clifford, F. (June 1, 1995) Babbitt OKs nuclear dump deal. *Los Angeles Times.*

[18] Gathright, A. (January 4, 1997) Scientists allege federal coverup of leaks, officials deny covering. *San Jose Mercury News.*

[19] Manning, M. (April 21, 1997) Cover-up charged over nuke dump leak. *Las Vegas Sun.*

[20] Clifford, F. (October 31, 1995) New data clouds Ward Valley plan. *Los Angeles Times.*

[21] Public Employees for Environmental Responsibility (PEER) (December 19, 1996) Press release.

[22] American Geological Institute (1998) Update and hearing summary on low-level nuclear waste disposal, October 23. (http://www.agiweb.org/legis105/lownuke.html)

[23] Congressman George Miller (June 6, 1997) Press release.

[24] US General Accounting Office (1997) Interior's review of the proposed Ward Valley waste site. GAO/T-RCED-97-212.

CHAPTER 11. WIPP

[1] Eliot, T. S. (1922) The Waste Land. *The Criterion.* First volume.

[2] Biello, D. (2009) Spent nuclear fuel: a trash heap deadly for 250,000 years or a renewable energy source? *Scientific American* January.

[3] WIPP Fact Sheet, How will future generations be warned? (http://www.wipp.energy.gov/fctshts/warned.pdf)

[4] The Editors of *Life* (1962) *The Epic of Man.* Golden Press.

[5] McCutcheon, C. (2002) *Nuclear Reactions: The Politics of Opening a Radioactive Waste Disposal Site.* University of New Mexico Press.

[6] National Research Council (1996) The Waste Isolation Pilot Plant: a potential solution for the disposal of transuranic waste. Washington, DC: National Academy Press.

[7] Sutcliffe, W. G. *et al.* (1995) A perspective on the dangers of plutonium. Lawrence Livermore National Laboratory, Report UCRL-JC-118825, CSTS-48-95.

[8] American Institute of Professional Geologists (1984) Radioactive waste: issues and answers. Arvada, Colorado.

[9] Stewart, R. B. and J. B. Stewart (2011) *Fuel Cycle to Nowhere: U.S. Law and Policy on Nuclear Waste.* Vanderbilt University Press.

[10] Lakey, L. T. *et al.* (1983) Management of transuranic wastes throughout the world. *Nuclear and Chemical Waste Management* 4: 35–46.

[11] Carter, L. J. (1987) *Nuclear Imperatives and Public Trust.* Washington, DC: Resources for the Future.

[12] Boffey, P. M. (1975) Radioactive waste site search gets into deep water. *Science* 190: 361.

[13] Carter, L. J. (1978) Trouble even in New Mexico for nuclear waste disposal. *Science* 199: 1050–1051.

[14] Metz, W. D. (1978) New review of nuclear waste disposal calls for early test in New Mexico. *Science* 199: 1422–1423.

[15] Carter, L. J. (1979) Carter says no to WIPP, but DOE may appeal. *Science* 206: 1287.

[16] Ford, D. (1984) *The Cult of the Atom*. Simon and Schuster, p. 46.

[17] Sun, M. (1982) Radwaste dump WIPPs up a controversy. *Science* 215: 1483–1486.

[18] Carter, L. J. (1983) WIPP goes ahead, amid controversy. *Science* 222: 1104–1106.

[19] Schneider, K. (August 30, 1992) Wasting away. *New York Times*.

[20] Andrus, C. D. and J. Connelly (1998) *Politics Western Style*. Sasquatch Books, p. 201.

[21] Schneider, K. (June 3, 1989) Nuclear waste dump faces another potential problem. *New York Times*.

[22] Mora, C. J. (2000) Sandia and the Waste Isolation Pilot Plant, 1974–1999. Presented at Southwest Oral History Association Meeting, April 28–30, 2000, Long Beach, CA.

[23] http://www.wipp.energy.gov/

[24] Wald, M. L. (May 14, 1998) Desert site approved for burying plutonium. *New York Times*.

CHAPTER 12. THE SEARCH FOR A GEOLOGIC REPOSITORY

[1] McCain, J. (May 23, 1991) In: *Addresses and Special Orders Held in the U.S. House of Representatives and the Senate, Presented in Honor of The Honorable Morris K. "Mo" Udall, A Representative from Arizona, (1993) One Hundred Second Congress, First Session*. Washington, DC: US Government Printing Office, p. 107. (http://www.library.arizona.edu/exhibits/udall/address/toc_sch.html)

[2] Carter, L. J. (1987) *Nuclear Imperatives and Public Trust*. Washington, DC: Resources for the Future.

[3] Ford, D. (1984) *The Cult of the Atom*. Simon and Schuster, p. 180.

[4] Yergin, D. (1991) *The Prize: The Epic Quest for Oil, Money, and Power*. Simon and Schuster.

[5] *Time* (September 29, 1980).

[6] *Time* (November 5, 1973) Changes in Dixyland.

[7] *Science* (1973) AEC shakes up nuclear safety research. 180: 934–935.

[8] Ray, D. L. (1973) The nation's energy future, a report to Richard M. Nixon, President of the United States. Washington, DC: US Atomic Energy Commission.

[9] Gillette, R. (1973) Energy R & D: under pressure, a national policy takes form. *Science* 182: 898–900.

[10] *Time* (February 4, 1974).

[11] US Congress, Office of Technology Assessment (1985) Managing the nation's commercial high-level radioactive waste. Report OTA-O-171. Washington, DC.

[12] Metlay, D. (1985) Radioactive waste management policymaking. In US Congress, Office of Technology Assessment, *Managing the Nation's Commercial High-Level Radioactive Waste*. Report OTA-O-171. Washington, DC, p. 223.

[13] Carter, L. J. (1979) Radioactive waste policy is in disarray. *Science* 206: 312–314.

[14] Carter, L. J. (1978) Nuclear wastes: the science of geologic disposal seen as weak. *Science* 200: 1135–1137.

[15] Bredehoeft, J. D. *et al.* (1978) Geologic disposal of high-level radioactive wastes –
Earth science perspectives. US Geological Survey Circular 779.

[16] Energy Research and Development Administration (1976) Alternatives for
managing wastes from reactors and post-fission operations in the LWR fuel
cycle. ERDA 76–43. Washington, DC.

[17] American Physical Society (1978) Report to the American Physical Society by
the study group on nuclear fuel cycles and waste management. *Reviews of
Modern Physics* 50(1), Part II: S139.

[18] US Department of Energy (1978) Report of the task force for review of nuclear
waste management. Washington, DC, February.

[19] Interagency Review Group (1979) Report to the President by the Interagency
Review Group on Nuclear Waste Management. NTIS Report TID-29442, p. 6.

[20] US Department of Energy (1980) Final environmental impact statement: management of commercially generated radioactive waste. DOE/EIS-0046F, Volume 1 of 3, p. iii.

[21] Carter, L. J. (1980) No veto for States on radwaste sites. *Science* 209: 788.

[22] Lee, K. N. (1980) A federalist strategy for nuclear waste management. *Science*
208: 679–684.

[23] http://www.udall.gov

[24] Lewis, M. (May 25, 1997) The subversive. *New York Times*.

[25] Gillette, R. (1975) Nuclear power: hard times and a questioning Congress.
Science 187: 1058–1062.

[26] *Time* (January 10, 1983) Too hot for the usual burial.

[27] Walker, J. S. (2009) *The Road to Yucca Mountain.* University of California Press,
p. 180.

[28] Ege, J. R. (1985) Maps showing distribution, thickness, and depth of salt
deposits of the United States. US Geological Survey Open-File Report 85–28.

CHAPTER 13. NEVADA WINS THE LOTTERY

[1] The Official Report, House of Commons (5th Series) (November 11, 1947)
vol. 444, cc. 206–207.

[2] Carter, L. J. (1987) *Nuclear Imperatives and Public Trust.* Washington, DC:
Resources for the Future.

[3] Crawley, J. W. (July 11, 2005) The day we nuked Mississippi. Waynesboro, VA:
News Virginian.

[4] *Time* (December 31, 1984) An unwelcome Christmas present.

[5] Marshall, E. (1986) Nuclear waste program faces political burial. *Science* 233:
835–836.

[6] US Department of Energy (1986) Recommendation by the Secretary of Energy
of candidate sites for site characterization for the first radioactive waste
repository. DOE/5-0048, May.

[7] US Congress, Office of Technology Assessment (1985) Managing the nation's
commercial high-level radioactive waste. Report OTA-O-171. Washington, DC.

[8] Gutekunst, P. (February 12, 1986) Explosive situation handled calmly. *Portland
Press Herald*.

[9] Maynard, J. (May 11, 1986) The story of a town. *New York Times*.

[10] Marshall, E. (1987) Thirty ways to temporize on waste. *Science* 237: 591–592.

[11] Holt, M. (2009) Nuclear waste disposal: alternatives to Yucca Mountain. Congressional Research Service. Washington, DC, p. 22.

[12] Marshall, E. (1988) Nevada wins the nuclear waste lottery. *Science* 239: 15.

CHAPTER 14. THE NEVADA TEST SITE

[1] Beano comic.
[2] Fehner, T. R. and F. G. Gosling (2000) Origins of the Nevada Test Site. US Department of Energy. Report DOE/MA-0518.
[3] Hacker, B. C. (1994) *Elements of Controversy*. University of California Press.
[4] Hewlett, R. G. and F. Duncan (1969) *A History of the United States Atomic Energy Commission, Vol. II: Atomic Shield, 1947–1952.* Pennsylvania State University Press.
[5] The Atomic Testing Museum (Las Vegas, NV).
[6] US Department of Energy, National Nuclear Security Administration (2011) News Nob. DOE/NV 774.
[7] US Department of Energy, National Nuclear Security Administration (2005) Atomic culture. DOE/NV 1042.
[8] US Department of Energy, National Nuclear Security Administration (2010) Miss Atom Bomb. DOE/NV 1024.
[9] Smith, M. (1991) Advertising the atom. In: M. J. Lacey (ed.), *Government and Environmental Politics: Essays on Historical Developments since World War Two.* The Johns Hopkins University Press, pp. 233–262.
[10] Seife, C. (2008) *Sun in a Bottle.* Viking.
[11] Vandegraft, D. L. (1993) Project Chariot: nuclear legacy of Cape Thompson. In *Proceedings of the U.S. Interagency Arctic Research Policy Committee Workshop on Arctic Communication, Session A: Native People's Concerns about Arctic Contamination II: Ecological Impacts, May 6, 1993.* Anchorage.
[12] Broad, W. J. (November 7, 1991) A Soviet company offers nuclear blasts for sale to anyone with cash. *New York Times.*
[13] US Department of Energy, National Nuclear Security Administration (2010) Sedan Crater. DOE/NV 712.
[14] Horgan, J. (1996) "Peaceful" nuclear explosions. *Scientific American* 274: 14–15.
[15] President J. F. Kennedy (May 25, 1961) "Man on the moon," Special Message to the Congress on Urgent National Needs.
[16] US Department of Energy (1996) Final environmental impact statement for the Nevada Test Site and off-site locations in the State of Nevada, Volume 1. DOE/EIS 0243.
[17] Laczniak, R. J. *et al.* (1996) Summary of hydrogeologic controls on ground-water flow at the Nevada Test Site, Nye County, Nevada. US Geological Survey Water-Resources Investigations Report 96–4109.
[18] Tompson, A. F. B. *et al.* (2002) On the evaluation of groundwater contamination from underground nuclear tests. *Environmental Geology* 42: 235–247.
[19] Lyon, J. L. *et al.* (1979) Childhood leukemias associated with fallout from nuclear testing. *New England Journal of Medicine* 300: 397–402.
[20] Simon, S. L. *et al.* (2006) Fallout from nuclear weapons tests and cancer risks. *American Scientist* 94: 48–57.

CHAPTER 15. YUCCA MOUNTAIN

[1] Tolkien, J. R. R. (1965) *The Lord of the Rings, Part One: The Fellowship of the Ring.* Ballantine Books, p. 349.
[2] Ekren, E. B. *et al.* (1974) Geologic and hydrologic considerations for various concepts of high-level radioactive waste disposal in conterminous United States. US Geological Survey Open-File Report 74–158.

[3] Winograd, I. J. (1972) Near surface storage of solidified high-level radioactive wastes in thick (400–2,000 foot) unsaturated zones in the southwest. *Geological Society of America, Abstracts with Programs* 4: 708.

[4] Winograd, I. J. (1974) Radioactive waste storage in the arid zone. *Eos, Transactions, American Geophysical Union* 55(10): 884–894.

[5] McKelvey, V. E. (July 9, 1976) Letter to Richard W. Roberts, Assistant Administrator for Nuclear Energy, ERDA.

[6] Hanks, T. C. *et al.* (1999) Yucca Mountain as a radioactive-waste repository. US Geological Survey Circular 1184.

[7] Swainston, H. W. (1991) The characterization of Yucca Mountain: the status of the controversy. *Federal Facilities Environmental Journal* Summer: 151–160.

[8] *60 Minutes* transcript, http://www.cbsnews.com/stories/2003/10/23/60minutes/main579696.shtml.

[9] Flynn, J. *et al.* (1995) *One Hundred Centuries of Solitude – Redirecting America's High-Level Nuclear Waste Policy.* Westview.

[10] Kittredge, W. (1988) In my backyard. *Harper's Magazine* October: 59.

[11] National Research Council (2006) Going the distance? The safe transport of spent nuclear fuel and high-level radioactive waste in the United States. Washington, DC: National Academies Press.

[12] Sproat, E. F. (September 24, 2008) Statement before the Committee on Commerce, Science, and Transportation, US Senate.

[13] Waymire, S. D. (2010) Yucca Mountain: the battle for national energy policy, Chapter 51. (www.yuccamountainexpose.com/Y51.htm)

[14] *Harper's Magazine* (March 1992) How to sell a nuclear dump.

[15] Rothstein, L. (1992) Nevadans dump dump ads. *Bulletin of the Atomic Scientists* May: 3–4.

[16] Flynn, J. *et al.* (1993) The Nevada Initiative: a risk communication fiasco. *Risk Analysis* 13: 497–502.

[17] *Las Vegas Sun* (December 30, 2000) New cries of outrage over DOE, Yucca relationship.

[18] Winograd, I. J. (1991) Yucca Mountain as a nuclear-waste repository – neither myth nor millennium. US Geological Survey Open-File Report 91–170.

[19] Roseboom, E. H., Jr. (1983) Disposal of high-level nuclear waste above the water table in arid regions. US Geological Survey Circular 903.

[20] Bruno, J. and R. C. Ewing (2006) Spent nuclear fuel. *Elements* 2: 343–349.

[21] Winograd, I. J. and E. H. Roseboom, Jr. (2008) Yucca Mountain revisited. *Science* 320: 1426–1427.

[22] Neff, E. (April 8, 2002) Guinn vetoes Yucca dump. *Las Vegas Sun.*

[23] Macfarlane, A. (May 28, 2003) Statement before Senate Energy and Water Development Subcommittee, Las Vegas, NV.

[24] Ewing, R. C. and A. Macfarlane (2002) Yucca Mountain. *Science* 296: 659–660.

[25] Comments and response to Ewing and Macfarlane: *Science* 296: 2333–2335 and http://www.sciencemag.org/cgi/eletters/296/5568/659?ck=nck#487.

[26] National Research Council (2003) One step at a time. Washington, DC: National Academies Press.

[27] Belcher, W. R. *et al.* (2006) Ground-water modeling of the Death Valley region, Nevada and California. US Geological Survey Fact Sheet 2006–3120.

CHAPTER 16. HOW *LONG* IS LONG?

[1] Irons, J. C. (1896) *Autobiographical Sketch of James Croll, with Memoir of his Life and Work.* London: Edward Stanford.

[2] Chapman, N. A. and I. G. McKinley (1987) *The Geological Disposal of Nuclear Waste*. John Wiley.

[3] The Editors of *Life* (1962) *The Epic of Man*. Golden Press.

[4] Wong, K. (2009) Twilight of the Neanderthals. *Scientific American* 301: 32–37.

[5] Huff, C. D. *et al.* (2010) Mobile elements reveal small population size in the ancient ancestors of *Homo sapiens*. *Proceedings of the National Academy of Sciences* 107: 2147–2152.

[6] Cameron, D. (1965) Goethe – discoverer of the ice age. *Journal of Glaciology* 5(41): 751–754.

[7] Weart, S. R. (2003) *The Discovery of Global Warming*. Harvard University Press.

[8] Imbrie, J. and K. P. Imbrie (1986) *Ice Ages: Solving the Mystery*. Harvard University Press.

[9] Fleming, J. R. (2006) James Croll in context: the encounter between climate dynamics and geology in the second half of the nineteenth century. *History of Meteorology* 3: 43–53.

[10] Macdougall, D. (2004) *Frozen Earth: The Once and Future Story of Ice Ages*. University of California Press.

[11] *Encyclopedia of Earth* (Milankovitch, Milutin, www.eoearth.org/article/ Milankovitch,_Milutin).

[12] Emiliani, C. (1955) Pleistocene temperatures. *Journal of Geology* 63: 538–578.

[13] Wertenbaker, W. (2000) William Maurice Ewing: pioneer explorer of the ocean floor and architect of Lamont. *GSA Today* October: 28–29.

[14] Bond, R. L. (1999) A core a day keeps "Doc" happy. In: L. P. Lippsett (ed.), *Lamont-Doherty Observatory of Columbia University, Twelve Perspectives on the First Fifty Years, 1949–1999*. (http://www.ldeo.columbia.edu/res/fac/CORE_REPOSITORY/ RHP5d.html)

[15] Broecker, W. S. *et al.* (1968) Milankovitch hypothesis supported by precise dating of coral reefs and deep-sea sediments. *Science* 159: 297–300.

[16] Hays, J. D. *et al.* (1976) Variations in the Earth's orbit: pacemaker of the ice ages. *Science* 194: 1121–1132.

[17] Hoffman, R. J. (1988) Chronology of diving activities and underground surveys in Devils Hole and Devils Hole Cave, Nye County, Nevada, 1950–86. US Geological Survey Open-File Report 88–93.

[18] Winograd, I. J. *et al.* (1988) A 250,000-year climatic record from Great Basin vein calcite: implications for Milankovitch theory. *Science* 242: 1275–1280.

[19] Winograd, I. J. *et al.* (1992) Continuous 500,000-year climate record from vein calcite in Devils Hole, Nevada. *Science* 258: 255–260.

[20] Winograd, I. J. *et al.* (2006) Devils Hole, Nevada, δ18O record extended to the mid-Holocene. *Quaternary Research* 66: 202–212.

[21] Broecker, W. S. (1992) Upset for Milankovitch theory. *Nature* 359: 779–780.

[22] Crowley, T. J. (2002) Cycles, cycles everywhere. *Science* 295: 1473–1474.

[23] Berger, A. and M. F. Loutre (2002) An exceptionally long interglacial ahead? *Science* 297: 1287–1288.

[24] Ruddiman, W. F. (2005) *Plows, Plagues, and Petroleum*. Princeton University Press.

CHAPTER 17. LEAVING ALMOST NO STONE UNTURNED

[1] Kerr, R. A. (1999) For radioactive waste from weapons, a home at last. *Science* 283: 1626–1628.

[2] Tetreault, S. and K. Rogers (June 4, 2008) DOE files to build Yucca. *Las Vegas Review-Journal*.

[3] Whipple, C. (2006) Performance assessment: what is it and why is it done? In: A. M. Macfarlane and R. C. Ewing (eds.), *Uncertainty Underground*. Cambridge, Massachusetts: MIT Press.

[4] Swift, P. (May 29, 2008) Presentation to Nuclear Waste Technical Review Board. (http://www.nwtrb.gov/meetings/2008/may/swift.pdf)

[5] Hanks, T. C. *et al.* (1999) Yucca Mountain as a radioactive-waste repository. US Geological Survey Circular 1184.

[6] Eckhardt, R. (1987) Stan Ulam, John von Neumann, and the Monte Carlo method. *Los Alamos Science* 15: 131–137.

[7] Federal Register (September 19, 1985) 50(182): 38066–38089.

[8] National Research Council (1995) Technical bases for Yucca Mountain standards. Washington, DC: National Academy Press.

[9] Nuclear Energy Institute vs. US Environmental Protection Agency (2004) US Court of Appeals for the District of Columbia Circuit, 373 F.3d 1251.

[10] National Council on Radiation Protection and Measurements (2009) Ionizing radiation exposure of the population of the United States. NCRP Report No. 160. (http://NCRPpublications.org)

[11] Tetreault, S. (August 10, 2005) Yucca radiation limits unveiled. *Las Vegas Review-Journal*.

[12] Kerr, R. A. (2000) Science and policy clash at Yucca Mountain. *Science* 288: 602.

[13] Peterson, P. F. *et al.* (2006) Nuclear waste and the distant future. *Issues in Science and Technology* Summer.

[14] Okrent, D. and L. Xing (1993) Future risk from a hypothesized RCRA site disposing of carcinogenic metals should a loss of societal memory occur. *Journal of Hazardous Materials* 34(3): 363–384.

[15] Isherwood, D. (1988) Nuclear waste disposal. *Science* 239: 1321–1322.

[16] Slovic, P. (1987) Perception of risk. *Science* 236: 280–285.

[17] Weart, S. (November 28, 1992) Fears, fantasies and fallout. *New Scientist*.

[18] Chauvet, J. M. *et al.* (1996) *Dawn of Art – The Chauvet Cave*. Harry N. Abrams, Inc.

[19] The Editors of *Life* (1962) *The Epic of Man*. Golden Press.

[20] Simmons, A. M. and J. S. Stuckless (2010) Analogues to features and processes of a high-level radioactive waste repository proposed for Yucca Mountain, Nevada. US Geological Survey Professional Paper 1779.

[21] Stuckless, J. S. (2000) Archaeological analogues for assessing the long-term performance of a mined geologic repository for high-level radioactive waste. US Geological Survey Open-File Report 00–181.

[22] Winograd, I. J. (1986) Archaeology and public perception of a transscientific problem – disposal of toxic wastes in the unsaturated zone. US Geological Survey Circular 990.

[23] Murphy, W. M. (1995) Natural analogs for Yucca Mountain. *Radwaste Magazine* 2(6): 44–50.

[24] Chu, M. S. Y. and J. R. Dyer (2003) Licensing, design, and construction of the Yucca Mountain repository. *The Bridge* 33: 18–25.

[25] Applegate, D. (1997) Yucca Mountain: no light at tunnel's end? *Geotimes* July: 13.

[26] *Las Vegas Sun* (June 2, 1998) Jury still out on science issues.

[27] Nuclear Waste Technical Review Board (1999) Moving beyond the Yucca Mountain Viability Assessment. (www.nwtrb.gov)

[28] Bodvarsson, G. S. (2006) Thermohydrologic effects and interactions. In: A. M. Macfarlane and R. C. Ewing (eds.), *Uncertainty Underground*. Cambridge, Massachusetts: MIT Press.

[29] Peterman, Z. E. (January 16, 2008) Presentation to Nuclear Waste Technical Review Board. (http://www.nwtrb.gov/meetings/2008/jan/peterman.pdf)

[30] Nuclear Waste Technical Review Board (October 21, 2003) Letter to Margaret S. Y. Chu. (http://www.nwtrb.gov/corr/mlc014016.pdf)

[31] Chu, M. S. Y. (October 27, 2003) Letter to Nuclear Waste Technical Review Board. (http://www.nwtrb.gov/corr/doe102703.pdf)

[32] Nuclear Waste Technical Review Board (September 25–26, 2006) Workshop on localized corrosion, Las Vegas, NV. (www.nwtrb.gov)

[33] Garrick, B. J. (April 22, 2008) Letter to Edward F. Sproat III, Director, Office of Civilian Radioactive Waste Management. (http://www.nwtrb.gov/corr/bjg087.pdf)

[34] Apted, M. J. *et al.* (2006) Evaluation of potential formation and impacts of deliquescent brines. In *Proceedings, 11th International High-level Radioactive Waste Management Conference.* La Grange Park, Illinois: American Nuclear Society, pp. 921–924.

[35] Ballinger, R. (May 29, 2008) Transcript, Nuclear Waste Technical Review Board Spring Meeting, p. 249. (http://www.nwtrb.gov/meetings/2008/may/08may29.pdf)

CHAPTER 18. SURPRISE

[1] Rumsfeld, D. (February 12, 2002) Department of Defense news briefing.

[2] Chamberlin, T. C. (1890) The method of multiple working hypotheses. *Science* 148: 754–759. (Reprinted in 1965.)

[3] Winchester, S. (2003) *Krakatoa.* Perennial, p. 74.

[4] National Research Council (1990) Rethinking high-level radioactive waste disposal: a position statement of the Board on Radioactive Waste Management. Washington, DC: National Academy Press.

[5] Flint, A. L. *et al.* (2001) Evolution of the conceptual model of unsaturated zone hydrology at Yucca Mountain, Nevada. *Journal of Hydrology* 247: 1–30.

[6] Interview with R. C. Ewing (2000) *Environmental Review: A Monthly Newsletter of Environmental Science and Policy* May.

[7] Scott, R. B. *et al.* 1983. Geologic character of tuffs in the unsaturated zone at Yucca Mountain, southern Nevada. In: J. W. Mercer *et al.* (eds.), *Role of the Unsaturated Zone in Radioactive and Hazardous Waste Disposal.* Ann Arbor Science, pp. 289–335.

[8] Roseboom, E. H., Jr. (1983) Disposal of high-level nuclear waste above the water table in arid regions. US Geological Survey Circular 903.

[9] Montazer, P. and W. E. Wilson (1984) Conceptual hydrologic model of flow in the unsaturated zone, Yucca Mountain, Nevada. US Geological Survey Water-Resources Investigations Report 84–4345.

[10] Metlay, D. (2000) From tin roof to torn wet blanket. In: D. Sarewitz *et al.* (eds.), *Prediction: Science, Decision Making and the Future of Nature.* Island Press, pp. 199–228.

[11] Bodvarsson, G. S. *et al.* (2003) Estimation of percolation flux from borehole temperature data at Yucca Mountain, Nevada. *Journal of Contaminant Hydrology* 62–63: 3–22.

[12] Fabryka-Martin, J. *et al.* (2006) Water and radionuclide transport in the unsaturated zone. In: A. M. Macfarlane and R. C. Ewing (eds.), *Uncertainty Underground.* Cambridge, Massachusetts: MIT Press.

[13] Paces, J. B. and R. C. Roback (2006) Chlorine-36 validation study at Yucca Mountain, Nevada. US Department of Energy Report TDR-NBS-HS-000017 Rev00.

[14] Cizdziel, J. V. *et al.* (2008) Recent measurements of 36Cl in Yucca Mountain rock, soil, and seepage. *Journal of Radioanalytical and Nuclear Chemistry* 275: 133–144.

[15] Dyer, J. R. (January 24, 2007) Presentation to Nuclear Waste Technical Review Board. (http://www.nwtrb.gov/meetings/2007/jan/dyer.pdf)

[16] Richardson, D. (January 25, 1999) Presentation to Nuclear Waste Technical Review Board. (http://www.nwtrb.gov/meetings/1999/jan/richardson.pdf)

[17] Ewing, R. C. and A. Macfarlane (2002) Yucca Mountain, *Science* 296: 659–660.

[18] Wald, M. L. (December 18, 2001) New suit filed against U.S. about nuclear waste dump. *New York Times.*

[19] Wald, M. L. (May 12, 2008) Joseph Egan, lawyer who fought nuclear waste site, is dead at 53. *New York Times.*

[20] Shoesmith, D. W. (2006) Waste package corrosion. In: A. M. Macfarlane and R. C. Ewing (eds.), *Uncertainty Underground.* Cambridge, Massachusetts: MIT Press.

[21] Nuclear Waste Technical Review Board (July 19, 2001) Meeting transcripts, p. 83. (http://www.nwtrb.gov/meetings/2001/july/01july19.pdf)

[22] Zumdahl, S. S. (1993) *Chemistry.* Third edition. D. C. Heath and Company, p. 945.

[23] Stahl, D. (2006) Drip shield and backfill. In: A. M. Macfarlane and R. C. Ewing (eds.), *Uncertainty Underground.* Cambridge, Massachusetts: MIT Press.

[24] Loux, R. (April 19, 2007) Letter to Dale Klein, Chairman, Nuclear Regulatory Commission. (http://www.state.nv.us/nucwaste/news2008/pdf/nv070419klein.pdf)

[25] US Department of Energy (2001) Yucca Mountain science and engineering report: technical information supporting site recommendation consideration. DOE/RW-0539. pp. 2–31.

[26] *Las Vegas Sun* (September 18, 1997) Plutonium movement through groundwater triggers questions.

[27] Buddemeier, R. W. and J. R. Hunt (1988) Transport of colloidal contaminants in groundwater: radionuclide migration at the Nevada Test Site. *Applied Geochemistry* 3: 535–548.

[28] Kersting, A. B. (2006) Colloidal transport of radionuclides. In: A. M. Macfarlane and R. C. Ewing (eds.), *Uncertainty Underground.* Cambridge, Massachusetts: MIT Press.

[29] Kersting, A. B. *et al.* (1999) Migration of plutonium in ground water at the Nevada Test Site. *Nature* 397: 56–59.

[30] Broad, W. J. (March 5, 1995) Scientists fear atomic explosion of buried waste. *New York Times.*

[31] Broad, W. J. (March 23, 1995) Theory on threat of blast at nuclear waste site gains support. *New York Times.*

[32] Taubes, G. (1995) Blowup at Yucca Mountain. *Science* 268: 1836–1839.

[33] Cowan, G. A. (1976) A natural fission reactor. *Scientific American* 235: 36–47.

[34] Meshik, A. P. (2009) The workings of an ancient nuclear reactor. *Scientific American* January.

[35] Bowman, C. D. and F. Venneri (1995) Nuclear waste storage at Yucca Mountain. *Science* 269: 906–907.

[36] Bowman, C. D. and F. Venneri (1996) *Science and Global Security* 5(3): 279–302.

[37] Taubes, G. (1996) Yucca blowup theory bombs, says study. *Science* 271: 1664.

[38] Nuclear Waste Technical Review Board (January 28, 2009) Transcripts of Winter Board Meeting, pp. 162–242. (http://www.nwtrb.gov/meetings/2009/jan/09jan28.pdf)

CHAPTER 19. SHAKE & BAKE

[1] Blakeslee, S. (July 4, 1992) Earthquake raises concern about nuclear waste dump. *New York Times.*

[2] Hill, D. P. *et al.* (1993) Seismicity remotely triggered by the magnitude 7.3 Landers, California, earthquake. *Science* 260: 1617–1623.

[3] Gilbert, G. K. (1875) Report on the geology of portions of Nevada, Utah, California, and Arizona examined in the years 1871–1872. US Geographical and Geological Surveys West of the 100th Meridian, v. 3, part 1.

[4] McPhee, J. (1981) *Basin and Range.* Farrar, Straus and Giroux, p. 49.

[5] Valentine, G. A. and F. V. Perry (2009) Volcanic risk assessment at Yucca Mountain, NV, USA: integration of geophysics, geology, and modeling. In: C. B. Connor, N. A. Chapman and L. G. Connor (eds.), *Volcanic and Tectonic Hazard Assessments for Nuclear Facilities.* Cambridge University Press, pp. 452–480.

[6] US Department of Energy (2001) Yucca Mountain science and engineering report: technical information supporting site recommendation consideration. DOE/RW-0539, pp. 4–378.

[7] Hinze, W. J. *et al.* (2008) Evaluating igneous activity at Yucca Mountain. *Eos, Transactions, American Geophysical Union* 89(4): 29–30.

[8] Electric Power Research Institute (2005) Potential igneous processes relevant to the Yucca Mountain repository: intrusive release scenario. EPRI Report 1011165. Palo Alto, CA.

[9] Crowe, B. M. *et al.* (2006) Volcanism: the continuing saga. In: A. M. Macfarlane and R. C. Ewing (eds.), *Uncertainty Underground.* Cambridge, Massachusetts: MIT Press.

[10] Kerr, R. A. (1996) A new way to ask the experts: rating radioactive waste risks. *Science* 274: 913–914.

[11] Connor, C. B. *et al.* (2000) Geologic factors controlling patterns of small-volume basaltic volcanism: application to a volcanic hazards assessment at Yucca Mountain, Nevada. *Journal of Geophysical Research* 105: 417–432.

[12] Macilwain, C. (2001) Out of sight, out of mind? *Nature* 412: 850–852.

[13] Smith, E. I. and D. L. Keenan (2005) Yucca Mountain could face greater volcanic threat. *Eos, Transactions, American Geophysical Union* 86(35): 317–324.

[14] Blakely, R. J. *et al.* (2000) Aeromagnetic survey of the Amargosa Desert, Nevada and California: a tool for understanding near-surface geology and hydrology. US Geological Survey Open-File Report 00–188.

[15] O'Leary, D. W. *et al.* (2002) Aeromagnetic expression of buried basaltic volcanoes near Yucca Mountain, Nevada. US Geological Survey Open-File Report 02–020.

[16] Swift, P. (May 29, 2008) Presentation to Nuclear Waste Technical Review Board. (http://www.nwtrb.gov/meetings/2008/may/swift.pdf)

[17] Hanks, T. C. *et al.* (1999) Yucca Mountain as a radioactive-waste repository. US Geological Survey Circular 1184.

[18] Spengler, R. W. *et al.* (1994) The Sundance fault: a newly recognized shear zone at Yucca Mountain, Nevada. US Geological Survey Open-File Report 94–49.

[19] US Department of Energy (2003) Studying the movement of rock and earthquakes. Fact Sheet. DOE/YMP-0344.

[20] Applegate, D. (2006) The mountain matters. In: A. M. Macfarlane and R. C. Ewing (eds.), *Uncertainty Underground.* Cambridge, Massachusetts: MIT Press.

[21] Stepp, J. C. *et al.* (2001) Probabilistic seismic hazard analyses for ground motions and fault displacement at Yucca Mountain, Nevada. *Earthquake Spectra* 17: 113–151.

[22] Hanks, T. C. *et al.* (2006) Report of the workshop on extreme ground motions at Yucca Mountain, August 23–25, 2004. US Geological Survey Open-File Report 2006–1277.

[23] Brune, J. N. *et al.* (2007) Gauging earthquake hazards with precariously balanced rocks. *American Scientist* 95: 36–43.

[24] Normile, D. (March 18, 2011) Devastating earthquake defied expectations. *Science* 331: 1375–1376.

[25] *The Associated Press* (September 24, 2007) Nuke dump structures moved after study.

[26] Rogers, K. (September 24, 2007) Yucca fault line might spring surprise. *Las Vegas Review-Journal.*

[27] Hoffman, R. J. (1988) Chronology of diving activities and underground surveys in Devils Hole and Devils Hole Cave, Nye County, Nevada, 1950–86. US Geological Survey Open-File Report 88–93.

CHAPTER 20. THE PROJECT GETS INTO HOT WATER

[1] Mencken, H. L. WMail ezine. Issue 20.

[2] Nuclear Waste Technical Review Board (September 10, 2002) Transcript of Fall 2002 Board Meeting, p. 17. (http://www.nwtrb.gov/meetings/2002/sept/02sept10.pdf)

[3] US Department of Energy (2001) Yucca Mountain science and engineering report: technical information supporting site recommendation consideration. DOE/RW-0539, pp. 4–405.

[4] Marshall, E. (1991) The geopolitics of nuclear waste. *Science* 251: 864–867.

[5] Szymanski, J. S. (1989) Conceptual considerations of the Yucca Mountain ground water system with special emphasis on the adequacy of this system to accommodate a high level nuclear waste repository. DOE internal report, Las Vegas, NV.

[6] Broad, W. J. (November 18, 1990) A mountain of trouble. *New York Times.*

[7] Dudley, W. W. *et al.* (1990) Unpublished letter from U.S. Geological Survey scientists to the editor of the *New York Times* magazine regarding William J. Broads' November 18, 1990 article on Yucca Mountain. US Geological Survey Open-File Report 91–58.

[8] Swainston, H. W. (2006) History of an issue: upwelling of water at Yucca Mountain. (http://www.nwtrb.gov/meetings/2006/feb/swainston-report.pdf)

[9] Archambeau, C. B. and N. J. Price (1991) An assessment of J. S. Szymanski's conceptual hydrotechtonic model, minority report of the Special DOE Review Panel. September, Unpublished report.

[10] Powers, D. W. *et al.* (1991) External peer review panel majority report. August, Unpublished report.

[11] National Research Council (1992) Ground water at Yucca Mountain: how high can it rise? Washington, DC: National Academy Press.

[12] Quade, J. and T. E. Cerling (1990) Stable isotopic evidence for a pedogenic origin of carbonates in Trench 14 near Yucca Mountain, Nevada. *Science* 250: 1549–1552.

[13] Stuckless, J. S. *et al.* (1991) U and Sr isotopes in ground water and calcite, Yucca Mountain, Nevada: evidence against upwelling water. *Science* 254: 551–554.

[14] Nuclear Waste Technical Review Board (1992) Sixth report to the US Congress and the US Secretary of Energy, December. (http://www.nwtrb.gov/reports/6report.pdf)

[15] Davies, J. B. and C. B. Archambeau (1997) Geohydrological models and earthquake effects at Yucca Mountain, Nevada. *Environmental Geology* 32: 23–35.

[16] Rojstaczer, S. (1999) Stress dependent permeability and its political consequences at Yucca Mountain. American Geophysical Union, Spring Meeting. Abstract No. U22A-03.

[17] Cohon, J. L. (July 24, 1998) Letter to L. H. Barrett, Acting Director, Office of Civilian Radioactive Waste Management. (www.nwtrb.gov)

[18] Roedder, E. (1984) Fluid inclusions. *Reviews in Mineralogy* 12.

[19] Wilson, N. S. F. and J. S. Cline (2006) Hot upwelling water: did it really invade Yucca Mountain? In: A. M. Macfarlane and R. C. Ewing (eds.), *Uncertainty Underground*. Cambridge, Massachusetts: MIT Press.

[20] Dublyansky, Y. *et al.* (1996) Fluid inclusions in calcite from the Yucca Mountain exploratory tunnel. In: P. E. Brown and S. G. Hagemann (eds.), *Program and Abstracts, Sixth Biennial Pan American Conference on Research on Fluid Inclusions, May 30 – June 1, 1996*. Madison, Wisconsin: University of Wisconsin Department of Geology, pp. 38–39.

[21] Johnson, J. (December 14, 1998) Yucca Mountain samples reheat scientists' row. *Chemical & Engineering News*.

[22] Paces, J. B. *et al.* (2001) Ages and origins of calcite and opal in the Exploratory Studies Facility Tunnel, Yucca Mountain, Nevada. US Geological Survey Water-Resources Investigations Report 01–4049.

[23] Nuclear Waste Technical Review Board (May 9, 2001) Transcript of Spring 2001 Board Meeting. (http://www.nwtrb.gov/meetings/2001/may/01may09.pdf)

[24] Wilson, N. S. F. *et al.* (2003) Origin, timing, and temperature of secondary calcite-silica minerals at Yucca Mountain. *Geochimica et Cosmochimica Acta* 67: 1145–1176.

[25] Dyer, J. R. (January 24, 2002) Letter to J. L. Cohon, Chairman, Nuclear Waste Technical Review Board. (http://www.nwtrb.gov/corr/doe12502.pdf)

[26] Hornberger, G. M. (May 8, 2006) Letter to H. W. Swainston. (http://www.nwtrb.gov/corr/dmd149.pdf)

[27] Neymark, L. A. *et al.* (2002) U-Pb ages of secondary silica at Yucca Mountain, Nevada: implications for the paleohydrology of the unsaturated zone. *Applied Geochemistry* 17: 709–734.

[28] Paces, J. B. *et al.* (2010) Limited hydrologic response to Pleistocene climate change in deep vadose zones – Yucca Mountain, Nevada. *Earth and Planetary Science Letters* 300: 287–298.

[29] Nuclear Waste Technical Review Board (2011) Technical advancements and issues associated with the permanent disposal of high-activity wastes: lessons learned from Yucca Mountain and other programs. A Report to Congress and the Secretary of Energy, June. (http://www.nwtrb.gov/reports/technical%20lessons.pdf)

[30] Fabryka-Martin, J. *et al.* (2006) Water and radionuclide transport in the unsaturated zone. In: A. M. Macfarlane and R. C. Ewing (eds.), *Uncertainty Underground*. Cambridge, Massachusetts: MIT Press.

[31] Wong, I. (2011) Advances made in evaluating seismic hazards at the Yucca Mountain site. *Geological Society of America Annual Meeting, Abstracts with Programs* 43: Paper No. 73–6.

[32] Coppersmith, K. (2011) Development of volcanic hazard analysis methodologies for Yucca Mountain, NV. *Geological Society of America Annual Meeting, Abstracts with Programs* 43: Paper No. 73–5.

[33] Simmons, A. M. and J. S. Stuckless (2010) Analogues to features and processes of a high-level radioactive waste repository proposed for Yucca Mountain, Nevada. US Geological Survey Professional Paper 1779.

[34] Birkholzer, J. T. (2011) Scientific issues related to geologic storage of carbon sequestration: learning from nuclear waste and Yucca Mountain R&D. *Geological Society of America Annual Meeting, Abstracts with Programs* 43: Paper No. 73–10.

CHAPTER 21. A NEW PRESIDENT, NEW POLICIES

[1] Shapiro, A. (June 30, 2010) Once a critic, Obama now embraces commissions. *National Public Radio.*

[2] *US Water News Online* (March 2009) Official says nuclear waste won't go to Nevada site.

[3] *Eos, Transactions, American Geophysical Union* (May 5, 2009) Obama indicates strong support for science.

[4] US Senate Committee on Environment & Public Works (April 29, 2009) Senators seek explanation of Obama's Yucca Mountain decision. Press release.

[5] American Geological Institute (2011) Yucca Mountain regulations move ahead slowly as alternatives explored *(2/09).* (http://www.agiweb.org/gap/legis111/nuclear.html)

[6] Rogers, K. (July 31, 2009) Reid declares Yucca victory. *Las Vegas Review-Journal.*

[7] US Department of Energy (March 3, 2010) Department of Energy files motion to withdraw Yucca Mountain license application. News release.

[8] US Nuclear Regulatory Commission, Atomic Safety and Licensing Board (ASLBP No. 09–892-HLW-CAB04. (www.state.nv.us/nucwaste/licensing/order100629deny.pdf)

[9] US House of Representatives, Committee on Science, Space, and Technology (2011) Yucca Mountain: the administration's impact on U.S. nuclear waste management policy. Report by the Majority Staff, June, pp. 24, 66–71.

[10] US Government Accountability Office (2011) Commercial nuclear waste: effects of a termination of the Yucca Mountain Repository Program and lessons learned. GAO-11-229.

[11] US Department of Energy, Office of Inspector General (2010) Need for enhanced surveillance during the Yucca Mountain project shut down. OAS-SR-10-01, July 21. (www.ig.energy.gov/documents/OAS-SR-10-01.pdf)

[12] Zacha, N. J. (2011) On nuclear waste, the DOE, and 1984 (the novel, not the year). *Radwaste Solutions* July–August: 4.

[13] American Geological Institute (2011) Showdown unfolds over Yucca Mountain saga. (http://www.agiweb.org/gap/legis111/nuclear.html)

[14] Cappiello, D. and M. Daly (June 12, 2011) NRC chief in hot seat for scrapping work on dump. *Associated Press.*

[15] O'Keefe, E. (December 13, 2011) White House: Nuclear Regulatory Commission feuded over "organizational issues." *Washington Post.*

[16] Tetreault, S. (March 22, 2011) Yucca Mountain fate argued in court. *Las Vegas Review-Journal.*

[17] Blue Ribbon Commission on America's Nuclear Future (2012) Report to the Secretary of Energy. Washington, DC.

[18] US Government Accountability Office (2011) DOE nuclear waste – better information needed on waste storage at DOE sites as a result of Yucca Mountain shutdown. GAO-11-230.

[19] Nuclear Waste Technical Review Board (2011) Experience gained from programs to manage high-level radioactive waste and spent nuclear fuel in the United States and other countries. A Report to Congress and the Secretary of Energy, April. (http://www.nwtrb.gov/reports/reports.html)

[20] Eriksson, L. G. (2010) Spent fuel disposal: success vs. failure. *Radwaste Solutions* January/February: 22–30.

[21] Black, R. (April 27, 2006) Finland buries its nuclear past. *BBC news.*

[22] Mörner, N.-A. (2001) In absurdum: long-term predictions and nuclear waste handling. *Engineering Geology* 61: 75–82.

[23] Butler, D. (2010) France digs deep for nuclear waste. *Nature* 466: 804–805.

[24] Tetreault, S. (March 9, 2012) Nye official gives consent to burying nuke waste. *Las Vegas Review-Journal.*

[25] *The Associated Press* (March 13, 2012) Sandoval: Despite what Nye says, he says no to Yucca.

[26] Stewart, R. B. and J. B. Stewart (2011) *Fuel Cycle to Nowhere: U.S. Law and Policy on Nuclear Waste.* Vanderbilt University Press, p. 293.

CHAPTER 22. NUCLEAR WASTE AND OUR ENERGY FUTURE

[1] Knowlton, D. R. (1939) *Unitization – Its Progress and Future, Drilling and Production Practice.* American Petroleum Institute Report 39: 630–635.

[2] Pacala, S. and R. Socolow (2004) Stabilization wedges: solving the climate problem for the next 50 years with current technologies. *Science* 305: 968–972.

[3] European Nuclear Society (2012) Nuclear power plants, worldwide. (http://www.euronuclear.org/welcome.htm)

[4] Nadis, S. (1996) The sub-seabed solution. *The Atlantic Monthly* 278: 28–39.

[5] *Las Vegas Sun* (February 7, 2010) Letter to the editor.

[6] National Research Council (1990) Rethinking high-level radioactive waste disposal: a position statement of the Board on Radioactive Waste Management. Washington, DC: National Academy Press.

[7] Wildavsky, A. (1979) No risk is the highest risk of all. *American Scientist* 67: 32–37.

[8] Holt, M. (2009) Nuclear waste disposal: alternatives to Yucca Mountain. Congressional Research Service. February, p. 11.

[9] Carter, L. J. (1987) *Nuclear Imperatives and Public Trust.* Washington, DC: Resources for the Future.

[10] Lilienthal, D. E. (1963) *Change, Hope, and the Bomb.* Princeton University Press, p. 64.

[11] Slovic, P. (1987) Perception of risk. *Science* 236: 280–285.

[12] Kristof, N. D. (March 30, 2008) "With a few more brains . . . " *New York Times.*

[13] Dalrymple, G. B. (1991) Good press for bad science. *Eos, Transactions, American Geophysical Union* 72(5).

[14] Robertson, J. B. (1997) Time to move ahead. *Geotimes* June: 21–23.

[15] Brinkley, D. (2001) Eisenhower the dove. *American Heritage* September: 58–65.

Index

359